T0223981

Lecture Notes in Computer Science 1478

Edited by G. Goos, J. Hartmanis and J. van Leeuwen

Springer

Berlin
Heidelberg
New York
Barcelona
Budapest
Hong Kong
London
Milan
Paris
Singapore
Tokyo

Moshe Sipper Daniel Mange
Andrés Pérez-Uribe (Eds.)

Evolvable Systems:
From Biology to Hardware

Second International Conference, ICES 98
Lausanne, Switzerland, September 23-25, 1998
Proceedings

Springer

Series Editors

Gerhard Goos, Karlsruhe University, Germany
Juris Hartmanis, Cornell University, NY, USA
Jan van Leeuwen, Utrecht University, The Netherlands

Volume Editors

Moshe Sipper
Daniel Manger
Andrés Pérez-Uribe
Logic Systems Laboratory, Swiss Federal Institute of Technology
IN-Ecublens, CH-1015 Lausanne, Switzerland
E-mail: {moshe.sipper,daniel.mange,andres.perez}@di.epfl.ch

Cataloging-in-Publication data applied for

Die Deutsche Bibliothek - CIP-Einheitsaufnahme

Evolvable systems : from biology to hardware ; second international
conference ; proceedings / ICES '98, Lausanne, Switzerland,
September 23 - 25, 1998. Moshe Sipper ... (ed.). - Berlin ; Heidelberg
; New York ; Barcelona ; Budapest ; Hong Kong ; London ; Milan ;
Paris ; Singapore ; Tokyo : Springer, 1998
  (Lecture notes in computer science ; Vol. 1478)
  ISBN 3-540-64954-9

Cover Photo: André Badertscher

CR Subject Classification (1991): B.6, B.7, F.1, I.6, I.2, J.2, J.3

ISSN 0302-9743
ISBN 3-540-64954-9 Springer-Verlag Berlin Heidelberg New York

Typesetting: Camera-ready by author
SPIN 10638732   06/3142 – 5 4 3 2 1 0   Printed on acid-free paper

Preface

The idea of evolving machines, whose origins can be traced to the cybernetics movement of the 1940s and the 1950s, has recently resurged in the form of the nascent field of bio-inspired systems and evolvable hardware. The inaugural workshop, *Towards Evolvable Hardware*, took place in Lausanne in October 1995, followed by the *First International Conference on Evolvable Systems: From Biology to Hardware (ICES96)*, held in Japan in October 1996.

These proceedings contain the papers presented at the *Second International Conference on Evolvable Systems: From Biology to Hardware (ICES98)*, which was hosted by the Swiss Federal Institute of Technology in Lausanne. The papers present the latest developments in a field that unites researchers who use biologically inspired concepts to implement real-world systems. This year's conference saw a wide range of topics, including: evolution of digital systems, evolution of analog systems, embryonic electronics, bio-inspired systems, artificial neural networks, adaptive robotics, adaptive hardware platforms, and molecular computing.

We thank the authors for their high-quality contributions and the members of the program committee for their invaluable help in the refereeing process. We also thank the International Latsis Foundation, the Swiss National Science Foundation, and the Swiss Federal Institute of Technology for financial support.

Lausanne, July 1998

Moshe Sipper
Daniel Mange
Andrés Pérez-Uribe

General Chair: Daniel Mange, *Swiss Federal Institute of Technology*

Program Chair: Moshe Sipper, *Swiss Federal Institute of Technology*

Conference Secretary: Andrés Pérez-Uribe, *Swiss Fed. Inst. of Technology*

International Steering Committee:

- Tetsuya Higuchi, *Electrotechnical Laboratory (ETL)*
- Hiroaki Kitano, *Sony Computer Science Laboratory*
- Daniel Mange, *Swiss Federal Institute of Technology*
- Moshe Sipper, *Swiss Federal Institute of Technology*
- Andrés Pérez-Uribe, *Swiss Federal Institute of Technology*

Program Committee:

- H. Adeli, *Ohio State U.*
- I. Aleksander, *Imperial College*
- D. Andre, *U. California, Berkeley*
- W. W. Armstrong, *U. Alberta*
- F. H. Bennett III, *Stanford U.*
- J. Cabestany, *Polit. Catalunya*
- M. Capcarrère, *EPFL, Switzerland*
- L. O. Chua, *U. California, Berkeley*
- C. Ciressan, *EPFL, Switzerland*
- R. J. Deaton, *U. Memphis*
- R. Dogaru, *U. California, Berkeley*
- B. Faltings, *EPFL, Switzerland*
- D. Floreano, *EPFL, Switzerland*
- T. C. Fogarty, *Napier U.*
- D. B. Fogel, *Natural Selection, Inc.*
- H. de Garis, *ATR-HIP, Japan*
- M. H. Garzon, *U. Memphis*
- E. Gelenbe, *Duke U.*
- W. Gerstner, *EPFL, Switzerland*
- R. W. Hartenstein, *Kaiserslautern U.*
- I. Harvey, *U. Sussex*
- H. Hemmi, *NTT, Japan*
- J.-C. Heudin, *Pôle Universitaire*
- L. Kang, *Wuhan U.*
- J. R. Koza, *Stanford U.*
- P. L. Luisi, *ETH Zentrum*
- B. Manderick, *Free U.*
- P. Marchal, *CSEM, Switzerland*
- J. J. Merelo, *U. Granada*
- J. F. Miller, *Napier U.*
- F. Mondada, *EPFL, Switzerland*
- J. M. Moreno, *Polit. Catalunya*
- P. Nussbaum, *CSEM, Switz.*
- C. Piguet, *CSEM, Switzerland*
- J. Reggia, *U. Maryland*
- E. Ruppin, *Tel Aviv U.*
- E. Sanchez, *EPFL, Switzerland*
- A. Stauffer, *EPFL, Switzerland*
- L. Steels, *Vrije U.*
- D. Thalmann, *EPFL, Switz.*
- A. Thompson, *U. Sussex*
- M. Tomassini, *U. Lausanne*
- G. Wendin, *Chalmers U.*
- L. Wolpert, *U. College London*
- X. Yao, *ADFA, Australia*
- J. Zahnd, *EPFL, Switzerland*

Table of Contents

Evolution of Digital Systems

Evolution of Analog Systems

Embryonic Electronics

Bio-inspired Systems

Artificial Neural Networks

Adaptive Robotics

Adaptive Hardware Platforms

Molecular Computing

A Gate-Level EHW Chip:
Implementing GA Operations
and Reconfigurable Hardware on a Single LSI

Isamu Kajitani[1] Tsutomu Hoshino[1] Daisuke Nishikawa[2] Hiroshi Yokoi[2]
Shougo Nakaya[3] Tsukasa Yamauchi[3] Takeshi Inuo[3] Nobuki Kajihara[3]
Masaya Iwata[4] Didier Keymeulen[4] Tetsuya Higuchi[4]

[1] University of Tsukuba, 1-1-1 Tennoudai, Tsukuba, Ibaraki, Japan
[2] Hokkaido University, North 13 West 8, Sapporo, Japan
[3] Adaptive Devices NEC Laboratory, Real World
Computing Partnership, Tokyo, Japan
[4] Electrotechinical Laboratory, 1-1-4 Umezono, Tsukuba, Ibaraki, Japan

Abstract. The advantage of Evolvable Hardware (EHW) over traditional hardware is its capacity for dynamic and autonomous adaptation, which is achieved through by Genetic Algorithms (GAs). In most
EHW implementations, these GAs are executed by software on a personal computer (PC) or workstation (WS). However, as a wider variety
of applications come to utilize EHW, this is not always practical. One solution is to have the GA operations carried out by the hardware itself, by
integrating these together with reconfigurable hardware logic like PLA
(Programmble Logic Array) or FPGA (Field Programmable Gate Array)
on to a single LSI chip. A compact and quickly reconfigurable EHW chip
like this could service as an off-the-shelf device for practical applications
that require on-line hardware reconfiguration. In this paper, we describe
an integrated EHW LSI chip that consists of GA hardware, reconfigurable hardware logic, a chromosome memory, a training data memory,
and a 16-bit CPU core (NEC V30). An application of this chip is also
described in a myoelectric artificial hand, which is operated by muscular
control signals. Although, work on using neural networks for this is being
carried out, this approach is not very promising due to the long learning
period required for neural networks. A simulation is presented showing
that not only is the EHW performance slightly better than with neural
networks, but that the learning time is considerably reduced.

1 Introduction

The advantage of Evolvable Hardware (EHW)[1] over traditional hardware is
its capacity for dynamic and autonomous adaptation. EHW can reconfigure its
structure dynamically (on-line) and autonomously, according to changes in task
requirements or in the environments in which the EHW is embedded. Usually,
this autonomous reconfiguration is carried out by Genetic Algorithm (GA)[2].
In most EHW implementations (excluding [3]), GA are executed by software on
PCs or WSs.

However, as a wider variety of applications come to utilize EHW, this is not always practical. One solution is to have the GA operations carried out by the hardware itself, by integrating these together with reconfigurable hardware logic like PLA (Programmable Logic Array) or FPGA (Field Programmable Gate Array) on to a single LSI chip. A compact and quickly reconfigurable EHW chip like this could service as an off-the-shelf device for practical applications which require on-line hardware reconfiguration.

In this paper, we describe an integrated EHW LSI chip which consists of GA hardware, reconfigurable hardware logic, a chromosome memory, a training data memory, and a 16-bit CPU core (NEC V30).

As the application of this chip in a myoelectric artificial hand is also described. The myoelectric artificial hand is operated by muscular control signals. However, it takes a long time, usually almost one month, before a disabled person is able to control the hand freely[4]. During this period, the disabled person has to adapt to the myoelectric hand. We hope to reverse this situation, by having the myoelectric hand adapt itself to the disabled person. Although, work on using neural networks for this is being carried out[4][5], this approach is not very promising due to the long learning period required for neural networks. A simulation is presented showing that not only is EHW performance slightly better than with neural network, but that the learning time is shorter by three order of magnitude.

This paper is organized as follows; In section 2, the design considerations for the EHW chip are presented, and then, in the following section, the architecture of EHW chip is described, with an evaluation of the implemented GAs in Section 4. In section 5, an application for the EHW chip in a myoelectric artificial hand is explained, with a summary in the final section.

2 Design considerations

In previous research on EHW, genetic operations are carried out with software on PCs or WSs. This makes it difficult to utilize EHW in situations which need circuits to be as small and light as possible. For example, the myoelectric artificial hand should be the same size as a human hand and weight less than 700 gram. Similar restrictions exist for autonomous mobile robots with EHW controllers[6]. One answer to these kinds of problems is to incorporate the EHW in a single LSI chip, as in Fig.1.

In the rest of this section, we review previous research on GA hardware, and examine the problems of implementing this effectively, before proposing a combination of suitable genetic operations to overcome these problems of hardware implementation.

2.1 Previous research on implementing genetic operators within the hardware

As shown in Table 1, some research on GA hardware have been done already, but most of this has been with the main aim of accelerating GA operators. That is,

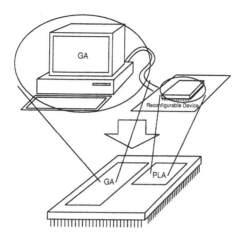

Fig. 1. The basic concept for incorporating EHW on a LSI chip.

these studies have simply incorporated GA operations on a FPGA in the hardware, and evaluated the effects on the speed of genetic operations. Furthermore, because these studies were only concerned with implementing simple GAs, most of them fail to consider the problems involved in the effective implementation of GA hardware. It is these problems that we discuss next.

Implementation Problems Here, we present three problems for the effective implementation of genetic operators.

1. Size of the memory for chromosomes
   The size of the memory needed for chromosomes increases in proportion to the number of individuals. Although, ideally the number of individuals should be small, if we are implementing them on a fixed size LSI, for effective genetic searches, a larger population is better.
2. Selection strategies
   Generational selection strategies which are used in simple GAs need extra memory to temporarily preserve selected individuals. Therefore, the selection strategy used should be a steady state one.
3. Random number generation for crossover points
   In the case of one point or multi-point crossover operations, we have to decide either one or a number of crossover points using random numbers. Although the maximum number of random numbers needed depends on the length of the chromosome, hardware implementation of random number generator can make only random bit strings in the form 2N (where N is the length of the bit string).

Based on these three problems, it is clear that to implement genetic operators on the hardware, they will have to meet the following three criteria.

Table 1. Previous work for GA hardware.

	generational or steady − state or grid − model	selection	random number generation	crossover	mutation
$Stephen D.Scott(HGA)$[7] $WashingtonUniversity$	steady − state	roulette	CA	1 − point	use
$PaulGraham(Splash2)$[8] $BrighamYoung$ $University$	generational	roulette	RNG	1 − point	use
$M.Salami(GAP)$[9] $VictoriaUniversity$	generational	roulette	CA	1 − point	use
$N.Yoshida(GAP)$[10][11] $KyushuUniversity$	generational steady − state	roulette and simplified tournament	CA	1 − point	use
$M.Goeke(evolware)$[3] $EPFL$	grid − model	grid − model	$LFSR$	uniform	not − use
$B.Shackleford(HGA)$[12] $MitsubishiElectric$ $Corporation$	generational	replace worst population	CA	1 − point	use

CA: Cellar Automaton. LFSR: Liner Feedback Shift Register.

1. An effective search method using a small population,
2. Steady state GAs,
3. A crossover operation, which does not require random numbers in the form 2N.

2.2 Suitable genetic operators for hardware implementation

In this section, we propose linking Elitist Recombination[13][14][15] and a uniform crossover[16], as an ideal combination capable of meeting the criteria identified in the previous section.

− Elitist Recombination
   The workflow for Elitist Recombination is as follow;
   1. Randomly select two individuals parents (Parent A, Parent B).
   2. Operate crossover and mutation to Parent A and Parent B, to make two children (Child A, Child B).
   3. Evaluate the two children.
   4. Select the two fittest individuals from amongst Parent A, Parent B, Child A, Child B.
   This method is basically an elite-strategy in which the fittest individual always survives. However, here selection only occurs within a family (i.e., the two parents and two children) and diversity can be easily kept in the population[13]. This method can therefore search effectively, even with a small number of individuals and thus meets the first criteria. Furthermore, because this method uses steady state GAs, it also satisfies the second.

– Uniform crossover

In a uniform crossover, there is no need to select crossover points, unlike with one or multi-point crossovers. Each allele will exchange its information with a fixed probability of 0.5. This method, therefore, meets the final criteria stated in the previous section. By combining these two genetic operators it is possible to meet all three of the criteria identified in the previous section.

3   The architecture of the EHW chip

The proposed EHW chip consists of seven functional blocks; GA UNIT, PLA, CPU, REGISTER FILE, RANDOM NUMBER GENERATOR, and two RAMs as shown in Fig.2. In the rest of this section the basic workflow of the EHW chip is explained, followed by a description of each functional unit.

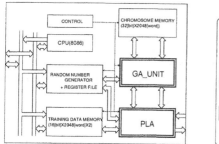 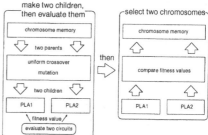

Fig. 2. A block diagram of the EHW chip. **Fig. 3.** The two phases of the EHW chip's work flow.

3.1   The workflow

The workflow of the EHW chip can be divided in two phases as shown in Fig.3. The first is the "make two children, and evaluate" phase, and the second is the "select two chromosomes" phase.

1. The "make two children, and evaluate" phase.
   – Select two chromosomes randomly (Parent A, Parent B).
   – Operate a uniform crossover and mutation on them, to make two new chromosomes (Child A, Child B).
   – Implement two circuits on two PLAs respectively by using the configuration bit strings of Child A, Child B.
   – Evaluate them.
2. The "select two chromosomes" phase.
   – Compare the fitness values for Parent A, Parent B, Child A, Child B.
   – Select the two fittest.

Note that when the chromosome for an EHW chip is very long (the maximum length is 2048bit) for parallel execution of crossover or mutation operators, then crossover and mutation is operated in units of 32 bits.

3.2 Details of each functional block

RANDOM NUMBER GENERATOR A parallel random number generator using cellar automata[17] was selected for implementation on the EHW chip, because this is very popular for GA hardware (see Table 1), and can produce 576 bits of random bit string every clock cycle.

Chromosome Memory A memory for the chromosomes of all the individuals. The maximum length for each chromosome is 2048 bits, and the maximum population is 32. Therefore, the size of this memory is 32[bit] × 2048[word]. This memory has two input/output ports, and two chromosomes can be read in units of 32 bits from the GA UNIT.

GA UNIT This block reads two chromosomes from the chromosome memory in units of 32 bits, then operates uniform crossover and mutation on them to make two segments (32 bits) of the chromosomes. Uniform crossover is operated using a random 32bit string. If a location in the random bit string has a value of '1', then the information corresponding to the same location in the two segments the chromosomes is exchanged. The two new segments of the chromosomes are then read from PLA.

PLA (Programmable Logic Array) There are two PLAs for parallel evaluation of two circuits. These blocks read two chromosomes from the GA UNIT in units of 32bits to implement the two circuits on the two PLAs respectively. The two circuits are then evaluated using training data. The maximum input width is 28bits, and the maximum output width is 8bits. Users can specify arbitrary input/output widths. The number of product term lines is 128 for each PLA.

Training data memory. (16[bit] times 2048[word] × 2) A pair of memory for the training data set. The training data set is used to evaluate the two circuits in the PLAs. Descriptions for the training data sets are shown in Section 5.

16bit CPU (8086 compatible,V30MX(NEC)) This CPU is used as the interface between outside and inside the chip. It can be used for to calculate fitness values for each circuit.

REGISTER FILE This is used for as the interface between the CPU and the other blocks. Users can specify chromosome length, mutation rate, maximum trial number and the input and output widths for each PLA.

CONTROL This makes control signals for each block.

In this work, we implemented an EHW chip on a LSI (QFP,304pins,33MHz). The number of gates for each block and the area rate for each block are shown in Table 2.

Table 2. The number of gates for each block, and their Area rates.

The name of block.	The number of gates.	area rates.(%)	The name of block.	The number of gates.	area rates.(%)
CONTROLS	7930	0.74	PLA	705300	66.7
GA UNIT	16118	1.52	CHROMOSOME-MEMORY	95816	9.06
RANDOM-NUMBER-GENERATOR, REGISTER FILE	103008	9.74	TRAINING-DATA-MEMORY	103676	9.76
			CPU	26248	2.48

3.3 Comparison of execution speed between the hardware implementation of GA operators and a GA program

To measure the effect of implementing the GA operations on to the hardware, we compared execution speed with a GA program. The GA program used the same algorithms implemented on the chip (i.e., the Elitist Recombination and the uniform crossover), and was executed on a Ultra Sparc 2(200MHz). The result shows with the GA program execution time for one set of crossover and mutation operation was $240[\mu]$, whereas, the operation time for the chip was just $3.97[\mu]$. This result shows that the execution speed for the hardware implementation of GA operations is about sixty-two times faster than that of the GA program on Ultra Sparc 2 (200MHz).

4 Evaluation of the implemented GAs

In this section, we describe a simulation to evaluate the adaptability of the implemented GAs, in quickly effecting a circuit synthesis. Specifically, a circuit synthesis with a three bit comparator circuit was simulated and the number of trials for evaluating candidate circuits were compared. The workflow for circuit synthesis with GAs is as follows.

1. Make a data set for evaluation of candidate circuits. (In this experiment, a truth table for the three bit comparator circuits was used.)
2. Implement each circuit on a PLA using configuration bit strings. Each chromosome represents one configuration bit string. Correct output rate for a training data set was treated as a fitness value.
3. Operate the genetic operations on the chromosomes to evolve the circuits.

4. Circuit synthesis terminated when the correct answer rate was 100

The result of the simulations showed that with the hardware implemented GAs, the three bit comparator circuits were synthesized after 32340.8 evaluations (average over ten simulations). In contrast, a simple GAs software (generational selection, one-point crossover), failed to synthesize the three bit comparator circuits in six of the simulations, that is, the circuit synthesis needed over 320000 evaluations. In the remaining four of simulations, the circuits were synthesized in 151128 evaluations (average over the four simulations). This result means that the speed of circuit synthesis with hardware implemented GAs is more than five times faster than that with simple GAs.

5 Synthesis of control circuits for a myoelectric artificial hand

5.1 A myoelectric artificial hand

In this paper, we present an application of the EHW chip for a control circuit in a myoelectric artificial hand. The hand can be controlled by myoelectric signals which are the muscular control signals. However, myoelectric signals vary from individual person to person. Accordingly, anybody wants to use a conventional myoelectric hands has to adapt to the controller through a long period of training (almost one month). To overcome this problem, research is being carried out on controllers that can adapt themselves to the characteristics of an individual person's myoelectric signals. Most of this research is using neural networks with back propagation (BP) learning. However, learning with back propagation also needs a great deal of time. Because EHW chips can adapt itself quickly, we propose to use them to control myoelectric hands.

5.2 Synthesizing control circuits of a myoelectric artificial hand

The myoelectric artificial hand, which has been used in this work (Fig.4), has six actions; open (A), grasp (B), supination (C), pronation (D), flection (E), and extension (F). The artificial hand is controlled by myoelectric signals, which are measured with sensors, as shown in Fig.5. Because there are three pairs of actions; (open, grasp), (supination, pronation), and (flection, extension), which can be controlled independently, the hand can perform two or three independent actions simultaneously. The control circuits for each pair of actions were synthesized with training data sets for each pair of actions, as follows.

1. Selection of ten thousand input patterns for the control circuits for each of the six actions respectively, as shown in Fig.6.
    - Detection of hand actions with a data glove and two angle sensors.
    - Classification of hand actions according to the six actions (A-F) using angular velocity.

 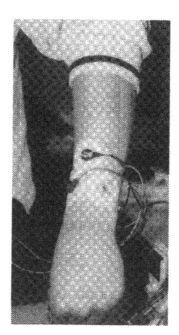

Fig. 4. A myoelectric artificial hand.　　**Fig. 5.** Sensors for myoelectric signals.

- Make frequency spectra of the myoelectric signals with FFT (Fast Fourier Transform), then select six frequency bands experimentally.
- Each frequency band was digitized into four bit binary numbers, giving twenty-four (four bits times six bands) bit input patterns for each circuit.
- Selection of ten thousand of these input patterns for each action.

2. Made training data set used to synthesize control circuits for each pair of actions, as follows.
   - Selection of two hundred input patterns from the input patterns selected for each action.
   - Each training data set for one pair of actions consisted of four hundred input patterns for a pair of actions, i.e., A, B or C,D or E,F, and eight hundred input patterns for no actions, as shown in Fig.6.

3. Evaluation of each circuit candidates with the training data sets. Specifically, the correct output rate for training input patterns was treated as fitness value for each circuit candidate.

4. Evolution of circuits with GAs. (The maximum number of evaluations was 10000 times.)

5. Evaluation of evolved circuits. The correct output rate of the evolved three control circuits for the remainder of the input patterns for each action (9800 times 6 actions = 58800) were calculated. (This rate is compared to the correct output rate for a neural network, which learned with BP in the next section.)

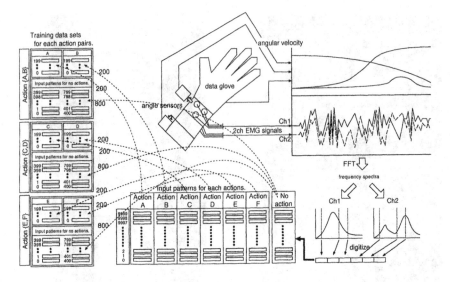

Fig. 6. How to make a training data set for each actions.

5.3 Evaluation of synthesized circuits

To evaluate the evolved control circuits for myoelectric hands, the correct answer rate calculated above was compared with that of neural networks, which is learned with the following way. Three neural networks to control each pair of actions respectively had six inputs ports for floating point numbers, which represented the six frequency bands selected above. They had sixty-four nodes in an intermediate layer and had two output ports for two bit binary numbers, respectively. These networks learned with BP using the training data sets, which were selected above. Then the correct output rate for the three networks for the remaining 58800 input patterns was calculated. The correct output rates for the evolved circuits and the learned neural networks are shown in Fig.7. For the evolved circuits, the average of each correct output rate was 85.0%, and for the learned neural networks, it was 80.0%. This means that the correct output rate for the evolved circuit is slightly better than that of the learned neural networks. Moreover, the speeds of evolution and learning differed greatly. Learning by the neural networks needed around three hours (Pentium-Pro, 200MHz) [1], but evolution of three control circuits needed only around 800 [ms] [2]. These results suggest that EHW chips are better suited as controllers than neural networks for myoelectric artificial hands.

---

[1] It takes three hours by off-line learning, while the latest on-line learning time with neural network is around ten minutes.

[2] 80.01 [μs] (for one evaluation) $\times$ 10000 evaluations, calculated from Verilog simulation.

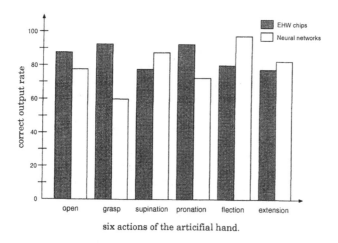
Fig. 7. The correct output rate of evolved circuits and learned neural networks.

6 Conclusions

Implementation of EHW on to a single LSI is essential for applications where circuit size is critical, such as the control circuits in myoelectric artificial hands, or in autonomous mobile robots. Therefore, we have made an EHW chip which implements reconfigurable hardware, hardware for genetic operators, memories and a 16-bit CPU core on a single LSI. The implemented genetic operators were selected after considering the problems of effective implementation. Execution speed for the incorporated GAs is sixty-two times faster than with a GA program on a Ultra Sparc 2 (200MHz), and their adaptation speed is over five times faster than that of simple GAs. The EHW chip can displace neural networks as the controllers for myoelectric artificial hands, because of its compactness and quick adaptability. In future work, we have to consider the size of PLAs and memories, because they occupy most of the area on the EHW chip (85.5%). In order to make them smaller, we would have to reduce the bit width of the inputs using other methods than FFT to make input patterns. The results presented in this paper is based on off-line learning. Currently, we are implementing the online learning of the EHW chip, which needs less training data set than current method.

7 Acknowledgements

This work is supported by MITI Real World Computing Project (RWCP). We thank Dr. Otsu and Dr. Ohmaki in Electrotechnical Laboratory, and Dr. Shimada in RWCP for their support.

References

1. T. Higuchi et al. Evolvable hardware with genetic learning: A first step towards building a darwin machine. In *Proc. of 2nd International Conference on the Simulation of Adaptive Behavior*, pages 417–424. MIP Press, 1993.

2. D.E. Goldberg. *Genetic Algorithms in Search, Optimization, and Machine Learning*. Addison-Wesley, 1989.

3. Maxime Geoke et al. Online autonomous evolware. In *Evolvable Systems: From Biology to Hardware*. Springer, 1996.

4. Masafumi Uchida et al. Control of a robot arm by myoelectric potential. In *Journal of Robotics and Mechatronics vol.5 no.3*, pages 259–265, 1993.

5. O. Fukuda et al. An emg controlled robotic manipulator using neural networks. In *IEEE ROMAN'97*, 1997.

6. D. Keymeulen et al. Robot learning using gate-level evolvable hardware. In *Sixth European Workshop on Learning Robots (EWLR-6)*, 1998.

7. Stephen D. Scott et al. Hga: A hardware-based genetic algorithm. In *Proc. of the 1995 ACM/SIGDA Third Int. Symposium on Field-Programmable Gate Arrays*, pages 53–59, 1995.

8. Paul Graham and Brent Nelson. Genetic algorithms in software and in hardware – a performance analysis of warkstation and custom computing machine implementation. In *Proc. of the IEEE symposium on FPGAs for Custom Computing Machines*, pages 216–225, 1996.

9. Mehrdad Salami. Multiple genetic algorithm processor for hardware optimization. In *Evolvable Systems : From Biology to Hardware*, pages 249–259, 1996.

10. Norihiko Yoshida et al. Gap: Generic vlsi processor for genetic algorithms. In *Second Int'l ICSC Symp. on Soft Computing*, pages 341–345, 1997.

11. N. Yoshida et al. Vlsi architecture for steady-state genetic algorithms (in japanese). Technical report, Research Reports on Information Science and Electrical Engineering of Kyushu University Vol.3 No.1, 1998.

12. Barry Shackleford et al. Hardware framework for accelerating the execution speed of a genetic algorithm. In *IEICE Trans. Electron Vol.E80-C NO.7*, pages 962–969, 1997.

13. Dirk Thierens and D.E. Goldberg. Elitist recombination: an integrated selection recombination ga. In *Proceedings of First IEEE conference on Evolutionary Computation*, pages 508–512, 1994.

14. Dirk Thierens. Selection schemes, elitist recombination and selection intensity. In *ICGA97*, pages 152–159, 1997.

15. C.H.M. and van Kemenade. Cross-competition between building blocks - propagating information to subsequent generations -. In *ICGA97*, pages 2–9, 1997.

16. G. Syswerda. Uniform crossover in genetic algorithms. In *ICGA89*, pages 2–9, 1989.

17. Peter D. Hortensius et al. Parallel random number generation for vlsi systems using cellular automata. In *IEEE Trans. on COMPUTERS vol.38 NO.10*, pages 1466–1473, 1989.

On the Automatic Design of Robust Electronics Through Artificial Evolution

Adrian Thompson

Centre for Computational Neuroscience and Robotics,
School of Cognitive & Computing Sciences, University of Sussex,
Brighton BN1 9QH, UK
adrianth@cogs.susx.ac.uk ; http://www.cogs.susx.ac.uk/users/adrianth/ade.html

Abstract. 'Unconstrained intrinsic hardware evolution' allows an evolutionary algorithm freedom to find the forms and processes natural to a reconfigurable VLSI medium. It has been shown to produce highly unconventional but extremely compact FPGA configurations for simple tasks, but these circuits are usually not robust enough to be useful: they malfunction if used on a slightly different FPGA, or at a different temperature. After defining an 'operational envelope' of robustness, the feasibility of performing fitness evaluations in widely varying physical conditions in order to provide a selection-pressure for robustness is demonstrated. Preliminary experimental results are encouraging.

1 Introduction

This paper describes first steps towards automatically designing robust electronic circuits using an evolutionary algorithm. Here, 'robust' means able to maintain satisfactory operation when certain variations in the circuit's environment or implementation occur. I will call the set of such variations with which a particular circuit must cope, its *operational envelope*. An operational envelope might include ranges of temperature, power-supply voltages, fabrication variations ('tolerances'), load conditions, and so on. The operational envelope necessary for a particular application will be assumed to be explicitly defined by a human analyst.

Evolutionary algorithms have been successfully used to design electronics previously, but either temporarily ignoring the need for robustness (eg. Thompson (1996), Koza, Bennett III, Andre, et al. (1997)), or constraining the available choice of components, component values, and interconnection topologies, such that all expressable circuits are guaranteed to be adequately robust (eg. Fogarty, Miller, and Thomson (1998), Salami, Iwata, and Higuchi (1997)). In contrast, the current work continues the development of 'unconstrained intrinsic hardware evolution', whereby evolution is given maximal freedom to exploit the full repertoire of behaviours that a physical reconfigurable VLSI device (such as an FPGA[1]) can produce. Conventional design principles are not enforced, in the hope that forms and processes that are 'natural' to the VLSI medium and the

---

[1] Field-Programmable Gate Array

evolutionary algorithm can result. Earlier experiments have shown that circuits so evolved can operate in very unusual ways, yet be impressively compact by putting to use more of the rich dynamics and interactions of the silicon components — possibly even including subtle aspects of semiconductor physics.

An example is shown in Fig. 1. This circuit was evolved using a standard genetic algorithm to discriminate between 1kHz and 10kHz audio tones: the fitness function was to maximise the difference in average output voltage as bursts of the two different inputs were applied. Of the 10×10 area of a Xilinx XC6216 FPGA provided (Xilinx, 1996), perfect behaviour was achieved using only 32 cells, as shown. There was no clock, and no external timing components, so to discriminate between input periods of 1ms and 0.1ms using so few components, each having timescales a factor of 10^5 shorter, is remarkable: a human designer would use more. It was shown that the rich dynamics and detailed physics of the VLSI medium were exploited to a high degree in achieving this efficiency. This experiment is built-upon later in the paper, and the reader is referred to Thompson (1997a) for full details and Harvey and Thompson (1997) for an evolution-theoretic analysis.

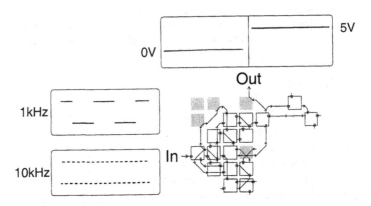

Fig. 1. The tone-discriminator evolved on a single FPGA. Only the functional part of the 10×10 region is shown. The waveforms are photographs of an oscilloscope screen monitoring the input and output.

Such early evolved 'unconstrained' circuits, though exciting and informative, are practically useless. Their operational envelope is inadequate for any real-world application, given the present norms and expectations of industry.[2] Most notably, they only work properly in a narrow temperature range, and only on the particular FPGA chip used to instantiate them for fitness evaluation during evolution. To be most useful, the configurations should work on any FPGA chip

---

[2] Although they are not necessarily useless as components in a less conventional architecture, such as certain artificial neural networks: Thanks to Ed Rietman for pointing this out.

of the same nominal specification (even though no two are absolutely identical), and preferably over a large temperature range. Then the user could purchase or licence the configuration (perhaps over the internet), obtain an appropriate FPGA from a local supplier, and use the evolved circuit in a product. Similar use of proprietary 'macros', 'cores' or 'intellectual property' (IP) as parts of a larger design on an FPGA is already well established in industry. Realistically useful behaviours have been successfully evolved (eg. the above 10×10 configuration was further evolved to distinguish between audio signals for the spoken words 'Go' and 'Stop'), but will remain lab curiosities until they can work reliably on other people's FPGAs of the same type.

This paper describes the beginning of an attempt to evolve such robust unconstrained circuits by physically exposing the evolving circuits to samples of the operational envelope during fitness evaluations. Robustness is as much part of the problem to be solved as the behavioural specification: a non-robust circuit simply will not work for part of its evaluation. Evolution is free to find a means to robustness tailored to the particular task, and to the physical resources available in the VLSI medium. Because the need for robustness is integrated with behavioural requirements, it can exert an influence at all stages during the evolution of the circuit (not just in a final 'implementation' step). Thus it is hoped the impressively effective exploitation of silicon properties seen previously will still be possible whilst meeting robustness requirements. Thompson (1997c) identifies possible mechanisms for robustness, drawing analogies with nature.

After clarifying the objectives, the main part of this paper is devoted to showing the technical feasibility of providing a selection pressure for robustness within intrinsic hardware evolution. Promising early results are then described. The modest aim of the paper is to share thoughts and experiences, to aid future research within the evolutionary electronics community.

2 Training, Testing, and Generalisation

It is not possible to test a circuit at every point within the operational envelope before assigning it a fitness. Considering the evolutionary algorithm as a machine learning method, during evolution the individuals are evaluated while exposed to a *training set* of conditions within the envelope, in the hope that the final result will generalise to work over the entire envelope defined for this task. Generalisation must then be tested by checking that the final circuit does work at points within the envelope not experienced during evolution.

A practical method for achieving adequate generalisation is the main goal of the project. For the current initial explorations, the training set consist of five combinations of conditions chosen to represent *extremes* of a usable operational envelope. Presumably the presence of some extremes in the training set is necessary for adequate generalisation, but it is not yet known if it is sufficient. Analogous difficulties arise in other application domains of evolutionary algorithms; see Jakobi (1997) for an interesting general framework.

It may be that aiming for an *adaptive* system, rather than a general one, would be more in harmony with an evolutionary approach.[3] For my particular project, this is a choice of viewpoint in interpreting the results, rather than a matter of experimental design. I have defined adequate behaviour at all points within an operational envelope as an engineering requirement, and a selection pressure towards this is provided in the least restrictive way possible. The circuits being evolved have internal state and rich dynamics, so do not necessarily display a constant behaviour over time. This means that although, in response to the selection pressure, evolution *could* produce a 'general' solution in the strict machine-learning sense, it could also produce a circuit which adapts to the current conditions through its own dynamics. If the time taken for this 'self-adjustment' were much shorter than the length of a fitness evaluation, then the adaptive circuit would be almost indistinguishable from a general one. Hence evolution is free to explore both avenues; for brevity I will use the term 'generalisation' to refer to both possible means of robust observed behaviour throughout this paper.

3   Questions to be Addressed

Of the many issues on the research agenda, the following are primary:

- What resources are needed for robustness? Are stable off-chip components required? Is it necessary to provide a stable oscillation as an extra input to the evolving circuits? Such a clock would be a resource, to be used in any way (or ignored) as appropriate, rather than an enforced constraint on the system's dynamics as in synchronous design.

- Can generalisation over an entire, practical, operational envelope be achieved through evolution in a relatively small number of different conditions?

- Given the experimental arrangement described herein, is the evolution of a robust circuit more difficult than of a fragile circuit evolved on just one FPGA? (An apparently harder task is not necessarily more difficult for evolution, depending on the pathways available for evolutionary change.) Does it make sense gradually to increase the diversity and span of the operational envelope during evolution, or should the population be exposed to a representation of the complete operational envelope right from the beginning?

- If robust circuits *are* evolved, how do they work? Do they look more like conventional circuits? Are they still considerably more efficient?

---

[3] Thanks to Dario Floreano for pointing this out.

4   Fault Tolerance and Yield

Most chips on a silicon wafer are thrown away after failing quality-control tests due to fabrication variations. Some of these 'fatal' variations can be seen as extreme cases of the variations between chips which do reach the market. If it proves to be possible to evolve circuits which tolerate unusually wide semiconductor variations, then the quality control of the chips to be used could be relaxed according the attained operational envelope, resulting in increased yield. This is speculation.

Similarly, some defects which develop during a chip's lifetime may fall within an evolved circuit's operational envelope, giving some degree of fault tolerance. Thompson (1997b) showed that some fault tolerance requirements can be explicitly included into the operational envelope specification as the goal of evolution.

5   Hardware Framework

This section describes the apparatus developed to support investigations into the evolution of robust electronics. A standard PC runs the evolutionary algorithm, and is host to the specialised hardware. There are five Xilinx XC6216 FPGA chips on which fitness evaluations can be performed simultaneously. One of these resides on a commercial PCI-bus card which plugs into the PC (Xilinx, Inc., 1998); a custom-made AT-bus card interfaces the other four. Test inputs for the FPGAs are given by thresholding the line output of the PC's standard sound card. There is an analogue integrator circuit for each FPGA, which can be used to measure the average output voltage over a period of time. The PC can read and reset the integrators via a commercial data-acquisition card, and use the measured values in a fitness function. The setup, jocularly named 'THE EVOLVATRON', is shown in Fig. 2. More details follow:

5.1   The Operational Envelope

The following parameters of the operational envelope are present:

Electronic surroundings. Four of the FPGAs are in separate metal boxes, configured by the PC via their serial interfaces. Most of the pins of these FPGAs are unused and unconnected. The FPGA on the PCI card is configured via its parallel interface, and nearly all of its pins are connected to support various features of the PCI card not used here.

Circuit position on FPGA. The 10×10 circuit being evolved is placed in a different position on each 64×64 FPGA. It is possible that adjacency to edges or corners of the array would affect the circuit, so the aim is to provide a variety of such conditions.

FPGA fabrication variations. Two of the devices are early engineering samples of the XC6216 made at a Yamaha foundry, and the remaining three are the current production version fabricated by Seiko.

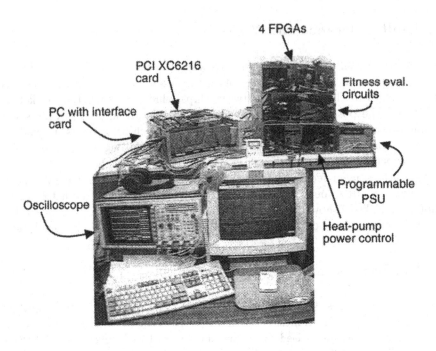

Fig. 2. Photograph of 'THE EVOLVATRON.' (Mk. I)

Packaging. Three of the devices are in 299-pin ceramic pin-grid-array packages ('PG299'). Another has a plastic 84-pin PLCC package ('PC84'). The remaining chip, on the PCI card, is surface-mounted in a 240-pin plastic PQFP-type package ('HQ240'). This package contains a grounded nickel-plated copper heatsink slug in very close proximity to the silicon die. See Xilinx (1996, chap. 10) for packaging details.

Temperature. Two of the FPGAs are at the ambient temperature of the lab (17–25°C). Another is maintained at a constant 60°C, and another at 12°C. Finally, a further chip has a thermal gradient of 15°C/cm maintained across the central region of the package in which the die is located. See 5.2 for details.

Power supply. The FPGA on the PCI card is running on the same power supply as the rest of the PC, as usual. Two of the other devices are powered by the 5V output of a separate typical computer switch-mode power supply unit (PSU), which is also running (from a different output) the numerous 12V fans involved in thermal regulation. The remaining two devices are powered by a precise programmable PSU,[4] with one chip at 4.80V and the other at 5.20V.

---

[4] Hewlett Packard HP E3631A.

Load. In this series of experiments, only one pin on each FPGA is configured as a user output from the evolving circuit. Each of these drives approximately 70cm of audio-grade cable with a co-axial shield,[5] before reaching an integrator (housed in a separate shielded box) used to measure average voltage. The integrator design used presents an almost purely resistive load impedance of approximately 350kΩ. To provide some variation, a 100kΩ resistor to 0V was connected to the output of one of the FPGAs close to the chip.

These factors are combined in the five FPGAs as shown in Fig. 3.

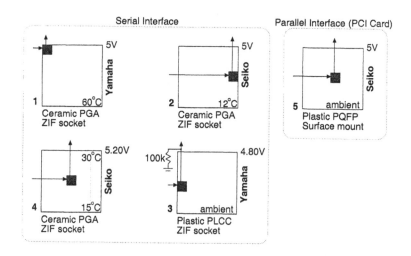

Fig. 3. Variations within the operational envelope: Evolvatron Mk. I

5.2 Thermal Control

It would be possible to vary the temperature of an FPGA during or in-between fitness trials. That was not adopted here in case the rapidly repeated thermal cycling would lead to premature chip-failure. Instead, the FPGAs are held at at different *but constant* temperatures.

Fig. 4 shows the present arrangement for the cold FPGA. A wafer-like Peltier-effect thermoelectric cooling device is used , which can pump heat from one face to the other at up to 68.8W (R.S. Components Ltd., 1988; Melcor, 1998). Its cold side is attached to the top face of the chip package, and the hot side to a heatsink. For low FPGA temperatures, good thermal insulation between chip and heatsink is required; this is aided by the copper spacing block. The FPGA package should also be insulated from the surroundings as much as possible to minimise 'heat leak.' For this reason, unused pins of the FPGA's socket are not

---

[5] This wiring will be eliminated in the Mk. II Evolvatron.

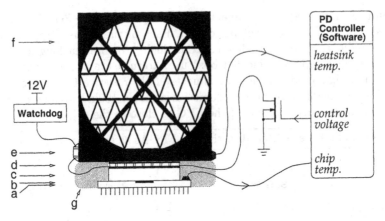

Fig. 4. Arrangement for the cold FPGA. The proportions of the main components are drawn to scale. (**a**) Ceramic FPGA package, (**b**) silicon die within, (**c**) copper spacing block, (**d**) Peltier-effect heat-pump, (**e**) thermal fuse, (**f**) large cuboid fan-cooled heatsink, (**g**) foam thermal insulation. The LM35CZ temperature sensors, and the fuse, are mounted in epoxy resin. Heat-transfer paste is applied at all thermal interfaces.

soldered into the circuit-board. To obtain low temperatures, the heatsink must be very effective: a 0.14°C/W part was used, with additional forced-air cooling.

The PC reads the value of a temperature sensor attached to the chip package, and in response provides an appropriate control voltage for the heat-pump. A standard proportional+derivative (PD) linear feedback control system is implemented in software, which attempts to hold the measured temperature at a given set-point. This was found to cope well with the various nonlinearities and time-delays present in the system.

If worked too hard (by attempting to cool too rapidly, or maintain too low a temperature for the current ambient temperature), the thermoelectric device enters a highly inefficient region of its operating characteristics, and converts a lot of electrical energy to heat. In this situation the chip temperature will rise, the PD controller will increase the pump power further, and thermal runaway will occur; the controller will not recover even if the original cause – such as an unusually large rise in lab temperature – is removed. For this reason, a rather conservative 'overdamped' controller is used. Before an evolutionary experiment, it spends 40 minutes gradually lowering the chip to the setpoint temperature, which is achieved without overshoot. Nevertheless, it is able to keep the chip package temperature to within ±1°C during the experiment (even as ambient temperature and FPGA power dissipation vary), with the state of the controller being updated every 30s. Evolving circuits can be prone to interference from nearby fluctuating signals, so the controller state is always updated *between*, not during, fitness evaluations.

The hot FPGA is thermostatically regulated in the same manner as the cold one described above, with the heat-pump connected in reverse. For the hot chip, it is less important to keep the thermoelectric device at an efficient operating point, so only a much smaller heatsink is needed. On a third FPGA, a constant thermal gradient is maintained by using a combination of heating and cooling pumps. By using a single large heatsink shared by the two pumps, there is a thermal cycle: one pump is heating the heatsink and the other cooling it. This makes it relatively easy to operate the thermoelectric devices efficiently at high power, so a surprising large thermal gradient can be maintained across the chip.

The three thermally regulated XC6216's were chosen in the Xilinx 299-pin pin-grid-array package. Unlike most IC packages, these are assembled 'Cavity Down', with the silicon die attached to the inside top of the package, for optimal heat transfer between the die and the upper surface of the case (Xilinx, 1996, chap. 10). Thus the thermal regulation apparatus will have a very direct influence on the temperature of the silicon.

With lab temperatures rising up to 25°C during the day, the set-point of the cold chip is currently set at 12°C during an evolutionary experiment. In the Mk. II Evolvatron (under development), the circuit-board with the cold FPGA will be in a plastic bag with some desiccant, and simply placed in a domestic freezer. This will allow lower chip temperatures even for hotter (summer-time) lab conditions.[6] The Peltier-cooling arrangement will be retained on another device for more precise temperature control for analysis purposes. For brief, imprecise, low-temperature tests (i.e. not during evolution), standard electronics freezer spray can also be used ($\sim -50°C$), or even liquid nitrogen ($-196°C$).

The hot FPGA is run at a case temperature of 60°C. A greater temperature could easily be maintained, but long-term time-to-failure must be considered, along with a margin for error. The FPGA with a thermal gradient is run with one of the temperature sensors at 30°C and the other at 15°C.[7] The silicon die is approximately 1cm^2, so these sensors are attached across the central region of the top surface of the case, with a separation of 1cm. The resulting thermal gradient of 15°C/cm is about the maximum that could reliably be obtained using the components chosen.

5.3 Safety

Fire-safety is a major design issue. An evolutionary machine needs to run unattended, 24 hours a day. In the Evolvatron, there are three lines of defence:

1. A temperature sensor on each heatsink causes the controller to shut down the pump completely and immediately if thermal runaway is detected.

---

[6] Another possibility would be the circulation of a refrigerated fluid through the heatsink.

[7] Other investigators are cautioned to beware condensation problems (corrosion, mould), especially in causing bad connections in socketed chips.

2. What if the computer running the PD controllers (and the evolutionary algorithm) crashes, due to a programming error, hardware failure, or operating-system freeze? To detect this condition, a re-triggerable monostable circuit is used as a watchdog. Every time the PD controllers' state is updated, the software causes a digital output signal from the PC to toggle. If the watchdog circuit has not detected a toggle within the last 100s (this time-constant is adjustable), then it trips-out a relay which cuts the power to all heat-pumps.

3. As a final resort, there is a thermal fuse electrically in series with each heat-pump, and attached to its heatsink. The fuses blow at 98°C.

5.4 Further Design Issues

All of the FPGAs except that on the PCI card are situated in separate housings to the PC, to allow for the large heatsinks, good airflow, etc., involved in thermal regulation. For maximum evolution speed, fitness evaluations need to take place on all FPGAs simultaneously. It has been observed in earlier experiments that some circuits arising in the intermediate stages of evolution oscillate wildly, potentially inducing interference on any wires connected to the FPGA. Evolved circuits have also sometimes shown susceptibility to picking up on the activity of FPGA pins to which they should have no direct access, such as configuration control signals. Therefore, all signals common to more than one FPGA are optically isolated to prevent evolved circuits (even multiple instantiations of the same configuration) from interacting during parallel fitness evaluations. The spatial separation of the PC and the FPGAs, along with the need for optical isolation of all common signals, makes the use of the XC6216's serial interface attractive, despite being slower than the parallel mechanism normally used.

The custom-made interface card in the PC accepts a 2Kbyte buffer of address+data pairs, and transmits them serially with appropriate timings. Using only 4 wires per FPGA, the interface allows configurations to be broadcast to any combination of FPGAs, including all or just one. The programmable clock generator of the PCI card is run at 6MHz to drive the serial configuration. The clock is only run for the minimum number of periods necessary for configuration, so that during fitness evaluations it is inactive and cannot interfere with the evolved circuits in any way. The configuration arrangement is arbitrarily expandable to more FPGAs, or to other members of the family (eg. XC6264), and is easily fast enough for the type of experiments planned.

Some of the FPGAs are mounted in zero insertion-force sockets to allow other devices to be easily substituted, to test for generalisation for example. Because of the potentially unruly behaviour of the circuits, careful attention is given to power-supply decoupling throughout. The host PC runs the GA and the PD thermocontrollers, manages configuration and fitness measurement, can control the temperature setpoints and programmable PSU voltages, and can measure current consumption on the programmable PSU and on the PCI card. This gives maximum flexibility in experimental design, as well as for the centralised logging of data for later analysis.

6   Results and Discussion

Taking the final population (generation 5000) of 50 individuals evolved for the tone-discrimination task on just one FPGA (see §1), evolution was continued with overall fitness now being the mean of the individual fitnesses measured on the five different FPGAs.[8] The other experimental details remained the same. FPGA 1 was the identical device used in the earlier experiment — but now at 60°C instead of ambient temperature — and the 10 × 10 circuit's position on that FPGA was unchanged. Fig. 5 shows the initial response to the new selection pressure for robustness.

FPGA

Fig. 5. Performance on the five FPGAs (Fig. 3) of the best overall individual in the population, at each generation. Intensity is proportional to performance, with dark black being near-perfect and white the worst possible.

Initially, there were some individuals in the population that performed better on FPGAs 2 and 5 than any did on the original chip at its newly elevated temperature. This is surprising as 2&5 are both from a different foundry, and 2 was at 12°C. Quickly, however, good performance was regained on the original chip at its new temperature. Within 300 generations, individuals emerged that had respectable performance on FPGAs 1,2 and 4. This is extremely promising, as these conditions include: {Yamaha silicon, 60°C}, {Seiko silicon, 12°C }, and {Seiko silicon, 15°C/cm gradient}. Most of the time, for any given FPGA there would be at least one individual in the population achieving fair performance on it; the difficult part is finding *single* individuals scoring well in *all* conditions. The role of population diversity appears worthy of further investigation.

At the time of writing, the experiment has not progressed beyond what is shown here, so it is too early to begin to answer the questions of §3, but the signs are encouraging. In particular, circuits coping with some extremities of the operational envelope have been found without having to increase the 10 × 10 area of FPGA made available to them — they are still extraordinarily compact. There is, however, a long way to go. This partially successful adaptation to operation in some extreme conditions was relatively rapid compared to the initial evolution

---

[8] Here, evolution time is dominated by the time taken for each fitness evaluation. Rather arbitrarily, an evaluation consisted of exposure to five half-second bursts of each tone (Thompson, 1997a), taking a total of 5s (whether evaluated on just one FPGA, or on several simultaneously). Hence, the original 5000 generations took around 5s×50×5000 ≃ 2 weeks, and the 300 generations reported here ≃ 21 hours. Possibly, much shorter evaluations could be made adequate.

of the behaviour on just one chip; it might be suspected that complete success will take much longer. The apparent ease of adapting a circuit (or population) to a *single* new set of conditions, although not the goal of this project, could conceivably be useful in some applications.

This paper has introduced the notion of unconstrained intrinsic evolution for robustness within an operational envelope. An experimental arrangement has been described in detail which permits a controlled and scientific exploration of this research avenue, to illustrate its feasibility and promise. Early results are positive: even if complete success should not be achieved, it is already clear that experiments within this framework will be illuminating.

This work is supported by EPSRC, with equipment donated by Xilinx and Hewlett Packard. Related work is supported by British Telecom and Motorola. Thanks to all.

Reference

Fogarty, T., Miller, J., & Thomson, P. (1998). Evolving digital logic circuits on Xilinx 6000 family FPGAs. In Chawdhry, Roy, & Pant (Eds.), *Soft Computing in Engineering Design and Manufacturing*, pp. 299–305. Springer.

Harvey, I., & Thompson, A. (1997). Through the labyrinth evolution finds a way: A silicon ridge. In Higuchi, T., & Iwata, M. (Eds.), *Proc. 1st Int. Conf. on Evolvable Systems (ICES'96)*, LNCS 1259, pp. 406–422. Springer.

Jakobi, N. (1997). Half-baked, ad-hoc and noisy: Minimal simulations for evolutionary robotics. In Husbands, P., & Harvey, I. (Eds.), *Proc. 4th Eur. Conf. on Artificial Life (ECAL'97)*, pp. 348–357. MIT Press.

Koza, J. R., Bennett III, F. H., Andre, D., et al. (1997). Automated synthesis of analog electrical circuits by means of genetic programming. *IEEE Trans. Evolutionary Computation*, *1*(2), 109–128.

Melcor, http://www.melcor.com/ (1998). *Thermoelectric Hardbook.*

R.S. Components Ltd., PO Box 99, Corby, Northants NN17 9RS, UK (1988). *Peltier effect heat pumps.* Datasheet 7562.

Salami, M., Iwata, M., & Higuchi, T. (1997). Lossless image compression by evolvable hardware. In Husbands, P., & Harvey, I. (Eds.), *Proc. 4th Eur. Conf. on Artificial Life (ECAL'97)*, pp. 407–416. MIT Press.

Thompson, A. (1996). Silicon evolution. In Koza et al. (Eds.), *Genetic Programming 1996: Proc. 1st Annual Conf. (GP96)*, pp. 444–452. MIT Press.

Thompson, A. (1997a). An evolved circuit, intrinsic in silicon, entwined with physics. In Higuchi, T., & Iwata, M. (Eds.), *Proc. 1st Int. Conf. on Evolvable Systems (ICES'96)*, LNCS 1259, pp. 390–405. Springer-Verlag.

Thompson, A. (1997b). Evolving inherently fault-tolerant systems. *Proc Instn Mechanical Engineers, Part I*, *211*, 365–371.

Thompson, A. (1997c). Temperature in natural and artificial systems. In Husbands, P., & Harvey, I. (Eds.), *Proc. 4th Eur. Conf. on Artificial Life (ECAL'97)*, pp. 388–397. MIT Press.

Xilinx (1996). *The Programmable Logic Data Book.* See http://www.xilinx.com.

Xilinx, Inc., http://www.xilinx.com/products/6200DS.htm (1998). *XC6200 Development System.* Datasheet V1.2.

Aspects of Digital Evolution: Geometry and Learning

Julian F. Miller[1] and Peter Thomson[1]

[1]Department of Computer Studies, Napier University, 219 Colinton Road, Edinburgh, EH14 1DJ, UK. Email: j.miller@dcs.napier.ac.uk, p.thomson@dcs.napier.ac.uk
Telephone: +44 (0)131 455 4305

Abstract. In this paper we present a new chromosome representation for evolving digital circuits. The representation is based very closely on the chip architecture of the Xilinx 6216 FPGA. We examine the effectiveness of evolving circuit functionality by using randomly chosen examples taken from the truth table. We consider the merits of a cell architecture in which functional cells alternate with routing cells and compare this with an architecture in which any cell can implement a function or be merely used for routing signals. It is noteworthy that the presence of elitism significantly improves the Genetic Algorithm performance.

1. Introduction

There is now a growing interest in the possibilities of designing electronic circuits using evolutionary techniques. [2, 3, 4, 5, 6, 7, 8, 10, 11, 12].

Arithmetic circuits are interesting examples to choose to use to examine the issue of evolving digital circuits [1, 9] since there are well known conventional designs with which the evolved solutions can be compared. Additionally arithmetic circuits are modular in nature so that there is practical interest in trying to evolve new efficient designs for the small building blocks. Indeed we showed in our earlier work that more efficient designs could indeed be produced using evolutionary techniques (e.g. the two-bit multiplier). We demonstrated how it was possible to evolve modular designs from which the general principle for building the larger system could be extracted (e.g. the two-bit ripple-carry adder). We also found that a modest increase in circuit complexity could mean that evolving the circuit became enormously more difficult (e.g. the three-bit multiplier). Previously all our attempts to evolve digital circuits have involved evaluating the fitness of chromosomes using the *complete* truth table. Clearly such a procedure can lead to very slow evolution if the size of the truth table is large. In this paper we set out to examine whether it is feasible to use randomly chosen examples to evaluate the fitness of the chromosome.

Recently we have begun to consider the relationship between the architectural platform used for circuit evolution and the ease with which circuit function can be evolved, and we examine a number of different configurations of functional cells and routing cells. In one scheme the rectangular array is divided into columns of functional cells alternating with routing cells, in another, the functional and routing cells make up a chequer-board pattern. These schemes are compared with one in which the GA decides whether a cell is to be used for function or routing. The results are presented in section 4.

All the experiments reported in this paper have involved a new chromosome representation that we have developed which closely follows the structure of the Xilinx 6216 FPGA. This differs from the previous structure on which we evolved arithmetic circuits which was modelled on less specific netlists. Although we have not yet employed the chip directly in the execution of our genetic algorithm, we are confident that the new representation will readily translate into the bit strings required to configure the device without violating the routing constraints.

2. A Chromosome Representation which Accurately Models the Xilinx 6216 Chip

Since we are considering only combinational designs (non-sequential) it is vital to allow only *feed-forward* circuits. To achieve this we chose a cell connection scheme in which inputs are fed into a cell which are East-going or North going only. If the cell is a Multiplexer (mux) then we allow the control input of the mux to arrive from the North (indicated by 'S'). This scheme is shown in Figure 1 below:

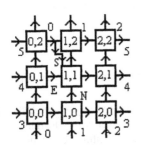

Fig. 1. The feed-forward connection of cells

The cells are numbered according to their column and row position with the origin at the bottom left hand corner of the array. All arrows pointing towards (outwards from) a cell represent inputs (outputs). The primary inputs connect to cells on the leftmost column and lowest row. The primary outputs exit the cells which are located on the topmost row and the rightmost column (in this case cells in column and row two). The cells are allowed to be one of the types shown in Table 1, where A and B represent 'E' and 'N' inputs and C represents 'S' input. If a cell at position (col, row) is a mux then the 'S' input is assumed to the 'E' output of the cell located at position (col-1, row+1). If the mux cell is located at column zero then we take the control input of the mux from the primary input located at position (-1, row+1). In such a scheme cells in the top row of the cellular array are not allowed to be multiplexers.

The chromosome has four parts; functional, routing, input, and output. The functional chromosome is a set of integers representing the possible cell types as indicated by gate type in Table 1. For N cells in the rectangular array there are N pairs (i,j), $i,j \in \{0,1,2\}$, of routing genes which make up the routing chromosome, where the first (second) element of the pair represents the North (East) output. If i or j equals 0 then the North (East) output is the connected to East (North) input, and if i or j equals 1 then the North (East) output is connected to the North (East) input If a routing gene is 2 then the corresponding cell output is a function of the cell inputs. Thus, the routing chromosome ignores the corresponding gene in the functional chromosome if its value is set to 0 or 1. The input chromosome has #rows + #columns elements. Each element can take any integer from 0 to #primary_inputs - 1. The output chromosome has #primary_outputs elements which represent the places in the cellular array from which the primary outputs are to be taken. To illustrate the interpretation of the chromosome consider the chromosome example (Table 2), which represents a chromosome for a 3x3 array of cells as seen in Figure 1. For the sake of argument imagine that the target function is a 1-bit adder with carry. This has three inputs A, B, Cin and two outputs S and Cout.

Table 1. Allowed functionalities of cells

Gate Type	Function	Gate Type	Function
-4	!A & !C + !B & C	6	A ^ B
-3	A & !C + !B & C	7	A | B
-2	!A & !C + B & C	8	!A & !B
-1	A & !C + B & C	9	!A ^ B
1	0	10	!B
2	1	11	A | !B
3	A & B	12	!A
4	A & !B	13	!A | B
5	!A & B	14	!A | !B

We read the chromosome in the following cell order (0,0) - (2,0), (0,1) - (2,1) and (0,2) - (2,2) as in Figure 1. The inputs and outputs are also read in as shown in Figure 1. Thus the cell at (1,1) is a multiplexer of type - 1 (see Table 1). Its outputs as indicated by (2,2) are routed out to North and East. The inputs on the bottom row are Cin, A, Cin, and along the left edge, A, B, B. The Outputs S and Cout are connected to the top row middle cell and top right hand corner cell (east output) respectively.

Table 2. Example chromosome for 3x3 array

Functional Part	Routing Part	Input part	Output part
2,9,11,12,-1,6,14,4,-3	0,1,2,1,1,2,2,2,2,2,0,0,1,1,1,0,2,2	2,0,0,2,1,1	5,1

In the genetic algorithm we used uniform crossover with tournament selection (size 2). Winners of tournaments are accepted with a probability of 0.7. The routing chromosomes are all initialised to 2; 0 and 1 are only introduced by mutation, this is found to be more effective than random initialisation. We use a fixed mutation rate which can be expressed as the percentage of all genes in the population to be randomly altered. The breeding rate represents the percentage of all chromosomes in the population which will be replaced by their offspring.

3. Aspects of the Evolution of Digital Circuits

3.1 How Does the Effectiveness of Learning Depend on the Number of Examples?

One of the fundamental problems with attempting to evolve digital circuits is that the size of the truth table grows exponentially with the number of inputs. This is not a problem if one is only trying to evolve quite simple circuits but would be a major difficulty if one were trying to evolve larger systems. The problem with digital circuits is that they are effectively useless even if a *single* output bit is incorrect. The degree of specification of the properties of an analogue circuit is not nearly so great as one is usually trying to obtain output characteristics to within a certain error. The question of interest is whether it is necessary to present all the examples in the truth table to every individual chromosome, or whether one can merely distribute all the examples over an entire run of the GA. Secondly if one does present an incomplete set of examples to each chromosome for fitness evaluation, how should this be done? We decided to carry out a number of experiments to investigate this. We fixed the total number of examples presented to the GA while we varied the population size, the number of examples, and the number of generations. The examples were generated randomly and we looked at two scenarios: (a) a set of random examples was generated for each new population and each chromosome fitness was evaluated using this, (b) as set of random examples was generated independently for each individual chromosome. Using an incomplete set of examples for fitness evaluation means that the evaluated fitness for the chromosome is unlikely to be the same as the fitness calculated using the entire truth table. We refer to the former value as the apparent fitness, and the latter value as the real fitness. Throughout the GA run the apparent fitness is used except where the number of examples happens to equal that in the truth table. We always evaluate the real fitness of the best solution in each GA run after it has terminated. It is only by doing this that we can determine the effectiveness of this approach.

3.2 Functionality and Routing

Our previous method of evolving pure netlists had been designed to limit the number of routing connections that each functional block could use to connect to its neighbours. We wondered whether it might make circuits easier to evolve using the new representation if this too was modified to limit routing and functionality. In other words, could the evolutionary process be forced, by limiting available resources, to converge more quickly to a desired 100% functionally correct solution. We conducted experiments in which the number of functional blocks and the number of routing blocks is pre-determined by the chosen geometry. In the linear chromosome, every odd numbered gene is chosen to represent a pure routing cell (we refer to this as a *differentiated* structure - an undifferentiated structure being one where there is no imposed difference between the cells). The functional cells were initialised to possess no routing, but could be mutated so that their outputs became routes. Figures 2 (a) and 2 (b) below show the arrangement which would result from the selection of 3 x 7 and 3 x 8 geometries respectively - where F represents a functional block and S is a switching block.

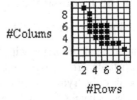

Fig. 2. Two differentiated structures with different geometries

Fig. 3. Geometry map for the experiments on the 2-bit multiplier

Thus if the number of columns in the chosen geometry is odd the functional and routing cells are arranged in a chequer board pattern while if the number of columns is even the entire columns of functional cells alternate with columns of routing cells. We carried out a large number of experiments with the geometries shown in Figure 3.

4. Results

In section 4.1, experiments were carried out on a 3-bit binary multiplier, whereas all remaining experiments have been carried out on a 2-bit binary multiplier. The GA was always run with a 100% breeding rate, a 1% mutation rate. The fitness of the chromosomes was calculated as the percentage of correct output bits.

4.1 Is Learning by Example Effective?

In Figure 4, parameters and results are presented for experiments in which a fixed number of randomly chosen examples (input combinations) were evaluated by every member of the population. It should be noted that in this set of experimental results, the total number of examples, which is equal to the product of the following quantities: number of randomly chosen examples, number of generations and population size. This is the total number of examples evaluated by the GA in each run. In figure 4 this is shown in multiples of 10^6, whilst the number of generations is presented in multiples of 10^3. We varied the number of random examples presented to each chromosome in the range 8 to 32. Our motivation here was to determine whether we could evolve the function using a subset of the total number of examples. We did not look at numbers of examples between 32 and 64 because we were attempting to investigate whether the circuit could be evolved whilst making a significant saving in terms of numbers of truth-table comparisons. Each experiment involved 10 runs of the GA. The standard deviation was calculated using the best solutions from each run. From all of these experiments, the first obvious conclusion is that evolution is much more successful at finding circuit design solutions when all possible examples (i.e. the entire truth table) is presented for fitness evaluation (Figure 4, result 2, GA with population size 20). It would appear that whenever a subset of examples is used the information is too sparse to allow the GA to properly converge upon the optimal solution (a circuit which is 100% functionally correct). This apparent lack of information with which the GA has to work is borne out by the fact that when the average final generation at which there was a design improvement (the Δ generation) is analysed, it is found that fitness evaluations based upon the full example set produce a Δ which was approximately half the total number of generations. The Δ of all other runs was a factor of 10-20 smaller, thus, when a small example subset is used, the GA finds it very difficult to improve. An extreme case of this is when only 8 examples (from a total possible of 64) was used. In this case, the GA always achieved 100% apparent correct functionality for the examples given, and the Δ value was extremely small (<10 generations in all cases).

Additionally, further evidence to support this point, regarding the lack of information available to the GA when evaluating fitness on a reduced sample set, is that in each run the algorithm always identifies the best available fitness. This is demonstrated by the fact that the standard deviation from the sample of these results was always zero. Thus, the GA always converged to the same fitness value. However, the standard deviation was not zero when the full truth table was used. It would seem that reducing the sample set to a subset of the entire truth table either makes the search space too discrete (by the removal of vital information), or that the chromosomes carry little memory about past generation performance because the basis for optimisation is apparently changing continually.

These results suggest that there is going to be very little hope that circuit designs can be directly evolved in this way by using sample subsets of the entire truth table, and that one must be prepared to accept that evolution needs perfect information regarding the circuit to be implemented. Indeed, it would seem that the greater the loss of information regarding the circuit to be evolved, the greater difficulty the GA has in performing well.

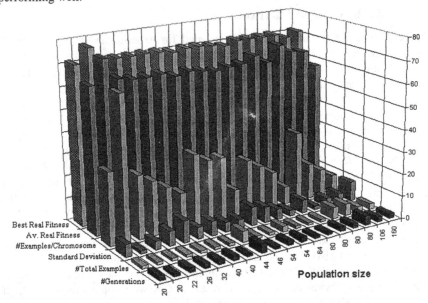

Fig. 4. Results for population based random examples.

In fact, it makes very little difference to the GA performance whether scenario (a) is used, where the same set of random examples is used for each population member, or scenario (b) is used, where a new random set of examples is generated for each chromosome within the population. Even increasing the total number of examples examined by the entire GA run does not produce a significant increase in performance (Figure 4, results 8, 15 and 16, population sizes 44 and 80). What seems to be much more important is the fact that the set examples being used has been reduced in the first place. The truly outstanding result in Figure 4 is result 2, with population size 22, where all examples from the entire truth table were used.

4.2 Choosing Cell Layout and Geometry

One of the potential problems with attempting to evolve circuits is the difficulty in routing the designs within the constraints of the implementation platform. This is because devices do not possess an infinite routing resource with which to inter-connect its functional blocks. In our earlier representation we attached no cost to routing whatsoever and this made it easier to evolve functionally correct designs. When we carried out 10 runs of the GA using a 4x4 geometry of cells we obtained an average fitness of 98.12 (including numerous 100% solutions). However using our new chromosome representation we obtained an average of 94.06 with no 100% solutions. We decided to investigate this by initialising the circuit cell structure with increased routing resources, and further allowing initialised functional blocks to mutate into routing cells (the differentiated model from above). This could be compared with a scheme in which the Genetic Algorithm decides for itself how to employ cells, either as functional cells, routing cells, or indeed, a mixture of both (undifferentiated model). Thus we carried out ten runs of the GA (with 10,000 generations and population size 50) with geometries ranging from 9 rows by 2 columns to 2 rows by 9 columns (see Figure 3). It should be emphasised that when the number of rows and columns is fixed, this represents the *maximum* number of cells which may be used. In reality, the circuits may use any number of cells up to and including this maximum, and so the evolutionary process is in effect choosing the size. We also looked at the effect of using elitism. The results for the undifferentiated model are shown in Figures 5(a) - 5(d) while those for the differentiated model are seen in Figures 6(a) - 6(d). The first immediate conclusion which can be drawn is that the use of elitism has a very considerable favourable impact on the quality of the solutions. Without elitism no 100% functional solutions were obtained with the experimental parameters. Indeed for some geometries 10 runs of 20,000 generations were carried out again no 100% functional solutions were obtained without elitism. The reason why in Figures 5(d) and 6(d) the graphs show "Number of 100% solutions" was because when elitism was used a great many cases had a best solution (of the ten runs) which was 100% or extremely close. When Figures 5 and 6 are compared with one another it appears to show that an undifferentiated cell structure is more effective than the differentiated pattern. However, one should be cautious about this conclusion because the number of functional cells which are available in a differentiated structure is approximately half that of the undifferentiated pattern. We know from our previous results that seven functional cells are the likely to be the theoretical minimum required to build the 2-bit multiplier, and many of the differentiated patterns do not possess a realistic number of actual functional cells to be able to find a 100% correct solution (i.e. the functional resources allocated are close to the theoretical minimum). Formerly, in our original chromosome representation, we were able to evolve many more 100% correct solutions because there was no effective cost incurred for routing. In other words, only functional cells were considered as bearing any actual cost. However, in this model, which is much closer to the actual Xilinx chip, cells have to be allocated to routing tasks and so represent a real resource cost. The more cells one gives over to pure routing functions - thereby reducing functionality, as in the differentiated cell approach - the more closely one operates to the resource margin. Therefore, a direct comparison of the undifferentiated and differentiated cell pattern is perhaps not entirely appropriate.

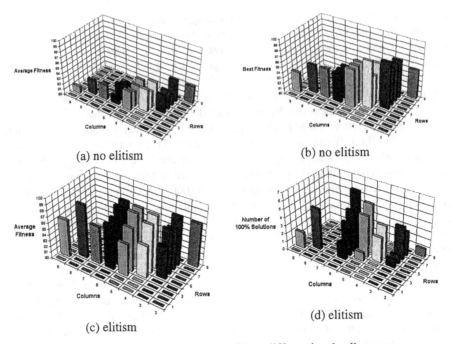

(a) no elitism

(b) no elitism

(c) elitism

(d) elitism

Fig. 5. Experimental results of runs with undifferentiated cell structure

Another feature which shows up the strong dependence of evolutionary success on functional and routing resources is depicted in Figure 6(d) where the number of 100% correct designs increases dramatically as resources are increased (by increasing cell geometry - with 6 x 6 as the most effective). What needs to be established is whether the differentiated cell pattern can become more effective than the undifferentiated approach when equal numbers of functional resource are allocated.

These results suggest that there needs to be a lot more work done to establish the relative importance of routing and functionality in attempting to evolve circuits. It is very tempting to concentrate purely on functionality when designing the GA for circuit evolution, and ignoring the inter-connectivity of cells and the available routing resources on the target device. There also needs to be a more open view of available architectures for evolution. It may be that the cell structure and connectivity arrangements currently employed by Xilinx are not best suited to the direct evolution of circuits, and that the entire process would benefit greatly from a more sympathetic internal architecture.

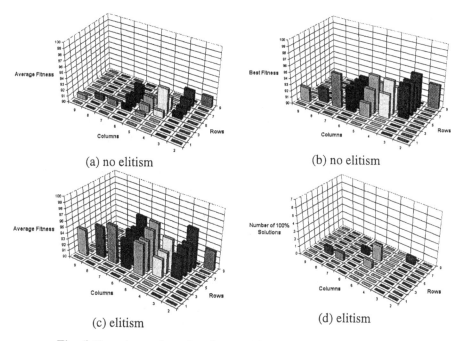

(a) no elitism

(b) no elitism

(c) elitism

(d) elitism

Fig. 6. Experimental results of runs with differentiated cell structure

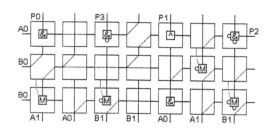

Fig. 7. Cell layout for fully evolved 2-bit multiplier

Figure 7 shows one of the 100% functionally correct evolved circuits. Each cell closely mimics the actual cells used on the Xilinx FPGA parts. As in our chromosome representation, the North and East (North and East going) connections to each cell are inputs, whereas the South and West going connections are the outputs. This means, therefore, that the cell at location 2,0 on the grid is a multiplexer (designated by the M), with one input B1 and the other A0 (both primary inputs on the truth table) routed from neighbouring cell (1,0). The control line for the multiplexer is passed down from the cell layer above and is actually the output of the multiplexer located at (0,0). The smaller square inside the cells indicates the cell's use as a functional logic gate, and the small circles on inputs and outputs of these denote logical inversion. The actual chromosome from which this circuit is generated is shown below:

-1 0 -3 0 3 0 -4 0 8 0 4 0 -3 0 3 0 4 0 6 0 8 (Functional)
2 2 1 0 2 0 1 1 2 2 1 0 2 0 0 0 0 1 1 1 0 1 1 0 2 0 1 0 2 2 1 1 2 2 0 0 2 2 1 0 2 2 (Routing)
0 1 2 2 1 0 2 3 3 1 (Inputs)
2 9 4 0 (Outputs)

Fig. 8. The chromosome which produced the circuit of Figure 9

Note that the zero entries in the functional part of the chromosome denote the absence of cell functionality. In other words, the routing part of the chromosome will never refer to these genes because these cells are only ever used as routing paths.

5. Conclusions

This paper has looked at a number of the issues that face researchers who are using evolutionary methods to create digital circuits directly. We are attempting to do this on a structure which is very closely modelled on that used by Xilinx on their 6216 FPGA part.

Experiments were conducted with regard to the evolution of a 2-bit multiplier circuit design on this new representation, and results gathered which examined: (a) the possibility of using truth table sample sets (as opposed to the entire truth table), and (b) the issue of providing the circuit representations with greater routing facilities (by differentiating cells as either functional or routers).

The first conclusion we may draw from our results is that it is extremely difficult to evolve functionally correct designs - even for a circuit as simple the 2-bit multiplier - using samples which are subsets of the truth table. In other words, it seems that the GA cannot use incomplete training data to derive the operational circuit, but requires the full truth table as a basis for fitness evaluation. There appears to be too great a loss of information concerning the circuit to be evolved for this method to succeed. This is clearly unfortunate as it means that the GA must be able to examine the entire truth table for a particular function - and this may be very large if the number of input variables is large. It is worth noting that the performance of the GA was not markedly different for the various runs involving different numbers of examples (where this number was less that the total number available in the full truth-table). This is surprising, but it may be due to the fact that without a very large percentage of the truth-table to work with, the GA is not much different from a random search. Clearly, this is a matter for further investigation.

We saw earlier that the number of cells used by the chromosome is variable but is bounded by the maximum number of cells available within a fixed geometry. When one considers allowing the maximum number of cells to be evolved, one would need to take into account the fact that initially smaller numbers of cells tend to produce higher fitnesses, although they eventually can be insufficient to realise the desired function. We feel that one needs to get the GA to reliably produce 100% functional solutions before one attempts to evolve the geometry.

Another conclusion is that using a GA which employs elitism is a considerable advantage in finding designs - using this chromosome representation - which are 100% functionally correct. Without elitism the GA struggled to find any fully correct solutions for what is essentially a very simple circuit, but with elitism the results were markedly improved. Further, there appears to be a strong dependency upon the cell resources given to the GA and its ability to deliver fully working solutions. We are intend to carry out further experiments in which we allocate equal functional resources to both differentiated and undifferentiated cell structures to ascertain if one technique is indeed more effective. Our original, more successful, chromosome representation employed effectively resource free routing, and we feel that an cell pattern or architecture which devotes more to routing must provide greater opportunity to evolve circuits which are functionally correct. It is possible that we could improve the performance of the GA by using the long-line routing that is incorporated into the Xilinx 6216 architecture.

References

[A] *Lecture Notes in Computer Science - Towards Evolvable Hardware*, Vol. 1062, Springer-Verlag, 1996.

[B] Higuchi T., Iwata M., and Liu W., (Editors), *Proceedings of The First International Conference on Evolvable Systems: From Biology to Hardware (ICES96), Lecture Notes in Computer Science*, Vol. 1259, Springer-Verlag, Heidelberg, 1997.

1. Fogarty T. C., Miller J. F., and Thomson P., "Evolving Digital Logic Circuits on Xilinx 6000 Family FPGAs" in Soft Computing in Engineering Design and Manufacturing, P.K. Chawdhry,R. Roy and R.K.Pant (eds),Springer-Verlag, London, pages 299-305,1998.

2. Goeke M., Sipper M., Mange D., Stauffer A., Sanchez E., and Tomassini M., "Online Autonomous Evolware", in [B], pp. 96 -106

3. Higuchi T., Iwata M., Kajitani I., Iba H., Hirao Y., Furuya T., and Manderick B., "Evolvable Hardware and Its Applications to Pattern Recognition and Fault-Tolerant Systems", in [A], pp. 118-135.

4. Hemmi H., Mizoguchi J., and Shimonara K., " Development and Evolution of Hardware Behaviours", in [A], pp. 250 - 265.

5. Iba H., Iwata M., and Higuchi T., Machine Learning Approach to Gate-Level Evolvable Hardware, in [B], pp. 327 - 343

6. Kitano H., "Morphogenesis of Evolvable Systems", in [A], pp. 99-107.

7. Koza J. R., *Genetic Programming*, The MIT Press, Cambridge, Mass., 1992.

8. Koza J. R., Andre D., Bennett III F. H., and Keane M. A., "Design of a High-Gain Operational Amplifier and Other Circuits by Means of Genetic Programming", in Evolutionary Programming VI, Lecture Notes in Computer Science, Vol. 1213, pp. 125 - 135, Springer-Verlag 1997.

9. Miller J. F., Thomson P., and Fogarty T. C., "Designing Electronic Circuits Using Evolutionary Algorithms. Arithmetic Circuits: A Case Study", in Genetic Algorithms and Evolution Strategies in Engineering and Computer Science: D. Quagliarella, J. Periaux, C. Poloni and G. Winter (eds), Wiley, 1997.

10. Sipper M., Sanchez E., Mange D., Tomassini M., Perez-Uribe A., and Stauffer A., "A Phylogenetic, Ontogenetic, and Epigenetic View of Bio-Inspired Hardware Systems", *IEEE Transactions on Evolutionary Computation*, Vol. 1, No 1., pp. 83-97.

11. Thompson A., "An evolved circuit, intrinsic in silicon, entwined with physics", in [B], pp. 390 - 405.

12. Zebulum R. S., Pacheco M. A., and Vellasco M., "Evolvable Systems in Hardware Design: Taxonomy, Survey and Applications", in [B], pp. 344 - 358.

Evolutionary Design of Hashing Function Circuits Using an FPGA

Ernesto Damiani[1], Valentino Liberali[2], and Andrea G. B. Tettamanzi[1]

[1] Università degli Studi di Milano, Polo Didattico e di Ricerca di Crema
Via Bramante 65, 26013 Crema, Italy
Phone: +39-373-898201; Fax: +39-373-898253
E-mail: edamiani@crema.unimi.it, tettaman@dsi.unimi.it
[2] Università di Pavia, Dipartimento di Elettronica
Via Ferrata 1, 27100 Pavia, Italy
Phone: +39-382-505226; Fax: +39-382-505677
E-mail: valent@ele.unipv.it

Abstract. An evolutionary algorithm is used to evolve a digital circuit which computes a simple hash function mapping a 16 bit address space into an 8 bit one. This circuit, based on FPGAs, is readily applicable to the design of *set-associative* cache memories. Possible use the evolutionary approach presented in the paper for *on-line* tuning of the function during cache operation is also discussed.

1   Introduction and Problem Statement

Integrated digital electronics is traditionally considered the ideal testbed for automated design techniques. Indeed, though to this date there is no known general technique for automatically designing VLSI digital circuits satisfying given specifications, considerable progress has been made to automate the design of some classes of digital systems.

Over the last decades, according to Moore's law, the number of active devices integrated into a single chip approximately doubled every 18 months, while the chip remained about the same size [13]. Such an exponential increase in circuit complexity leads to a demand for new design solutions. A great deal of effort has been devoted by the research community to the development of suitable design automation tools.

Evolutionary algorithms are a broad class of optimization methods inspired by Biology, that build on the key concept of Darwinian evolution [3,5]. After some successful applications of evolutionary algorithms to physical design (partitioning, placement and routing), now the same techniques are being considered also for structural design. This has given rise to an innovative field of research, called *evolvable hardware* [8,20,18]. Recently, evolutionary design automation has been applied to digital electronic design [15]. It has been proposed for analog circuits as well, with encouraging results [12].

The main concepts and issues relevant to evolutionary algorithms can be found, among others, in [17,7,4,14,11,1]; [9,6,16] are more of a historical interest.

The main motivation for the work presented in this paper is the understanding of how an evolutionary algorithm could be applied to the synthesis in hardware of a digital non-linear function. In this paper, circuits are evolved to compute a simple hashing function. A hashing function is a many-to-one mapping of a key set into an address set, having the design property of uniform filling of the co-domain. Such a function can be used, for example, in set-associative cache memories, and in hardware cryptographic systems.

Based on the hardware architecture described in Section 2, an evolutionary algorithm (described in Section 3) is devised to evolve the hashing function. Section 4 presents experimental results, while Section 5 envisages an application of the circuit to the design of set-associative cache memories.

2 Hardware Environment and Behavioural Description

The example presented in this paper is a hashing circuit that maps a 16-bit address space into an 8-bit index space as uniformly as possible.

The hashing function is implemented using an FPGA. Regularity and modularity of FPGAs, as well as their easy-to-use development tools, make them very attractive to implement evolvable digital circuits [15].

The Xilinx SRAM-based reconfigurable FPGAs have been chosen as the reference architecture for this application [21].

2.1 Architecture of FPGA Implementation

We consider the FPGA device XC3020, made of an array of 8×8 configurable logic blocks (CLBs) [10]. Each CLB has one output bit and four input bits. More recent FPGA families (e.g. XC4000) offer more design flexibility, both in the number and in the complexity of blocks. However, the simplicity of the XC3020 makes it the ideal candidate for an evolutionary study. Moreover, later FPGA families are fully compatible with this FPGA device.

Since the hashing function does not require any state information and therefore can be computed through a combinational circuit, the registers of the FPGA are not used. Our design employs only the combinational logic of each CLB, as illustrated in Figure 1.

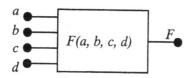

Fig. 1. The elementary cell of the circuit.

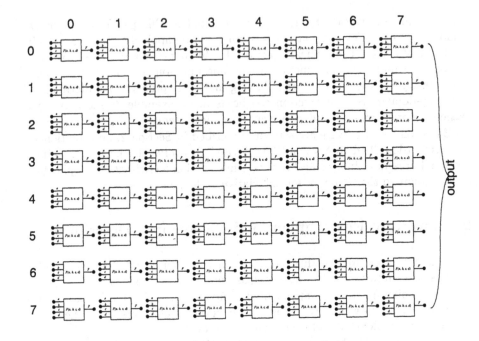

Fig. 2. The reference architecture adopted for the hashing circuits.

The complete combinational circuit is illustrated in Figure 2. For each CLB, each inputs can be any bit of the input key or can be the output of a previous cell located in the same row. In this way, each row processes the same input data independently. The signal flow is from left to right, and there is no feedback loop. The 16 inputs are transported on the horizontal long lines and distributed to the cells using both vertical general purpose interconnects and long lines. Connections between cells within the same row are carried out by horizontal general purpose interconnects. These conventions do not guarantee that all individuals generated by the algorithm will be routable using the FPGA hardware resources; however, in practice the individuals that cannot be routed using this technique are effectively filtered out by assigning zero fitness to them.

The output bits of the 8 rightmost blocks are the output of the circuit, i.e. the index computed by the hashing function.

2.2 Behavioural Description

The modularity of the circuit allows us to specify the behaviour simply by describing each CLB. For each block, we must specify its four inputs, which can come from input lines, from previous cells, or can be grounded (i. e. bound to zero), and how they are combined to give the output.

No global signal or supervision mechanism is used.

3 The Evolutionary Algorithm

In order to make hashing circuits evolve, a very simple generational evolutionary algorithm using fitness-proportionate selection with elitism and linear scaling of the fitness has been implemented, whose details are given in the following subsections.

3.1 Encoding

An encoding was adopted that satisfies the constraints described in Section 2 by construction. Each cell is encoded through five 16-bit unsigned integers; the first four of them define the four inputs to the cell, which can be either input lines of the circuit or outputs of previous cells in the same row. Even though five bits per input would be enough for coding 24 possible inputs, the choice of using a full 16-bit word is dictated by performance considerations, all the more so because the genotype already fits in 5 kbit of memory, which is negligible for state-of-the-art computers. The fifth integer encodes the truth table of the combinational network the cell implements.

Each of the integers encoding for the cell inputs is decoded by taking the rest of the division of its value by the number of legal inputs to the cell, i.e. the sixteen inputs lines of the circuit, the output lines of the previous cells in the same row and the ground line. This gives 17 possibilities for the first cell up to 24 for the last cell in a row.

The null (ground) input allows the algorithm to evolve the number of inputs to each cell, as well as its logical function, while maintaining a fixed-length encoding.

As an example, the cell in Figure 3 could be encoded as follows:

$$(400, 821, 37, 399, 54990). \tag{1}$$

The first four numbers encode the four inputs. Being the cell in column 3, its inputs can come from 20 possible sources, namely, one of the 16 inputs, outputs of cells $(0,0)$, $(0, 1)$, $(0,2)$, and the ground. Thus, 400 (mod 20) = 0, indicating i_0 as the first input; 821 (mod 20) = 1, yielding i_1, and so on. The last number, $54990 = (1101011011001110)_2$, is a straightforward encoding of the truth table defining $F(0,3)$. The reader should be aware that in general there are many alternative ways of encoding for the same cell. For instance, in this case there are 256 distinct ways of encoding just the same logical function, since one of the four inputs has been grounded.

Overall, a circuit is represented by a string of 5,120 bits, interpreted as a vector of $64 \times 5 = 320$ 16-bit integers. This is the genotype corresponding to a candidate solution.

3.2 Mutation

Whenever an individual is replicated to be inserted in the new generation, each bit of its genotype is flipped with the same probability p_{mut} and independently of the others.

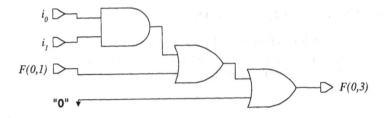

Fig. 3. Equivalent circuit diagram of a sample cell, imagined to be located in $(0, 3)$, as decoded from Expression 1.

There are some considerations to be done with respect to the encoding of cells:

1. for the truth table part, this mutation induces a small perturbation that does not disrupt the semantics of the genotype in that flipping a bit in this part amounts to modifying a single entry of the truth table; the resulting function is identical to the former one except for the value associated to a single input;

2. for the input selection part, flipping one or more bits amounts to replacing the old input with a new one at random, at least to a first approximation. In fact, it could be shown that codes corresponding to least significant input lines have a slightly higher probability of being generated. However, the bias introduced is negligible;

3. thanks to the design of the encoding, the offspring resulting from mutation are always legal; at most, they will not be routable (and therefore unfit for reproduction).

The mutation rate has to be chosen carefully, since it can dramatically affect the overall algorithm performance. Indeed, if the mutation rate is too low, the evolution is exceedingly slow; if it is too high, the algorithm tends to behave as a random search.

We have experimentally verified that the algorithm performance is good when the ratio $r = f^*/\bar{f}$, where f^* is the fitness of the best individual and $\bar{f}$ is the average fitness of the population, is close to $R = \frac{3}{2}$.

The mutation rate p_{mut} is dynamically adjusted in order to maintain the ratio r as close as possible to its optimal value R. The updating rule is:

$$p_{\text{mut}} \leftarrow \frac{R}{r} p_{\text{mut}}. \tag{2}$$

If the distance between the best and the average circuit increases, p_{mut} is accordingly lowered, while if the average circuit is catching up with the best, p_{mut} is increased to allow for more diversity in the population.

3.3 Recombination

For recombination, each 16-bit unsigned integer is treated as an atomic unit of the genotype.

Each such unit is inherited from either parent with $\frac{1}{2}$ probability (uniform crossover). This ensures that recombination will preserve the CLBs, which are the basic building blocks of a circuit.

3.4 Objective Function and Fitness

A "good" hashing function should map a set of M input keys into a set of indices $i = 1, \ldots, N$ in the most uniform way as possible. In order to assess the quality of the generated circuits, a set of keys is randomly generated over the set $\{1, \ldots, 2^{16}\}$ of possible key values. These keys are offered as inputs to each circuit and the number of hits h_i for each index i is calculated. The fitness of a circuit is then given by the following formula:

$$f = \frac{1}{1 + \frac{1}{N} \sum_{i=1}^{N} (h_i - \bar{h})^2},\tag{3}$$

where $\bar{h} = M/N$ is the expected number of hits per index in the ideal case of uniform key distribution.

An alternative measure of fitness could be based on the entropy of the key distribution generated by the circuit, i.e.

$$f = MH = \sum_{i=1}^{N} h_i \ln \frac{M}{h_i} = M \ln M - \sum_{i=1}^{N} h_i \ln h_i.\tag{4}$$

However, it was experimentally verified that such a fitness reduces the algorithm performance, for it does not sufficiently reward the best circuits, and the additional reward for a better circuit greatly diminishes as a high-entropy solution is approached.

In order to evaluate the results, it is interesting to study the probabilistic behaviour of a "random" hashing function. By "random" we mean that the probability that a random key is mapped to any index is independent of the index and uniform over the N indices. The number of hits for all indices $i = 1, \ldots, N$, would have binomial distribution with probability

$$\Pr[h_i = h] = \binom{M}{h} \frac{(N-1)^{M-h}}{N^M}.\tag{5}$$

Therefore, the mean is $E[h_i] = \bar{h} = M/N$ and the variance is $\mathrm{var}[h_i] = \frac{M(N-1)}{N^2}$.

4 Experimental Results

Experiments have been carried out with $N = 256$ and $M = 4,096$ out of the 65,536 possible keys. This gives an average number of hits per index $\bar{h} = 16$, regardless of the hashing function employed.

A population size of 100 was used, with crossover rate $p_{cross} = \frac{1}{2}$.

According to (3), the standard deviation σ of hits per index for a hashing circuit having fitness f is

$$\sigma = \sqrt{\frac{1}{f} - 1}. \tag{6}$$

For the random hashing function, the distribution of hits per index given in (5) would have mean $\bar{h} = 16$ and variance 15.9375, or standard deviation $\sigma_{rnd} = 3.9922$. This can serve as a term for comparison to assess the effectiveness of the evolved solutions.

Figure 4 shows a typical circuit synthesized by the algorithm described in Section 3. It has been obtained after 257 generations, and it generates a key distribution with a standard deviation of $\sigma = 3.0375$ hits per index, which is considerably better than the σ_{rnd} of the random hashing function. Repetition of the experiments always leads to similar results. This means that the evolved circuit is actually fitted to the actual set of keys used for evaluating fitness.

Fig. 4. The best circuit at Generation 257, having fitness 0.097785.

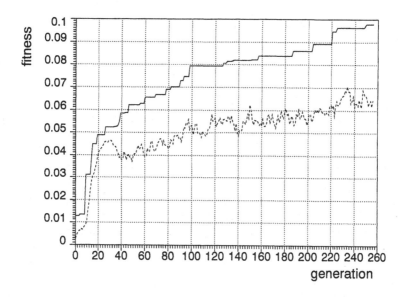

Fig. 5. Average and best fitness during a sample run of the evolutionary algorithm.

It has been observed that the evolution proceeds in three successive phases:

1. in the first few generations, the best combinational functions are selected for blocks;
2. in the second phase, the topology of circuits evolves in such a way that each input bit has a path through the output;
3. in the last phase, the evolution process is slower, and tends to increase the connectivity within circuit blocks.

As shown in Figure 5, the first phase is characterised by a steep increase of fitness and by a high rate of change in the population. In subsequent phases, one can observe the typical punctuated equilibria where abrupt changes are interleaved with long periods of stasis.

5 Applications and Future Work

This Section outlines a possible application for the evolutionary technique described in this paper: the adaptive design of cache memories. Commonly available cache memories are either *fully* or *set-associative* [19]. A fully associative cache memory comprises several *block frames* holding copies of main memory blocks. The cache also contains a table, holding the address of the main memory block associated to each block frame.

If a linear 32-bit address space is used for main memory and a block size of 64 kbyte is adopted, the table holds a set of 16-bit block addresses or *tags*, each coupled with the corresponding block frame identifier. Fully associative memories operate as follows: before the CPU performs an access to main memory, a fast combinational network simultaneously compares the 16 most significant bits of the desired address with all the block tags stored in the table. If a matching is found, the cache block frame is accessed instead of the main memory block. Otherwise, main memory is accessed and the missing block is copied in cache. The capacity of fully associative cache memories is severely limited by the topological complexity of the combinational network used to perform simultaneous comparisons. For this reason, large cache memories used in current microcomputer architectures are usually set-associative. A typical set-associative cache maps 16-bit block identifiers to a smaller set of 256 tags. This is usually done by performing associative comparison on the most significant 8 bits of the block frame identifier to select a set of blocks. The desired block is then located via linear search, usually without excessive performance degradation [2].

The circuit described in previous sections can be readily used to achieve uniform mapping of block identifiers to tags, as an alternative to associative comparison. In this case, each identifier can be regarded as a hash key identifying a set of 256 cache blocks which must be searched in order to locate the desired one. However, it is interesting to remark that if actual memory accesses are not evenly distributed across blocks, as it is often the case, uniform mapping of blocks into tags is not the best choice for set-associative cache memories. Indeed, the cost of linear search would be decreased by letting seldom consulted blocks to share a tag and countersigning frequently accessed blocks with a unique tag. The "best" mapping between tags and block sets should be dynamically searched while the circuit operates, through a second evolutionary phase of *on-line* optimisation. We intend to explore this issue in a future paper.

6 Discussion

This paper has described the evolutionary design of a digital circuit to compute a non-linear mapping suitable for hashing and associative memories. The feasibility of the approach has been demonstrated. Further research is in progress, in order to obtain a full implementation in hardware of the described algorithm, much in the spirit of the evolvable hardware approach.

The circuit layout proposed above has some advantages, above all, a regular structure allowing a straightforward encoding. However, it is neither the only possible, nor perhaps the best. As a matter of fact, it could be interesting to allow the connections between cells in a column-by-column fashion, instead of a within-row fashion. Furthermore, since the solutions obtained in Section 4 correspond to small depth functions, smaller portions of the 8×8 array could be used to implement the solution. By the way, this approach would lead to an overall layout reminiscent of multi-layer perceptrons.

Another interesting possibility (which has not been yet taken advantage of) is to use an actual FPGA for fitness evaluation. Given that each fitness evaluation implies running M software simulations of the same circuit, it is evident that hardware fitness evaluation becomes more and more attractive as M increases.

Acknowledgments

This work was partially supported by M.U.R.S.T. 60% funds.

The authors would like to thank Prof. Gianni Degli Antoni for his support and criticism, and are also grateful to the anonymous referees for their valuable suggestions and observations.

References

1. T. Bäck. *Evolutionary algorithms in theory and practice.* Oxford University Press, Oxford, UK, 1996.
2. W. Burkardt. Locality aspects and cache memory utility in microcomputers, *Euromicro Journal,* vol. 26, 1989.
3. C. Darwin. *On the Origin of Species by Means of Natural Selection.* John Murray, 1859.
4. L. Davis. *Handbook of Genetic Algorithms.* Van Nostrand Reinhold, New York, NY, 1991.
5. R. Dawkins. *The blind Watchmaker.* Norton, 1987.
6. L. J. Fogel, A. J. Owens, and M. J. Walsh. *Artificial Intelligence through Simulated Evolution.* John Wiley & Sons, New York, NY, 1966.
7. D. E. Goldberg. *Genetic Algorithms in Search, Optimization & Machine Learning.* Addison-Wesley, Reading, MA, 1989.
8. T. Higuchi, T. Niwa, T. Tanaka, H. Iba, H. De Garis, T. Furuya. Evolving hardware with genetic learning: a first step towards building a Darwin machine, in *Proc. of the 2nd Int. Conf. on the Simulation of Adaptive Behavior (SAB92).* The MIT Press, Cambridge, MA, 1993.
9. J. H. Holland. *Adaptation in Natural and Artificial Systems.* The University of Michigan Press, Ann Arbor, MI, 1975.
10. J. H. Jenkins. *Designing with FPGAs and CPLDs.* Prentice-Hall, Englewoou Cliffs, NJ, 1993.
11. J. R. Koza. *Genetic Programming: on the programming of computers by means of natural selection.* The MIT Press, Cambridge, MA, 1993.
12. J. Koza, F. Bennett, D. Andre, M. Keane. Evolution using genetic programming of a low distortion 96 decibel operational amplifier, in *Proc. ACM Symp. on Applied Computing (SAC '97),* San José, CA, Feb.-March 1997.
13. M. S. Malone. *The Microprocessor: A Biography.* TELOS (The Electronic Library of Science, Springer-Verlag), Santa Clara, CA, 1995.
14. Z. Michalewicz. *Genetic Algorithms + Data Structures = Evolution Programs.* Springer-Verlag, Berlin, 1992.
15. J. F. Miller, P. Thomson, T. Fogarty. Designing electronic circuits using evolutionary algorithms. Arithmetic circuits: a case study, in D. Quagliarella, J. Périaux, C. Poloni, G. Winter (Eds.), *Genetic Algorithms and Evolution Strategies in Engineering and Computer Science.* Wiley, Chichester, UK, 1998.
16. I. Rechenberg. *Evolutionsstrategie: Optimierung technischer Systeme nach Prinzipien der biologischen Evolution.* Fromman-Holzboog Verlag, Stuttgart, 1973.

17. H.-P. Schwefel. *Numerical optimization of computer models*. Wiley, Chichester, UK, 1981.
18. M. Sipper, E. Sanchez, D. Mange, M. Tomassini, A. Pérez-Uribe, A. Stauffer. A phylogenetic, ontogenetic, and epigenetic view of bio-inspired hardware systems, *IEEE Trans. Evolutionary Computation*, vol. 1, 1997.
19. A. Smith. Cache memory design: an art evolves, *IEEE Spectrum*, vol. 24, 1987.
20. A. Thompson. Silicon evolution, in *Proc. of Genetic Programming 1996*, Palo Alto, CA, 1996.
21. Xilinx, Inc. *The Programmable Logic Data Book*. San José, CA, 1996.

A New Research Tool for Intrinsic Hardware Evolution

Paul Layzell

Centre for Computational Neuroscience and Robotics,
School of Cognitive and Computing Sciences,
University of Sussex, Brighton BN1 9QH, UK.
paulla@cogs.susx.ac.uk

Abstract. The study of intrinsic hardware evolution relies heavily on commercial FPGA devices which can be configured in real time to produce physical electronic circuits. Use of these devices presents certain drawbacks to the researcher desirous of studying fundamental principles underlying hardware evolution, since he has no control over the architecture or type of basic configurable element. Furthermore, analysis of evolved circuits is difficult as only external pins of FPGAs are accessible to test equipment. After discussing current issues arising in intrinsic hardware evolution, this paper presents a new test platform designed specifically to tackle them, together with experimental results exemplifying its use. The results include the first circuits to be evolved intrinsically at the transistor level.

1 Introduction

In recent years, evolutionary algorithms (EAs) have been applied to the design of electronic circuitry with significant results being attained using both computer simulations and physical hardware. The possibility of the latter is largely due to the advent of re-configurable Field Programmable Gate Array (FPGA) chips consisting of many small circuit elements which can be configured almost instantaneously to produce a huge variety of different circuits. Any particular configuration can be specified by a genotype, allowing FPGA circuits to be evolved with basic EAs such as genetic algorithms [8]. It is tempting to think therefore, that FPGAs are the *ideal* medium for research of intrinsic[1] hardware evolution (IHE), however, there are several reasons why this may not be the case. Firstly, there is very little choice at present over the type of basic configurable element employed by commercial devices, the vast majority using digital configurable logic blocks (CLBs). This can be particularly restrictive if analogue functions are required. While evolution is capable of producing analogue behaviours from digital CLBs, such circuits may suffer from a number of undesirable characteristics such as dependence on temperature and lack of portability [9]. Analogue FPGAs using operational amplifiers as basic elements are

---

[1] The term 'intrinsic' is used to indicate that fitness evaluations are performed on the physical circuits, rather than in simulation [5]

starting to become available (for example the Motorola/Pilkington FPAA [3]), but we do not yet know exactly what the most appropriate basic element for IHE might be. Depending on the task it could be extremely basic, such as a transistor; some higher level multi-functional unit; or combinations of different components including passive resistors or capacitors. Secondly, the interconnection highways between circuit elements in FPGAs are designed with the conventional, modular circuit design methodology in mind. Evolution may be able to exploit a different system of interconnections or architecture - perhaps a more arbitrary one - much more effectively, but once again not enough is known at this stage to specify exactly what that system might be. Finally many circuits which have evolved on an FPGA to produce some desired behaviour are extremely difficult to analyse, since it is not possible to access individual circuit elements with test equipment, and computer simulations cannot practically reproduce all of the physical properties of silicon that may be being exploited. This paper presents a new configurable testbed which allows the user to investigate IHE with any type of basic element, and a variety of interconnection architectures. Each element can be accessed with test equipment at the component level, facilitating analysis of evolved circuits. The motivation behind this research is to improve evolutionary electronics as an engineering technique, and in particular provide a tool to aid exploration of fundamental principles underlying IHE. After briefly exploring some fundamentals and benefits of the field (Section 2), difficulties of current approaches are examined (Section 3). Next, the testbed's operation and its place within hardware evolution is described (Section 4). Finally, experimental results show how it can be employed to aid further research into the principal difficulties (Sections 5 and 6).

2   Motivation: Hardware Design Using Evolutionary Techniques

IHE design typically comprises a computer running an EA, and an FPGA or other configurable device on which individual genotypes are instantiated as physical electronic circuits. Fitness evaluations can be the measurement of the evolved circuit's performance, behaving in real time according to the laws of semiconductor physics. At no stage in this process are we concerned with the topology or other details of the circuit within the FPGA, only its behaviour. This means that evolution is free to explore all possible configurations and hence behaviours in the repertoire of the FPGA. There is no need to constrain evolution to work only within the relatively small subset of circuits that can be designed, analysed, or simulated by human designers. Hence, when allowed to exploit richer topologies, dynamical behaviours and complex aspects of semiconductor physics, evolution can produce circuits which are beyond the scope of conventional design methodologies. This offers significant benefits to electronic engineering. For example, by evolving a circuit on a small corner of a Xilinx XC6216 to discriminate between two square wave inputs of 1kHz and 10kHz, Thompson was able to show that evolution is capable of producing functionality in the absence of components that would be deemed necessary under conventional design maxims, as the circuit had no access to the counter/timers or RC networks normally associated with this task [9]. This has significant implications for VLSI where components such as capacitors, resistors and inductors are difficult to implement in small size. In another experiment, a robot controller was evolved to keep moving in the centre of an arena. The robot was equipped with two ultrasonic transducers and propelled by two wheels powered by d.c. motors, however the hardware normally associated with these facilities (reflection timers and p.w.m. controllers) were denied. Using 'genetic latches' evolution was able to choose which part of the circuit, if any, was synchronised to a clock, and the resulting circuit using a mixture of synchronous/asynchronous control was able to achieve the required behaviour using only 32-bits of RAM and a few flip-flops [10]. Hence complex behaviours can be achieved with smaller component count than would normally be expected of the conventional approach. The relative insensitivity of certain types of EA to mutation can be exploited to produce systems which are inherently robust to noise and certain types of fault [11]. This is of particular significance to areas such as the space industry, where manufacture of 'rad-hard' devices has decreased in recent years and alternative methods of radiation tolerance are sought [4]. As research progresses, other potential benefits are likely to be realised. Circuits produced using evolution may reveal electronic properties not yet exploited by conventional designers. Analysis of such circuits may lead to such properties being used to create new and innovative designs in the conventional manner. Also, the human requirement to the evolutionary approach is to produce a behavioural specification, which can be achieved with little or no knowledge of electronics, thereby opening up the field of hardware design to non-specialists. For detailed discussions of target domains and current research in IHE, the reader is referred to [7],[14].

3   Difficulties and Issues Arising in the Evolutionary Approach

Along with the benefits of unconstrained IHE come a number of difficulties and uncertain issues whose resolution will be an important factor in its progression as an engineering tool. Principle issues are portability, temperature-dependence, scalability, and analysis. In conventionally designed FPGA circuits, digital functions are normally co-ordinated by a global clock so that factors such as unequal gate propagation times do not affect operation. Where no clock is present, the digital elements should be treated as high-gain analogue devices whose gain, slew rate, and other parameters cannot be considered uniform. Exploiting the analogue aspects of a digital device increases the scope of possible behaviours attainable with it, but also affects the *portability* of the circuit - a circuit evolved on one FPGA may fail to work when instantiated on another of the same batch. Furthermore, many analogue parameters are temperature dependent, so that even on the chip upon which it was evolved, a circuit's operational temperature range may be small compared to that expected of commercial devices. Both of these effects have already been observed on evolved circuits, and a number of proposals exist to tackle them [12]. A commonly levelled criticism of much evolutionary research is that success can only be achieved with small genotypes and thus, in the case of hardware evolution, small circuits. Large genotypes correspond to large search spaces, and it is often perceived that as search space increases in size, more computational time is required to evolve useful behaviours. In the case of IHE, genotype length can be effectively shortened by adjusting basic element functionality, changing the genotype-phenotype mapping in software, and/or adjusting the physical architecture of interconnections between basic elements. In both cases the goal is to arrive at a search space rich in useful behaviours, and sparse in trivial, non-useful ones. It should be noted however, that smaller search spaces do not always lead to less computational expense. Recent research in *neutral networks* - pathways through search space along which a population can travel with single mutations, but no change in fitness - suggests that good solutions can be found faster in large search spaces with highly percolating neutral networks than in smaller ones without them [1]. Research in this area applied to IHE is currently being carried out with the testbed and will be the subject of a future publication. Furthermore, the scalability issue is dependent not only on genotype length, but also on defining the evaluation procedure and fitness function. Fortunately this aspect applies to the EA field as a whole, and a good deal of research has already been carried out, for example [2],[6],[13]. Evolved circuits often comprise numerous feedback paths making their analysis very difficult. One viewpoint is that analysis should be confined to the circuit's external behaviour, whilst acknowledging the internal circuit as a 'black box'. Determining exactly how for example the tone discriminator works may well be impossible, due to difficulty of measurement and the inability of current software models to simulate it. Time spent on analysis of such circuits may well be time wasted if the results are no more than speculative. On the other hand, much could be gained by electronic engineering in knowing exactly what properties are being exploited by evolution, and at what scale. Are intrinsically evolved FPGA circuits capable of exploiting the medium down to the molecular level [12], as has been found in biological systems, or would this result in ludicrous dependency on for example temperature? The question is also of importance to those trying to establish fundamental differences between silicon and organic material as evolutionary media.

4 The Evolvable Motherboard - A Testbed for IHE

The Evolvable Motherboard was conceived to address the issues presented in the previous section, and to help build a framework for an FPGA ideally suited to IHE. Fig. 4.1 is a simplified representation. It is essentially a diagonal matrix of analogue switches, connected to up to 6 plug-in daughterboards, which contain the desired basic elements for evolution. The diagram shows transistors and operational amplifiers, but any component type from capacitors to logic functions may be exploited for experimentation. Each daughterboard takes up to eight lines on the switch matrix, plus a further eight connections to allow for various power lines and I/O, which may be required by certain components on them. The matrix is designed to provide the minimum number of switches necessary so that every combination of interconnection between basic elements can be configured. In total there are approximately 1500 switches, giving a search space of 10^{420} possible circuits. Most of these circuits will be useless, since with this configuration of switches, there are many combinations which result in every wire effectively connected to every other wire. However the software interface together with the switch arrangement allows many different interconnection architectures (for example the local and global bus system used commonly by FPGAs) to be investigated. Also, via software, the board can be subdivided (for example into 2 matrices each containing three plug-in boards) to allow for mapping individuals to two sets of components thereby allowing the impact of manufacturing tolerance to be assessed. The motherboard incorporates additional connectors so that several can be daisy-chained together, should the need arise. Programming is achieved via an interface card plugged into a host PC's ISA bus. The switches can therefore be configured by direct writes to the PC's internal I/O ports, meaning that genotypes can be instantiated in hardware in a very short time (<1ms).

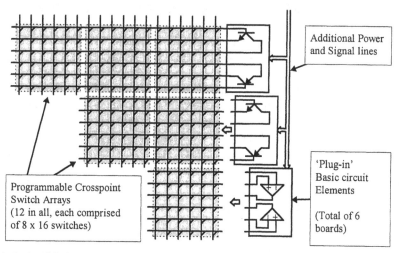

Fig. 4.1. A simplified representation of the Evolvable Motherboard

An important function of the motherboard is to allow any point of the circuit to be accessible for measurement. If evolution exploits the components allowed it in an unexpected way, it should be simpler to deduce which properties are being used than is the case for an FPGA. Note however that the analogue switches are themselves semiconductor devices, contained within integrated circuits. These switches behave like low-value resistors, but they also have other properties which may be exploited making analysis difficult. Should this circumstance arise, a circuit can be configured with various different switches to determine whether subtle properties of particular switches are indeed being exploited. Of the many circuit configurations possible, some of them could lead to the motherboard's self-destruction (for example if the power lines become shorted together). Each switch can only handle a relatively small current (up to 30mA) and if this current is exceeded, the switch will blow-up. To prevent self-destruction, the power lines are routed such that maximum switch current cannot be exceeded as long as the power supply does not exceed 3 Volts. Unfortunately the problem does not end there - certain devices that could be used as basic elements may be destroyed with much smaller currents - however this can be compensated for by incorporating additional resistors in the plug-in daughterboards.

5   Experimentation and Results

To illustrate the capabilities of the motherboard, and to highlight its potential as a tool for investigating important issues current in IHE, three experiments involving the evolution of a NOT gate are presented in this section, using bipolar transistors as the evolutionary building block. Whilst in itself a fairly trivial function, the NOT gate is quite a difficult test because as with many digital functions, evolution can easily tend to a local optimum which does not achieve the required behaviour. The following experiments are also significant in that they are the first ever circuits to be evolved intrinsically at the transistor level. A generational GA was used with single-point crossover, rank-based selection, and elitism. To evaluate the circuits, a series of 100 test inputs containing 50 '1s' and 50 '0s' (Logical Highs and Lows, respectively) was applied sequentially in random order. For each test input, the output voltage was measured five times with random delays between measurements, and summed. Fitness f was scored according to equation 1, where t signifies the test input number, S_L and S_H the set of Low and High test inputs respectively, and v_t ($t = 1,2,..,100$) the summed output voltage in millivolts of the circuit corresponding to test number t. The power supply was set at 2.8V, thus ranges of fitness from -140,000 to 140,000 are expected. Results shown are typical of those obtained from several runs.

$$f = \frac{1}{5}\left\{ \left(\sum_{t \in S_L} v_t \right) - \left(\sum_{t \in S_H} v_t \right) \right\} \tag{1}$$

5.1   A Hand-Seeded NOT Gate

The first experiment involves seeding the initial population with a hand-designed circuit to determine how evolution could improve on an existing prototype. An initial population of 50 individuals was created randomly with genotypes containing 3% '1's, and biased mutation rates averaging .008 (1 → 0) and 0.0015 (0 → 1) mutations per bit. Each genotype bit represents a single switch on the motherboard. A single individual of the initial random population was replaced by the 'poor' NOT gate circuit shown in Fig. 5.2. This circuit conforms to the NOT function in that its output corresponding to a '0' input is of slightly higher voltage than that corresponding to a '1' input, however this difference is too small to be of any practical use. In electronic parlance, its swing is not great enough to cross the digital logic threshold. The experiment was carried out on a small portion of the motherboard, allowing just two transistors. Over 470 generations the elite fitness score increased from 32000 to 97000. Fig. 5.2 *(centre)* shows the best individual. In essence the circuit is the same as the seeded one, but evolution has accomplished increased performance by exploiting the resistance of the programmable switches, effectively increasing the resistance between power and collector, and decreasing it between emitter and ground. To achieve this, evolution has had to exploit a fundamental law concerning resistance: that resistors placed in series *increases* the combined resistance, but placing them in parallel *decreases* it. Referring to Fig. 5.1, the portion of the motherboard used for this experiment allows up to eight paths between external connections and the transistors. The evolved circuit used three for connecting input, output and +2.8V, and dedicated all remaining paths to the 0V connection to achieve parallel resistance. This has important implications concerning interconnection architecture. Evolution may not have been able to make such good use of parallel resistance with a more restrictive interconnection system.

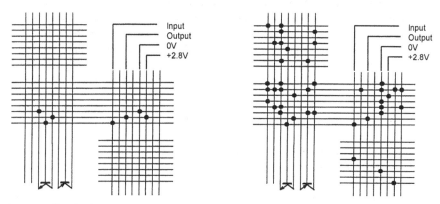

Fig. 5.1. Circuits instantiated on the Evolvable Motherboard. The black dots represent closed switches. *(Left)* Original hand seeded prototype.*(Right)* Fittest individual after 570 generations

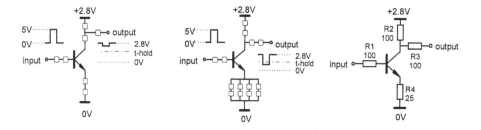

Fig. 5.2. Circuit diagrams. *(Left)* Hand-designed prototype NOT gate. The boxes represent Motherboard switches. *(Centre)* Fittest individual after 570 generations. *(Right)* Circuit representation replacing switches with resistors of equivalent value

5.2 Evolving a NOT Gate From Scratch

This experiment uses the same mutation rates and portion of the motherboard as for the first. Starting from a random population, evolution continued for many generations with no result, but eventually after 2000 generations a circuit of fitness consistently greater than zero arose, after which only a few more generations were required to produce fitness scores of around 133,000 - significantly better than had been achieved with the first experiment. Fig. 5.3 *(left)* shows the circuit. Although near-perfect behaviour was observed on the oscilloscope, it is not at all clear what is going on at first glance. Evolution has chosen *not* to make use of the 0V line, implying that the circuit should not work at all. In fact a path to 0V is achieved via the 10M input impedance of the oscilloscope (Ro on the diagram). TR2 is operating in reverse mode, and for best operation, R7 should be small compared to Ro, which is evidently why evolution has exploited the oscilloscope in this manner. If the scope is unplugged, the circuit ceases to work. The experiment has also been successfully run without using an oscilloscope to monitor it. In either case eventual fitness is always much higher than that achieved for the hand-seeded design. Imposing the human-designed local optimum has constrained evolution to explore within a limited region of the search space.

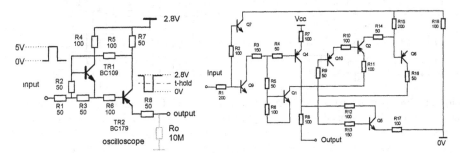

Fig. 5.3. NOT gates evolved from scratch. *(Left)* using small portion of motherboard. *(Right)* using full complement, and alternative connection architecture

5.3 An Alternative Interconnection Architecture

2000 generations may seem excessively many before the first stable prototype solution is found, but even with the small portion of the motherboard used, the search space is very large at 10^{77} possible circuits. However, the vast majority of possible circuits give constant output because they short together the pins of the active components. The final experiment investigates an alternative architecture intended to prevent such shorting, by allowing only a limited number of up to 3 switches per motherboard row to be set. The experiment uses the whole complement of the motherboard, with ten transistors on a 48 x 48 wire matrix. Thus, as well as exploring the new encoding, the experiment will give an indication of whether or not evolution will exploit all of the resources open to it, as ten transistors would normally be considered an excessive amount for a NOT gate. Fig. 5.3 shows the fittest evolved circuit after 600 generations, the first prototype gate having evolved after only 200 generations. Mutation rate was 0.008 mutations per bit. The circuit uses eight out of the ten transistors, each of these when unplugged resulting in circuit failure. Certain aspects of this evolved circuit run counter to intuition, in particular the necessity of so many transistors. The circuit performs no better than the previous one, neither does it exhibit any apparent tolerance to mutations or faults. Indeed one would expect that a circuit using as few transistors as possible would emerge, since mutation should have fewer deleterious effects on such a circuit. The evolved topology may be related to a phenomenon suggested by a follow-up experiment performed with the population of 50 individuals in the generation that produced this circuit. The experiment consisted of removing one of the transistors from the motherboard, and re-evaluating all individuals. No matter which transistor was removed, at least one individual from the population scored higher than 75% of maximum possible fitness. The population is fault/mutation tolerant while the elite individual is not. Thus circuit sub-structures which are only being used as 'wires' in this circuit may have more active roles in other individuals within the population. Research of this effect is currently being pursued, and results will be reported in the near future. If 'populational' fault tolerance is true of evolved circuits in general, rather than being due to the encoding or the search space characteristics for this one task, then there are implications for the portability issue: Evolved circuits may be made to be portable by a manufacturing process which implements a population, rather than a single individual.

6   Conclusion

The experiments of section 5 show that the evolvable motherboard can be used to investigate many important issues arising in current IHE research, including analysis; fault-tolerance; genotype encoding; portability; basic elements, and evolved topologies. The genotype-phenotype mappings of section 5 are two of many that are possible, due to the array of switches allowing any combination of pin connections. Mapping is analogous to the architecture of some configurable device, and analysis of various mappings is likely to lead to a picture of the sort of architecture that an FPGA ideally suited to evolution should have. While section 5 includes suggestions to account for the phenomena observed and their possible implications, it should be noted that they are not, nor are they intended to be, conclusive. The experiments are presented primarily to illustrate the capabilities of the evolvable motherboard as a research tool. Research continues to reveal positive and negative characteristics which seem to be shared by generic classes of evolved circuits. Whatever the benefits they offer, one principal characteristic is that they are very difficult to analyse due to the non-standard ways in which components are exploited and numerous feedback loops. Whilst analysis of very large complex evolved circuits should probably not be attempted at least until convenient methods become available, a methodical approach must be taken if this field is to progress as a credible engineering tool. Thus research on simple, albeit unimpressive circuits continues to play a major role in the development of IHE. Analysis of such circuits is far from impractical, and is likely to contribute to the understanding of the properties that evolution can and cannot exploit, and why.

Acknowledgements: This research was funded by British Telecommunications (BT). Thanks also to Phil Husbands, Inman Harvey, and especially to Adrian Thompson for his comments on numerous drafts of this paper.

References

1. Barnett, L.: Tangled Webs. Evolutionary Dynamics on Fitness Landscapes with Neutrality. MSc. Dissertation, COGS, University of Sussex. (1997)
2. Blickle, T., Teich, J., Thicle, L,: System-Level Synthesis Using Evolutionary Algorithms. Computer Engineering and Communication Networks Lab (TIK), Swiss Federal Institute of Technology (ETH), TIK-Report, Nr. 16, April, (1996)
3. Bratt, S. & Macbeth, I.: Design and Implementation of a Field Programmable Analogue Array. Technical report, ACM 0-89791-773-1. (1996)
4. Castro, H, Gough, M.P.,"Dynamic Reconfigurability with Fault Tolerance", ENGGRP, School of Engineering, University of Sussex. (1995)
5. De Garis,H.: Growing an Artificial Brain with a Million Neural Net Modules Inside a Trillion Cell Cellular Automaton Machine. Proc. 4th International Symposium On Micro Machine and Computer Science. (1993). 211-214
6. Harvey, I., Husbands, P., Cliff, D: Seeing the light: Artificial Evolution, Real Vision: In Cliff, D. et al. (Eds.), From Animals to Animats 3: Proc of 3rd Int. Conference on simulation of adaptive behaviour,. MIT Press. (1994) 392-401
7. Layzell, P.: The 'Evolvable Motherboard' A Test Platform for the research of intrinsic Hardware Evolution. Technical Report CSRP479. COGS, University of Sussex (1998)
8. E. Sanchez,: Field Programmable Gate Array (FPGA) Circuits. Lecture Notes in Computer Science -Towards Evolvable Hardware, Vol. 1062. Springer-Verlag, (1996) 1-18
9. Thompson, A.: An evolved circuit, intrinsic in silicon, entwined with physics. International Conference on Evolvable Systems (ICES96), Vol 1259. Springer LNCS. (1997) 390-405
10. Thompson, A.: Evolving electronic robot controllers that exploit hardware resources. In Moran, F., et al. (Eds.), Advances in Artificial Life: Proc. 3rd Eur. Conf on Artificial Life (ECAL95), Vol.929 of LNAI.. Springer-Verlag. (1995) 640-656
11. Thompson, A.: Evolving fault tolerant systems. In Proc. 1st IEE/IEEEE Int. Conf. On Genetic Algorithms in Engineering Systems: Innovations and Applications (GALESIA '95),. IEE Conf. Publication No. 414. (1995) 524-529
12. Thompson, A.: Temperature in Natural and Artificial Systems. In Husbands, P., & Harvey, I., (Eds.) Proc, Fourth European Conference on Artificial Life, MIT Press. (1997) 388-397
13. Worden, R.: A Speed Limit for Evolution. J. of Theoretical Biology 176. (1995) 137 - 152
14. Zebulum, R., Pacheco, M. A., Vellasco, M.: Evolvable Systems in Hardware Design:Taxonomy, Survey and Application", International Conference on Evolvable Systems: from Biology to Hardware. Japan. (1996)

A Divide-and-Conquer Approach to Evolvable Hardware

Jim Torresen

Department of Informatics, University of Oslo, PO Box 1080 Blindern
N-0316 Oslo, Norway
E-mail: jimtoer@idi.ntnu.no

Abstract. Evolvable Hardware (EHW) has been proposed as a new method for designing systems for complex real world applications. One of the problems has been that only small systems have been evolvable. This paper indicates some of the aspects in biological systems that are important for evolving complex systems. Further, a divide-and-conquer scheme is proposed, where a system is evolved by evolving smaller subsystems. Experiments show that the number of generations required for evolution by the new method can be substantially reduced compared to evolving a system directly. However, there is no lack of performance in the final system.

1   Introduction

Evolvable hardware (EHW) has been introduced as a target architecture for complex system design based on evolution. So far a very limited number of real applications have been proved to be solvable by this new scheme. There are several reasons for this. One is the problem of evolving systems based on a long chromosome string. The problem has been tried solved by using variable length chromosome [1]. Another option, called functional level evolution, is to evolve at a higher level than gate level [2]. Most work is based on fixed functions. However, there has been work in Genetic Programming for *evolving* the functions [3]. The method is called Automatically Defined Functions (ADF) and is used in software evolution.

Both gate and function level of evolution have been applied to real applications. Simulations of data compression using function level evolution indicates performance comparable to other compression methods like JPEG compression [4]. The scheme is designed for implementation in a custom ASIC device. A function based FPGA has been proposed for applications like ATM cell scheduling [5] and adaptive equalizer in digital mobile communication [6]. Except for the few real problems studied, there is a larger range of small and non-real problems, see [7, 8].

This paper presents some concepts from biological systems and how they can be applied into architectures for evolvable hardware to be used for real applications.

The next section introduces the aspects from biological systems influencing on evolvable hardware. Section 3 presents a new scheme for evolving hardware.

Results from experiments are given in Section 4, which are followed by conclusions in Section 5.

2 A Framework for Evolvable Hardware

2.1 The Inspiration from Nature

The idea behind evolutionary schemes is to make models of biological systems and mechanisms. From nature the following two laws seem to be present:

1. The *Law of Evolution*. Biological systems develop and change during generations by combination and mutation of genes in chromosomes. In this way, new behavior arises and the most competitive individuals in the given environment survive and develop further. Another expression for this law is *phylogeny* [9].

2. The *Law of Learning*. All individuals undergo learning through its lifetime. In this way, it learns to better survive in its environment. This law is also referred to as *epigenesis* [9].

The two laws are concerning different aspects of life and should be distinguished, when studying artificial evolution. Most work on genetic algorithms are inspired by the Law of Evolution, while artificial neural networks are considering learning. Biological memory is still not fully understood. However, a part of the memory is in the biological neural networks. An interesting approach to design artificial evolutionary systems would be to combine the two laws into one system. One approach to this is Evolutionary Artificial Neural Networks (EANNs), which adapt their *architectures* through simulated evolution and their *weights* through learning (training) [10]. Neural networks are based on *local* learning, while the evolutionary approach is not [11].

Concerning FPGAs, this approach could be implemented by evolving a configuration bit string into a network of cells. After evolution their connections (weights) become trained.

Multi-Environment Training. So far artificial systems have mainly been trained by a single training set or in limited environment. This is in contrast to biological organisms trained to operate in different and changing environments. When training in a limited environment, the system with maximum fitness is selected. This fitness value does not have to represent the system with the best generalization [11]. It would be an interesting approach to investigate if an artificial system trained to operate in one environment could be combined with a similar system trained in other environments. If the resulting system could be able to contain knowledge about the different environments it would be able to operate in various environments *without* retraining. A possible implementation could be by including a soft switching mechanism between control systems, when the system enters a new environment. This would be an interesting approach for e.g. a vacuum cleaner moving from one room to another.

Online Adaptation in Real-Time. Research on evolutionary methods have been based on one-time learning. However, there is a wish of being able to design on-line adaptable evolvable hardware. The hardware would then have to be able to reconfigure its configuration dynamically and autonomously, when operating in its environment [12]. To overcome the problem of long evolution time, local learning should be investigated. Further, a time switching approach between learning and performing could be investigated.

An Appropriate Technology for Biological Modeling. Reconfigurable hardware like FPGAs are slower than microprocessors. However, a highly interconnection ratio is possible. These aspects are similar to those in the biological neural networks, where the neurons are massively interconnected. However, their speed of operation is slow. This indicates that reconfigurable technology should be an interesting approach for solving problems, where the biological brain is superior to ordinary computers.

2.2   The Basic Building Blocks and Their Interconnections

An evolutionary system would have to be based on some kind of basic unit. The model of a multi-cellular living organism would be to use a cell, which is able to reproduce itself by duplicating the full description of the complete organism. This corresponds to developing an individual (phenotype) from a chromosome (genotype). One cell could then be used to start reproduction to be carried out until the full organism is populated with cells [13]. Then, a specialization phase starts, where each cell interprets one piece of the chromosome depending on the location in the system. The major problem of this approach – named embryological development, is the tremendous amount of memory required within each cell for storing the complete chromosome. Thus, simpler logic gates or higher level functions have been used as the basic building blocks for evolution.

Most research on evolvable hardware is based on gate level evolution. There are several reasons for this. One is that the evolved circuit can be partly inspected by studying the evolved Boolean expressions. However, then the complexity of the evolved circuit is limited. Functional level evolution has been proposed as a way to increase the complexity. However, so far experiments have only been undertaken using ASIC chips. In the work presented in this paper, it is of interest to study the use of commercial FPGAs to higher than gate level evolution. Table 1 lists the possible combinations of low level building blocks and their interconnections.

Combining flip-flops[1] and feed-back connections have not been applied in EHW. One of the reasons for this is probably the difficulties and time consuming fitness evaluation of each individual. However, including flip-flops could be required to solve more complex problems by EHW.

---

[1] Flips-flops are essentially logic gates with feed-back connections, but are here listed as a special building blocks as they are in FPGAs.

Feed-back.-conn.	Flip-flops	Logic Gates	Applic.	Comments
No	No	No	-	
No	No	Yes	+	Applied for evolution in PLD and FPGA.
No	Yes	No	-	
No	Yes	Yes	+?	Could be applied to evolve delayed data networks.
Yes	No	No	-	
Yes	No	Yes	+	Applied by Thompson [14]
Yes	Yes	No	-	
Yes	Yes	Yes	+?	This would imply evolving state machines.

Table 1. Possible basic evolvable units and interconnections. "-" indicates a not applicable combination. "+" means that the combination has already been applied, while "+?" indicates a possible useful combination.

2.3   Local Learning

The internal operation of FPGAs available today is given by the configuration bit string loaded from an external source. This is in conflict to the wish that the operation of each cell in a device should only be based on local interactions with no global control. This is not a problem during normal operation, but rather during evolution. To make an FPGA device evolvable based on local interaction, it is possible to arrange sets of FPGA cells into subsets of cells. Each subset should be able to be evolved based on local interaction. In this scheme, the configuration bit string will not be changed during evolution, but rather the content of the internal registers. The application of local interactions would make local learning possible. The main challenge of this approach would be how to implement the genetic operators inside the FPGA.

Cellular Automata (CA) has been introduced to design systems based on local interactions. An experimental system based on FPGAs has been built for CA and local learning [9, 15]. The states of the cells in the system are evolved to oscillate between all zeros and all ones on successive time steps like a swarm of fireflies.

The long computation time has been one of the major problems for genetic algorithms, which is present also for evolvable hardware [14]. Evolving a 10x10 corner — a configuration string of 1800 bits, of a 64 x 64 array XC6216 FPGA required 2 – 3 weeks. The circuit was evolved to discriminate between two tones. The long evolution time even for a small circuit indicates that alternative evolution schemes should be investigated. The inherent parallelism in programmable hardware should be applied not only for operation, but also for the evolution.

The basic units in FPGAs are logic gates and 1-bit registers. Neural networks, on the other hand, have been shown to acquire at least 16 bits floating point values for inputs and weights for the best performance. Thus, further investigations should be conducted to find the optimal macro-cell size compared to the network size for evolution at a higher level than gate level. A large or complex cell structure would lead to a smaller number of such cells within a single FPGA device. Thus, the network will be small compared to using a simple macro-cell.

3 A New Approach to Hardware Evolution

Evolving systems for real applications of average complexity seem to be a near unobtainable task by the computer hardware available today. As mentioned earlier, several approaches have been suggested to overcome the problems, including variable chromosome length and function level evolution.

In this section, a scheme based on the principles described in the previous sections will be investigated. The method introduces a new approach to evolution called *increased complexity evolution*. The idea is to evolve a system gradually as a kind of divide-and-conquer method. Evolution is first undertaken individually on a large number of simple cells. The evolved functions are the basic blocks used in further evolution or assembly of a larger and more complex system. This may continue until a final system is at a sufficient level of complexity.

The main advantage of this method is that evolution is not performed in one operation on the complete evolvable hardware unit, but rather in a bottom-up way. The chromosome length can be limited to allow faster evolution. The problem of the approach would be how to define the fitness functions for the lower level sub-systems. However, for some applications it is possible to *partition* each training vector. Further, low level training vectors – e.g. speech *phoneme* recognition, can be used in the first evolution, followed by a higher level evolution using the evolved first level systems – e.g. to do *word* recognition. In this paper, the principles will be illustrated by a character recognition system. The evolution will be based on gate level, but can easily be changed to function level. The target EHW is an array of logic gates similar to that found in the Xilinx XC6200 and illustrated in Figure 1.

The array consists of n number of layers of gates from input to output. Except for layer 1, the Logic Gate (LG) is either a *Buffer, Inverter, AND* or *OR* gate. In layer 1, only *Buffer* and *Inverter* gates are available. Each gate's two inputs[2] in layer l is connected to the outputs of two gates in layer $l-1$. The function of each gate and its two inputs are determined by evolution. The evolution is undertaken off-line using software simulation. However, since no feed-back connections are used and the number of gates between the input and output is limited to n, the real performance should equal the simulation. Any spikes could be removed using registers on the output gates.

The application to be used in the experiments are the problem of recognizing characters of 5 x 6 pixels size, where each pixel can be 0 or 1. Each of the pixels is connected to *one* input gate. In addition to the 30 pixel inputs, two extra inputs are included as a bias of value 0 and 1, respectively. The output of the gate array

---

[2] *Buffer* and *Inverter* gates have only one input.

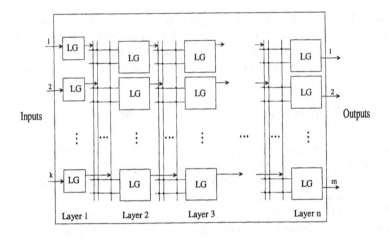

Fig. 1. An array of gates. "LG" indicates logic gate and can be *Buffer Inverter*, *AND* or *OR* gate.

consists of one output gate for each character the system is trained to recognize. During recognition, the output gate corresponding to the input character should be 1, while the other outputs should be 0. Thus, if the training set consists of m *different* characters, the gate array should consist of m output gates.

The aspect of *increased complexity evolution* is introduced by the way the evolution is undertaken. We compare evolving a system directly to evolving sub-systems. In the former case, the system is evolved to classify all training vectors in the training set. In the latter case, an evolved sub-system is able to classify a *subset* of the training characters. That is, each subsystem input *all* the 30 input pixels and *all* training vectors are applied during fitness evaluation. However, each subsystem has a limited number of *output* gates. In this way, the sub-systems are evolved without lack of generalization. The benefit is that each gate array is smaller and thus, should easier become evolved to perform the correct operation. For this application, the integration of sub-systems are straightforward by running them in parallel. For more complex applications, like speech recognition, a next level of evolution could be applied, where the sub-systems are the basic block in the evolution. The number of gate layers in the array is flexible and different numbers will be tested in the experiments. A large number would allow for more complex logic expressions. However, the chromosome becomes longer and obtaining a correctly evolved circuit could be more difficult.

Except for the output layer in the gate array, 32 gates are applied in each layer in the array. Thus, the *complete* system will be larger as the number of sub-systems increases. The main motivation for this work is to allow for evolution of complex systems and limiting the number of gates is not regarded as an important topic. The reason for this is that the main problem of today's research seems to be the lack of evolutionary schemes overcoming the chromosome length problem, more than the lack of large gate arrays.

The basic Goldberg style of GA was applied for the evolution with a population size of 50. For each new generation an entirely new population of individuals is generated. The mutation rate is 0.001. For each test, ten circuits were evolved and the circuit requiring the least number of generations was picked as the best.

4 Results

This section describes the results of evolving the character recognition system. Two separate experiments have been conducted. One with four characters and one with eight characters. A larger number of characters was also tested, but it was impossible to evolve a complete system in one operation.

Type of system	3 gate layers in array	4 gate layers in array
4 characters (A,B,C,D)	274	101
2 characters (A,B)	1	7
2 characters (C,D)	35	9

Table 2. The result of evolving a circuit for classifying four patterns, for three and four layers of gates in the array.

First, the experiment using four characters was undertaken. Table 2 shows the required number of generations used for obtaining a circuit that correctly classifies the patterns in the training set. In average, over ten times as many generations are required for evolving the system for recognizing four characters compared for each of the two sub-systems. Each of the sub-systems recognizes two characters. When other characters in the training set are input, the system is evolved to output the value 0 on each output.

The chromosome length is not very different for the two systems – 846 and 822 bits for the 4 and 2 character system[3], respectively. However, the number of outputs influencing the fitness evaluation is half the number for the complete system. Thus, a correct circuit can easier be found. In average, less number of generations is required for systems using four layers of gates compared to three. One layer of gates requires 384 chromosome bits.

Table 3 shows the results for the same kind of experimental setup, but with 8 patterns to be classified. As for 4 patterns, the number of generations required for evolving a classification system increases with the number of patterns. However

---

[3] This is for systems with four layers of gates.

Type of system	3 gate layers in array	4 gate layers in array
8 characters (A,...,H)	2709	3514
4 characters (A,B,C,D)	697	941
4 characters (E,F,G,H)	776	179
2 characters (A,B)	275	49
2 characters (C,D)	93	63
2 characters (E,F)	133	75
2 characters (G,H)	4	38

Table 3. The result of evolving a circuit for classifying eight patterns, for three and four layers of gates in the array.

as mentioned earlier, the final systems still has the same classification ability. It is interesting to observe in this experiment that for the systems evolved to classify many patterns, a larger number of generations is required for four layers of gates compared to three layers of gates. This is in contrast to the first experiment and may indicate that a shorter chromosome string is beneficial for a larger training set.

In this work, no separate test set has been used and thus, no concern about the overall generalisation and noise robustness has been included. The main goal has been to show that the evolution time can be reduced by dividing the systems into smaller sub-systems.

Both experiments show that the evolution time can be reduced by dividing the problem into sub-tasks to be evolved separately. This shows that the *increased complexity evolution* method should be promising for evolving complex systems for real applications. The scheme should, in future work, be tested for more complex applications and by using higher level functions.

5   Conclusions

This paper has introduced some aspects of biological systems to be applied for evolvable hardware. A scheme, called *increased complexity evolution*, is introduced. The method is based on sub-system evolution for the design of complex systems. A character recognition application is used to present one implementation of the scheme. It is shown that the number of generations can be drastically reduced by evolving sub-systems instead of a complete system.

Acknowledgements: The author would like to thank the group leader Tetsuya Higuchi and the researchers in the Evolvable Systems Laboratory, Electrotechnical Laboratory, Japan for inspiring discussions and fruitful comments on my preliminary work, during my visit there in November 1997.

References

1. M. Iwata et al. A pattern recognition system using evolvable hardware. In *Proc. of Parallel Problem Solving from Nature IV (PPSN IV)*. Springer Verlag, LNCS 1141, September 1996.

2. M. Murakawa et al. Hardware evolution at function level. In *Proc. of Parallel Problem Solving from Nature IV (PPSNIV)*. Springer Verlag, LNCS 1141, September 1996.

3. J. R. Koza. *Genetic Programming II: Automatic Discovery of Reusable Programs*. The MIT Press, 1994.

4. M. Salami et al. Lossless image compression by evolvable hardware. In *Proc. of 4th European Conf. on Artificial Life (ECAL97)*. MIT Press, 1997.

5. W. Liu et al. Atm cell scheduling by function level evolvable hardware. In T. Higuchi et al., editors, *Evolvable Systems: From Biology to Hardware. First Int. Conf., ICES 96*, pages 180 – 192. Springer-Verlag, 1997. Lecture Notes in Computer Science, vol. 1259.

6. M. Murakawa et al. Evolvable hardware for generalized neural networks. In *Proc. of Fifteenth Int. Joint Conf. on AI (IJCAI-97)*. Morgan Kaufmann Publishers, 1997.

7. J. Torresen. Evolvable hardware — A short introduction. In *Proc. of International Conference On Neural Information Processing (ICONIP'97, Dunedin, New Zealand)*. Springer-Verlag, November 1997.

8. J. Torresen. Evolvable hardware — The coming hardware design method? In N. Kasabov and R. Kozma, editors, *Neuro-fuzzy tools and techniques for Information Processing*. Physica-Verlag (Springer-Verlag), 1998. To appear.

9. M. Sipper et al. A phylogenetic, ontogenetic, and epigenetic view of bio-inspired hardware systems. *IEEE Trans. on Evolutionary Computation*, 1(1):83–97, April 1997.

10. X. Yao and Y. Liu. Evolutionary artificial neural networks that learn and generalize well. In *Proc. of IEEE Int. Conf. on Neural Networks, Washington DC*. IEEE Press, 1996.

11. X. Yao and T. Higuchi. Promises and challenges of evolvable hardware. In T. Higuchi et al., editors, *Evolvable Systems: From Biology to Hardware. First Int. Conf., ICES 96*. Springer-Verlag, 1997. Lecture Notes in Computer Science, vol. 1259.

12. T. Higuchi et al. Evolvable hardware and its applications to pattern recognition and fault-tolerant systems. In E. Sanchez and M. Tomassini, editors, *Towards Evolvable Hardware: The evolutionary Engineering Approach*. Springer-Verlag, 1996. Lecture Notes in Computer Science, vol. 1062.

13. P. Marchal et al. Embryological development on silicon. In R. Brooks and P. Maes, editors, *Artificial Life IV*, pages 371–376. MIT Press, 1994.

14. A. Thompson. An evolved circuit, intrinsic in silicon, entwined with physics. In T. Higuchi et al., editors, *Evolvable Systems: From Biology to Hardware. First Int. Conf., ICES 96*. Springer-Verlag, 1997. Lecture Notes in Computer Science, vol. 1259.

15. M. Sipper. *Evolution of Parallel Cellular Machines: The Cellular Programming Approach*. Springer-Verlag, 1997. Lecture Notes in Computer Science, vol. 1194.

Evolution of Astable Multivibrators *in Silico*

Lorenz Huelsbergen[1] and Edward Rietman[1] and Robert Slous[2]

[1] Bell Laboratories, Lucent Technologies***
[2] Xilinx Inc.†

Abstract. We use evolutionary search to find automatically electronic circuits that toggle an output line at, or close to, a given target frequency. Reconfigurable hardware in the form of field-programmable gate arrays—as opposed to circuit simulation—computes the fitness of a circuit which guides the evolutionary search. We find empirically that oscillating circuits can be evolved that closely approximate some of the supplied target frequencies. Our evolved oscillators alias a harmonic of the target frequency to satisfy the fitness goal. Frequencies of the evolved oscillators were sensitive to temperature and to the physical piece of silicon in which they operate. We posit that such sensitivities may have negative implications for demanding applications of reconfigurable hardware and positive implications for adaptive computing.

1 Introduction

Complex entities—biological and artificial, for example—are in part governed by systolic processes. In many animals, hearts beat at (perhaps variable) frequencies to distribute fluids. Circuits in machines coordinate information flow in step with a local or global clock. Biological oscillation arose from primitive components (molecules) in an environment (physics) via the process of natural selection. Mechanical oscillation arose in computers, and in many other machines, through human design. We seek to understand if evolution can be harnessed to design pieces of computing machines. Toward this end, we are conducting experiments to "evolve" circuits—oscillators (astable multivibrators) in particular—from primitive logic components. Our goals are twofold: The exploration of the capabilities of *in Silico* evolution and the investigation of whether computational circuits based on oscillators can thus be constructed.

The genetic algorithm (GA) [4] is a form of *evolutionary search*. GAs have been shown to perform well as a general optimization technique across a broad range of domains (see Goldberg [2] for examples). The GA maintains a population of individuals (*bit strings*) over a series of *generations*. The initial population is random. Using an externally supplied *fitness function* (environment), the GA selects promising individuals for the next generation. Some such selected individuals are then paired and, with random substrings interchanged, placed in the next generation.

Evolutionary search—most recently in the form of GA-based *genetic programming* (GP)—has been used to evolve computer software (*e.g.*, [6]). In this

*** {lorenz,ear}@bell-labs.com
† robert.slous@xilinx.com

context, the bit string comprising an individual is interpreted as a (perhaps variable length) sequence of instructions written in a computer language. Distance, in some metric space, between the result of evaluating a GP individual and the desired target result constitutes an individual's fitness. It is well known that digital software and hardware are computationally equivalent. This suggests that application of software evolution techniques may also be fruitful in a hardware realm [8, 3, 1, 12]. With evolvable computational structures, the programming onus shifts from providing an algorithm (circuit or program) for solving the task at hand to crafting a function that assigns an accurate fitness measure to partial and complete solutions. Search—taking the form of a GA for this paper—can then automatically perform algorithm discovery.

The recent experiments of Thompson [8] in particular demonstrate that reconfigurable logic in the form of field-programmable gate arrays (FPGAs) can serve as a viable substrate for gate-level hardware evolution. Thompson evolved discriminator circuits that, when presented with one of two possible input tones (frequencies), would correctly classify their input. Our system for *in Silico* evolution is similar to Thompson's. Our study however concerns circuits with fundamentally different characteristics than input-sensitive frequency discriminators— we are evolving computational components, namely oscillators, that function as stand-alone clocks.

Our result is the automatic generation of oscillators at specified *target frequencies* from primitive electronic components. Given only a target frequency f, our system can produce—from logic gates (such as "not") and wires to connect them—a circuit whose single output oscillates between the logic states low and high with frequency f (or a harmonic multiple thereof). Note that the circuit undergoing evolution receives no input in general and no clock signal in particular. It completely synthesizes oscillation.

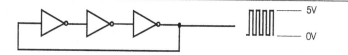

Fig. 1.: A manually designed ring oscillator constructed from three inverters. The output oscillates due to gate and signal propagation delays. A genetic algorithm can find circuits with similar behavior—but not necessarily of similar structure—that oscillate at predefined target frequencies.

One route to such oscillation is exploitation of gate and signal propagation delays along with feedback. The manually designed ring oscillator of Figure 1 illustrates this basic principle. This oscillator's frequency depends on the speed of the substrate's implementation technology.[5] Note that the oscillator's frequency may be reduced by inserting additional inverters into the ring. It is difficult, even for skilled designers, to craft feedback circuits with specified characteristics (*e.g.*, frequency) solely from logic gates. In practice, therefore, oscillators are usually constructed from (relatively expensive) analog components.

---

[5] Gates in contemporaneous off-the-shelf FPGAs typically switch in nanoseconds.

Though most often deployed in digital circuits, transistors are analog amplifiers. Since gates are constructed from transistors, circuit evolution is free to exploit their analog behavior to satisfy its goal. Furthermore, since electronic components are physical devices that operate electromagnetically, effects such as cross coupling are also available to evolution. As we will discuss (§5), and Thompson considers at length [10, 9], exploitation of such "features" poses new engineering concerns that must be addressed. On the other hand, we observe functional portability of evolved circuits from one piece of silicon to another which discounts the wide-spread exploitation of such chip-specific peculiarities.

We report results of *in Silico* oscillator evolution for ten target frequencies in three cell-array sizes (6x8, 8x8, and 16x16). Considering all three cell-array sizes, our system discovered quite accurate oscillators—over 97% of their pulses correct—for five of the ten frequencies and required only a small number of GA runs. Our empirical results are statistically reproducible. Random search confirms that the directed search of the GA in the space of oscillator circuits is effective; that is, the evolutionary search is directed and does not "blindly stumble" upon solutions.

2 Related Work

Although all experiments on hardware evolution are recent, the idea [12, 1] spans at least two decades. To date a few groups have used software simulation (digital and analog) to provide circuit fitness measures [3, 5]. Aside from the work of this paper, one other experiment [8] used reconfigurable logic to directly, in real time, to evolve circuits *in Silico*. We now provide further details on the prior work.

Wolfram [12] and Atmar [1] are among the first to describe frameworks based on (evolutionary) search to create computational hardware. Wolfram posits that it may be possible to find a realization of simple combinational circuits using genetic search and, thereby, to mitigate the complexities of hardware design. In his thesis, Atmar considers hardware that admits rapid evolution of finite state machines. Since the design and construction of such hardware was relatively expensive at that time, Atmar simulated his approach in software. Though slow, the simulation was applied with some success to the task of character recognition.

The most significant experimental result to date is that of Thompson [8]. He used a reconfigurable hardware device (FPGA) to evolve a binary frequency discriminator that classifies its input signal into one of two frequencies. His search was also GA based and obtained, as we do here, its fitness measures in real time from the FPGA. Thompson identified the phenomenon of evolution exploiting the device's analog nature (*i.e.*, the physics of its immediate environment) in finding solution circuits [9] (see also Section 5 below). Earlier, Thompson *et al.* [11] simulated oscillator evolution in a software model of an FPGA. They report evolving pulse trains with frequencies much lower than speeds of the simulated logic gates, but not at the target frequency. Because of simulation overheads, circuit evaluation times were limited to 10ms; our experience reveals that times an order of magnitude longer are necessary to yield stable oscillators in practice. Our work differs from Thompson *et al.*'s [11] in three respects: (1) our evolved oscillators function in an actual reconfigurable device; (2) their frequencies are very close to a multiple of the target frequency; and, (3) they are stable over time (*i.e.*, for many hours).

Others have coupled GAs to electronic and electrical circuit simulators as sources of fitness measures. Koza *et al.* (*e.g.*, [5]) coopted genetic programming of software to evolve instead descriptions of analog electrical circuits. They however do not report on the conversion of evolved descriptions to actual hardware. It has not been demonstrated that analog circuits evolved via simulation are implementable in practice. *In Silico* evolution, on the other hand, assures that solution circuits function in at least the device (and environment) in which they originally evolved. Higuchi *et al.* (*e.g.*, [3]) similarly use simulation of, in their case, digital circuits to guide evolution. They do test and verify their evolved circuits in an FPGA, but only after evolution is complete. *In Silico* evolution—Thompson's [8] and our approach—can speed circuit evolution since evaluation is in hardware proper. The *in Silico* approach also provides evolution the ability to exploit analog circuit characteristics whereas digital simulation abstracts this physics. Higuchi *et al.* [3] describe various means by which hardware evolution can be extended; by performing the genetic search in hardware instead of software, for example.

3 System for *in Silico* Evolution

Our custom system for *in Silico* gate-level evolution consists of a hardware component coupled with a GA implementation. We first describe the hardware and then the software; further details of the hardware are available elsewhere [7].

3.1 Hardware: FPGA and PC

The hardware consists of two pieces: an FPGA used exclusively for assigning fitness values to circuits and a general purpose computer (PC) that services the FPGA and executes the search algorithm. Figure 2 is a block diagram of this system.

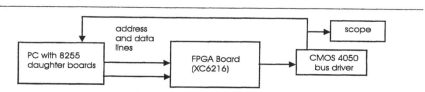

Fig. 2.: A system for *in Silico* evolution. The PC downloads configurations (circuits) into the FPGA for fitness evaluation. It also samples and records the FPGA's output. The bus driver insures that the FPGA output is forced either completely low or high. Peripheral IO (PIO) devices consitute the PC/FPGA interface.

An FPGA is an electronic device (usually packaged as a single chip) that can be configured to function as an arbitrary digital circuit.[6] Reconfigurability can be achieved by associating memory cells with logic gates and interconnections.

---

[6] The FPGA's size limits the digital circuits it may realize.

The states of the configuration cells govern a gate's specific logic function or an interconnection's source and destinations. That is, a circuit for embodiment in an FPGA can be described by a bit string that contains information on the type of gates a circuit requires and how they are to be connected. The genetic algorithm described in the next section will search for bit strings describing circuits that induce a target behavior (*i.e.*, oscillation) in the FPGA.

FPGAs generally consist of a collection of *cells* along with one or more levels of hierarchical interconnect among the cells. Additionally, FPGAs contain IO blocks (IOBs) that enable the configured circuit to communicate with external devices.

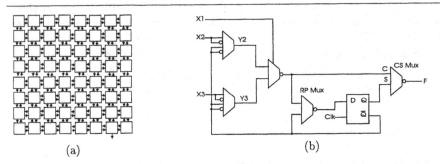

(a) (b)

Fig. 3.: Relevant portions of the XC6216 FPGA architecture. Sample logic-cell layout (a) used for the experiments of this paper. Nearest-neighbor connections and the single bit-wide output line (lower right) are depicted. Digital per-cell logic (b) consists of multiplexors controlled via configuration bits.

Figure 3 contains high and low level views of the Xilinx XC6216 architecture relevant to our experiments. Figure 3a depicts a sub-array of the FPGA's cells and enabled connections (nearest-neighbor). Note the bit-wide output line emanating from a cell on the bottom row. The oscillator experiments of this paper were conducted in such 6x8, 8x8, and 16x16 arrays with single outputs. The output of a single cell was routed to a XC6216 IOB which transmitted the signal to the PC as well as to an oscilloscope for monitoring. Note that—among other points discussed below—we utilize only the lowest level of inter-cell communication and that the FPGA circuit receives no input (but may produce a time varying output signal).

The Xilinx XC6216 FPGA [13] was chosen for our implementation for a number of reasons:

1. it may be reconfigured indefinitely,
2. it may be partially reconfigured,
3. its design insures that an invalid circuit will not harm it,
4. its specification is non-proprietary.

Point 1 allows its use in repeated fitness evaluation which is the central component of a GA.[7] Point 2 guarantees that configuration time (a non-negligible cost

---

[7] Our chip has been reconfigured millions of times.

in fitness computation) is linear in the size of the circuit being loaded. Point 3 tolerates random circuit configurations which are prevalent in the circuit population maintained by the GA. Point 4 permits bit-level circuit description while bypassing conventional circuit entry tools.

Figure 3b is a schematic of the digital logic within a single XC6216 cell. Multiplexors, controlled via configuration bits, implement common logic functions. Wires into a cell (from its nearest neighbors) constitute its $X1$, $X2$, and $X3$ inputs which, by appropriate configuration of the multiplexors, can compute all $2:1$ and some $3:1$ boolean functions for the cell's output, F. A one-bit register is also available per cell. However, we disabled all registers for this paper's experiments. (Evolution is free to construct register structures by appropriately connecting multiple cells.) Configuration of a single XC6216 cell nominally requires three bytes of information.

It is important to note that although the cell function is described in terms of digital logic, the cell is constructed from gates which are in turn built from analog components (transistors). Thus, a circuit constructed in an unconventional manner may exhibit analog behavior that may extend to the global (circuit) level. GA fitness evaluation is an unconventional FPGA application and we make no attempt to detect or restrict configuration strings describing such circuits.

3.2 Software: Search Algorithms

This section describes the implementation of the genetic algorithm used in the oscillator experiments and also recounts the random-search algorithm used to verify the efficacy of *in Silico* evolution. We defer elaboration of the various constants (*e.g.*, population and sample sizes) to Section 4.

Evaluation Function. Fitness assignment requires an assessment of an individual's performance on some particular task. This is done via an *evaluation function*:

$$\mathcal{E}_O : (I, f) \to S \tag{1}$$

$\mathcal{E}_O$ maps an n-bit configuration string (individual) I and a frequency f to an m element *sample vector* S. This vector contains m output values (each either low or high) of the FPGA's output line at $1/f$ second intervals and is used in assigning individual I's fitness.

We note that in our implementation the frequency f may only be instantiated to values that can be readily synthesized in software by the PC (see §4).

Fitness Function. A circuit I's fitness is computed using only the sample vector S returned by the evaluation function (Equation 1). Lower fitness values are better. The fitness is given by:

$$\mathcal{F}_o(S) \equiv \sum_{i=1}^{m} S_i \oplus (i \bmod 2) \tag{2}$$

where S_i is the i-th element of the sample and $\oplus$ denotes "exclusive or." The fitness, therefore, is the number of "missed" pulses. A missed pulse occurs when an even (odd) sample element does not match a low (high) target pulse.

Figure 4 illustrates the fitness computation process. The top pulse (of sample frequency f) indicates the points in time when samples are taken. The FPGA's

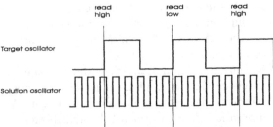

Fig. 4.: Juxtapostion of the target oscillator (at sampling frequency f) and a possible solution oscillator at a harmonic of f.

output is always sampled at the beginning of a high pulse and alternating samples are taken to be high followed by low, *etc.* Note that an oscillator of a harmonic multiple of f may satisfy the desired criteria.

Search Methods. Here, we describe details of the genetic and random search.

Genetic Search. We use *tournament selection* with *elitism* as the GA's population-selection mechanism. *Crossover* and *mutation* are its genetic operators.

Population selection, for the construction of successive generations, is performed via k-tournament selection. Let P be a population (set) of N individuals (configuration strings). To select a single individual from P, tournament selection examines k individuals in P and selects the one with best fitness. This mechanism is used to generate candidate pairs for crossover, for example.

We use a recombination operator that performs two-point crossover. Crossover of two n-bit configuration strings I_a and I_b first selects a subsequence of bits[8] starting at a random point $0 \leq p_a < n$ in configuration I_a. The length $k > 0$ of the subsequence is chosen randomly such that $p_a + k < n$. A random point p_b, $0 \leq p_b + k < n$, is then chosen in configuration I_b. Finally, the k bits in I_a starting at p_a are interchanged with the k bits in I_b starting at p_b.

Crossover in a population P is performed by first selecting a subset $P' \subseteq P$ of configurations from the population (using tournament selection). A configuration is randomly selected for P' with probability $Prob_{crossover}$. The configurations in P' are then randomly paired and the crossover operator is applied to each pair to produce a replacement pair.

Bits in every configuration string in P are mutated (negated) with probability $Prob_{mutate}$ during generation transitions.

Random Search. Random search randomly generates an individual I, evaluates I and computes its fitness, and (optionally) records I's fitness as the best seen if I improves on the current best fitness. This process continues until a sufficient number of global solutions are found or until the number of configuration evaluations exceeds a predetermined threshold.

4 Experiments

For three array sizes—6x8, 8x8, and 16x16—we performed oscillator evolution experiments for ten target frequencies in a Xilinx XC6216 FPGA [13]. Each of

---

[8] To simplify implementation, we select points at byte boundaries. We do not expect that this simplification fundamentally affects our results.

the thirty experiments consisted of ten randomly seeded GA runs. Table 1 lists the frequencies and summarizes the results. Note that the evolved oscillators are close to harmonics of the target frequency. The highest target frequency is the one at which the PC can most rapidly sample the FPGA by polling the IO cards mapped to the PC's port. The nine lower frequency targets were obtained by successively inserting blocks of 100 null operations (NOPs) into this polling loop. (See §5 below for alternatives to software sampling.) Actual frequency determination was then done with an external oscilloscope attached to the bus driver output.

f	f'	GA	Random		f	f'	GA	Random		f	f'	GA	Random
88.4		4949	–		88.4		4381	–		88.4		6033	–
66.3		8298	–		66.3	1185	466	9734		66.3	909	575	9931
55.3		5527	–		55.3	929	213	9099		55.3	925	52	8313
47.4		5475	–		47.4		6032	–		47.4		6920	–
40.1	925	149	9943		40.1		5025	–		40.1		4134	–
35.2		6808	–		35.2		7066	–		35.2		8194	–
31.5		8808	–		31.5		8772	–		31.5		7523	–
28.4		6475	–		28.4	909	341	9905		28.4	943	552	9886
25.9		6947	–		25.9		6535	–		25.9		6491	–
23.7	925	220	9475		23.7		8016	–		23.7		8533	–

6x8 Cell Array	8x8 Cell Array	16x16 Cell Array

Table 1.: Results of evolving oscillators for ten target frequencies in three cell-array sizes. The target frequency f and the evolved frequecy f' are in kHz. (f' is given only for experiments that produced fitness values less than 1000.) The table lists the best fitness (as the number of missed pulses out of 2×10^4) found by the GA and, for targets that gave oscillators, by a random search.

GA parameters were set as follows: population size $N = 512$, $Prob_{crossover} = 25\%$, $Prob_{mutate} = 0.01\%$. The probability settings mean that a quarter of every generation was generated through crossover (the other three quarters being directly selected) and that one bit (randomly selected) in every 10^4 bits in the population was negated during transition into a new generation. Binary tournament selection was used. A GA run was deemed complete when the fitness of the population's best individual did not improve for 50 consecutive generations (stasis). We have no reason to believe these settings to be in any way "optimal" for the problem under investigation; alternate settings have not been tried.

The length of the configuration string was 576 bytes of which 1920 bits (41.6%) actively controlled the FPGA. To simplify the problem, the remaining bits were masked to enable only nearest-neighbor interconnections and to disable cell registers. We chose to disable these features to simplify the problem and hence do not yet have data on whether their inclusion affects the solutions.

In a fitness evaluation, the FPGA's output line was sampled $m = 2 \times 10^4$ times at the target rate. We arrived at this sample size by noting that shorter samples (e.g, $m = 1 \times 10^3$) produced unstable oscillators. That is, oscillators evolved using a significantly shorter sample sequence decayed after producing pulses only for the duration of the sampling. We find that the GA can encourage oscillator stability by integrating over longer sample times. In light of these findings, it would be interesting to examine the stability of Thompson *et al.*'s [11] oscillators evolved under software simulation.

For each target frequency f, Table 1 records the best GA individual fitness (fewest missed pulses) found for the three array sizes over ten runs. For individuals with fitness less than 1000 we list the oscillating frequency f'. The fitness of the best individual found via random search is also given. Our results are statistically reproducible. This means that in a fixed number of runs, evolutionary searches will find circuits with comparable fitness even though the searches' random seeds differ.

The GA found oscillators of good, but not perfect, accuracy. Solutions always appeared within the first 300 generations. Note that all evolved oscillators are close to a harmonic of the target frequency. The oscillators found in the smallest array (6x8) are qualitatively different than those found in the larger (8x8 and 16x16) arrays in that they were found at different frequencies. In particular, the evolved 6x8 oscillator for target 40.1kHz has a frequency similar to the 16x16 evolved oscillator for the 55.3kHz target; yet no oscillator was found in the 6x8 cell space for the latter target. That is, the 6x8 oscillator for 40.1kHz should also be an oscillator for 55.3kHz in the same cell space; yet it was not found for target 55.3kHz. Restriction of resources apparently enables discovery of certain oscillators but also inhibits discovery of others. A possible explanation is that the restricted cell space blocks some evolutionary pathways that are available with more resources.

The evolved oscillator circuits require further study to discern their internal structure. It is straightforward to construct—from the configuration bit strings— a schematic of the cells' functions and the wires that connect them. However, it is unlikely that a digital implementation (with different layout, for example) constructed from a disassembled schematic will yield identical, or even similar, behavior. In other words, it is possible that the evolved circuits exploit parallel interactions between gates used in an asynchronous uncontrolled fashion. In the oscillator experiments we observe that the the frequency of the oscillator circuits could be governed by such effects but not their general function (oscillation); we conclude this because the circuits portably oscillate in different silicon pieces, but at different frequencies (see §5).

Figure 5a depicts an oscilloscope output of the 929kHz 8x8 oscillator found during the search at target frequency $f = 55.3$kHz. The voltage on the FPGA output line was captured *before* the signal passed through the bus driver (Figure 2). Capture of the output before the driver yields a sawtooth function that oscillates between TTL threshold voltage levels. Note that the circuit's gates are exploiting threshold levels to achieve oscillation. The signal *after* the bus driver approaches the form of a square wave (not shown). Figure 5b is the power spectrum for this sawtooth signal; it indicates that the signal lies mainly about 900kHz, but also contains a few high-frequency components. In other words, the oscillator is not spectrally pure, but quite close.

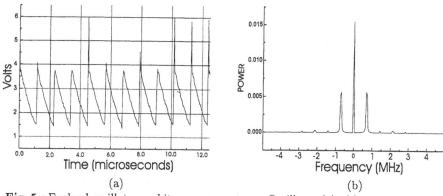

Fig. 5.: Evolved oscillator and its power spectrum. Oscillator (a) of frequency 929 kHz, a harmonic of the target frequency $f = 55.3$ kHz; its pulses coincide with 98% of f's pulses. The power spectrum (b) indicates that the signal's dominant components lie at the measured harmonic frequency.

To assure that evolutionary search was indeed approaching solutions and not randomly "stumbling" upon them, we conducted 10^6 random-search evaluations (§3.2) for each experiment that yielded a circuit with GA fitness below 1000. The best fitness produced by random search never reached 9000. (In comparison, ten GA runs for a given frequency required approximately 3.5×10^5 runs.) Note that a circuit that holds its output either low or high will trivially score a fitness of $m/2$ (where m is the number of pulse samples). Random search—even when allotted more evaluations than the GA—has not located a circuit whose output was in the correct state more than 52% of the time. Furthermore, evolutionary search always found circuits closer to the goal while requiring fewer evaluations.

5 Discussion

We conclude this paper with a discussion of the results and of some engineering issues raised by *in Silico* evolution.

Evolutionary search discovered oscillators for some, but not all, presented target frequencies. It is desirable, however, to develop methods that can evolve oscillation for more—if not for most—frequencies. We believe oscillator circuits were not found for some frequencies for two reasons: (1) oscillation at the target frequency is not possible given the available resources and (2) the length of the search was insufficient. Comparison of array sizes 6x8, 8x8, and 16x16 was done to begin investigation of the first point. Since the size of the search space grows exponentially in the size of the hardware array, searches in larger arrays will likely be more computationally expensive. (For the 8x8 array the search space already consists of 2^{1920} possible configuration strings, some of which describe semantically equivalent circuits; this size increases by a factor of 16 for the 16x16 array.) Conducting similar experiments with larger populations or greater stasis settings would expand the search which—in exchange for longer run times— could turn up other oscillating circuits

Our method of sampling the FPGA with a software polling loop running on a PC introduced some indeterminacy—such as operating system interrupts—into the fitness measurement process. Future systems for oscillator evolution could employ dedicated frequency generators and microcontrollers to better control the sample intervals and to provide a larger and finer spectrum of target frequencies. Software sampling did not however pose a significant obstacle to evolving some fairly accurate oscillators; better sampling technique may yield more accurate oscillators or even oscillators for frequencies so far found untenable by evolutionary search.

It is not at all understood *how* the evolved circuits function. For example, relative to the speed of the FPGA's gates (nanosecond transition times), the evolved oscillators are of rather low frequency. To explore this point, we manually constructed the largest (63 inverters), and hence slowest, ring oscillator that fits in the 8x8 cell space. (Figure 1 depicts a similar three-inverter ring.) This oscillator has a frequency of 3560kHz which is still much higher than the frequencies of the evolved oscillators in all three array sizes. Our evolved circuits must therefore be using mechanisms other than simple ring oscillation to attain their behavior. Disassembly of the circuits into digital schematics may provide some clues, but it is unlikely that this will be sufficient to fully understand their behavior.

Evolution operates with respect to an environment. *In Silico* circuit evolution occurs not only in the silicon of the reconfigurable logic device, but also in the environment where that device is located. Like Thompson [10], we have observed that temperature affects the evolved circuits. Cooling the FPGA increases the frequency of oscillation in our case. The oscillators of this paper were evolved at "room temperature." Good temperature control seems paramount to evolving more precise circuits. Thompson [10] is investigating solutions using simultaneous evolution in multiple environments to increase circuit applicability and robustness. Other factors, such as the position of the circuit within the FPGA's full 64x64 array, or the silicon wafer from which the particular FPGA was cut, may have an effect on the portability of evolved circuits. We verified that it is possible to use an oscillator evolved in FPGA chip A in another chip B (of the same variety)—albeit with a slight change in frequency.

The influence of factors such as temperature and silicon quality on circuit evolution is an impediment to the rigorous adherence to specification (of frequency for example) required by conventional system design. However, evolved circuits are potentially quite adaptive. For example, the evolved oscillators function as thermometers that could control the device over a temperature range. Evolution is an adaptive process: it could potentially salvage defective chips by evolving circuitry to avoid—or even harness—the defects.

6   Summary

We have described a system for using evolutionary search to find electronic circuits that approximate or meet a given specification—in particular, to find oscillators at various target frequencies. Real-time fitness measures, obtained from a reconfigurable FPGA, guide the genetic algorithm that performs the search. For five of ten target frequencies, the system was able to construct circuits that oscillated at a harmonic close to the target frequency. Automated search algorithms operating in the space of circuits can be harnessed as a powerful new aid in the design and construction of certain electronic hardware.

Acknowledgments

Thanks to Bob Frye for valuable insights into this work. The anonymous referees, Brian Kernighan, and Adrian Thompson supplied useful comments.

References

1. J. W. Atmar. *Speculation on the Evolution of Intelligence and its Possible Realization in Machine Form.* PhD thesis, New Mexico State University, April 1976.
2. D. E. Goldberg. *Genetic Algorithms in Search, Optimization and Machine Learning.* Addison-Wesley, 1989.
3. T. Higuchi, H. Iba, and B. Manderick. Evolvable hardware with genetic learning. In H. Kitano and J. A. Hendler, editors, *Massively Parallel Artificial Intelligence*, pages 399–421. MIT Press, 1994.
4. J. Holland. *Adapation in Natural and Artifical Systems.* University of Michigan Press, 1975.
5. J. R. Koza, F. H. Bennett III, D. Andre, and M. A. Keane. Automated WYWIWYG design of both the topology and component values of analog electrical circuits using genetic programming. In *Proceedings of the First Conference on Genetic Programming*, pages 123–131. MIT Press, July 1996.
6. J. R. Koza, K. Deb, M. Dorigo, D. B. Fogel, M. Garzon, H. Iba, and R. L. Riolo, editors. *Proceedings of the Second Genetic Programming Conference.* Morgan Kaufmann, July 1997.
7. E. Rietman, R. Slous, H. Hemmi, H. de Garis, and K. Shimohara. Building a machine for evolution *in silico*. In M. Sugisaka, editor, *Proceedings of the Third International Symposium on Artificial Life and Robotics*, pages 186–189, January 1998.
8. A. Thompson. Silicon evolution. In J. Koza, editor, *Proceedings of the First Conference on Genetic Programming*, pages 444–452. MIT Press, July 1996.
9. A. Thompson. An evolved circuit, intrinsic in silicon, entwined with physics. In T. Higuchi and M. Iwata, editors, *First Int. Conference on Evolvable Systems: from Biology to Hardware (ICES96)*, pages 390–405. Springer Verlag LNCS 1259, 1997.
10. A. Thompson. Temperature in natural and artificial systems. In P. Husbands and I. Harvey, editors, *Fourth International Conference on Artificial Life*, pages 388–397. MIT Press, 1997.
11. A. Thompson, I. Harvey, and P. Husbands. Unconstrained evolution and hard consequences. In E. Sanchez and M. Tomassini, editors, *Towards Evolvable Hardware: The Evolutionary Engineering Approach*, pages 136–165. Springer–Verlag, 1996.
12. S. Wolfram. Approaches to complexity engineering. *Physica D*, 22:385–399, 1986.
13. Xilinx Inc. *The Programmable Logic Data Book. XC6200 Advanced product specification V1.0*, 1996. http://www.xilinx.com.

Some Aspects of an Evolvable Hardware Approach for Multiple-Valued Combinational Circuit Design

Tatiana Kalganova, Julian F. Miller and Terence C. Fogarty

Dept. of Computing, Napier University, 219 Colinton Road,
Edinburgh, UK, EH14 1DJ. E-mail: t.kalganova,
j.miller,t.fogarty@dcs.napier.ac.uk Ph: +44 (0)131 455 4304/5

Abstract. In this paper a gate-level evolvable hardware technique for designing multiple-valued (MV) combinational circuits is proposed for the first time. In comparison with the decomposition techniques used for synthesis of combinational circuits previously employed, this new approach is easily adapted for the different types of MV gates associated with operations corresponding to different algebra types and can include other more complex logical expressions (e.g. single-control MV multiplexer called T-gate). The technique is based on evolving the functionality and connectivity of a rectangular array of logic cells. The experimental results show how the success of genetic algorithm depends on the number of columns, the number of rows in circuit structure and levels-back parameter (the number of columns to the left of current cell to which cell input may be connected). We show that the choice of the set of MV gates used radically affects the chances of successful evolution (in terms of number of 100% functional solutions found).

1 Introduction

Evolvable Hardware extends the concepts of Genetic Algorithms to the evolution of electronic circuits. A central idea of this is that each possible electronic circuit can be represented as a chromosome in an evolutionary process in which the standard genetic operations over the circuits, such as initialization, recombination, elitism, selection are carried out. The evolving circuits may be evaluated using software simulation models [4, 7, 8, 12], or in some cases implemented directly in hardware [2, 11]. A number of investigations have been carried out for synthesis of binary logic circuits [4, 7].

Multiple-Valued (MV) Logic refers to the adoption of logic systems having more than two levels [3, 9]. It is generally felt that MV logic allows circuits to have increased functionality with a reduction in wiring density.

In this paper we present a method for the synthesis of combinational MV circuits. This approach is an extension of evolvable hardware method for binary logic circuits proposed in [1, 10]. The first attempts to evolve MV arithmetical combinational circuits have been discussed in [5, 6] and here we present more detailed results and further discussion about the most suitable parameters used in this approach. We examine the most effective geometry which allows the evolution of 100% functional circuits for a 3-valued one bit adder with carry logic function. We present the results of a number of experiments which show that the number of rows which allows most fully functional solutions depends on the number of outputs in logic function, and that the optimal number of columns depends strongly on the levels-back parameter. The levels-back parameter defines the number of columns to the left of current cell to which cell input may be connected. An analysis of different sets of MV gates shows that there is a particularly effective set of MV gates which allows us to evolve three times as many 100% functional solutions as any other set. A notable feature of this paper is that it shows how the chosen geometry of circuit design has a strong influence on the GA performance.

2 The Evolvable Hardware Method for Combinational MV Circuits

The method proposed here is based on evolving combinational MV networks employing a rectangular array of the logic cells. The logic cells in this array are uncommitted and can be removed from the network if they prove to be redundant. Let $X=\{x_1, x_2, ...,x_n\}$ and $Y=\{y_1, y_2, ... y_m\}$ be input and output variables of implemented r-valued function respectively. The number of variables in X is denoted by n and the number of functions in Y is defined by m. The inputs that are made available are logic constants "0", "1", ...,"r-1", where r is the radix, all primary inputs x_1, x_2, ..., x_n and the unary operators acting on the primary inputs, for instance, complement of all primary inputs $\bar{x}_1$, $\bar{x}_2$, ..., $\bar{x}_n$. To illustrate this let us consider a 3-input 3-output 3-valued logic function which will be implemented on a 4 x 4 array of 3-valued logic cells with two inputs and one output (Figure 1). Input encoding is carried out as follows. The number of logic constants encoded as 0, 1, 2 respectively is 3. The set of primary inputs $X=\{x_1, x_2, x_3\}$ are labeled with 3, 4 and 5 output numbers respectively. The inverted inputs $\bar{x}_1$, $\bar{x}_2$, $\bar{x}_3$ are labeled as 6, 7 and 8 correspondingly.

Each cell in the array is labeled with an output number. The first cell having an output is labeled $(2n + r)$. These are numbered connection points, which are important for the network representation of the chromosome (discussed below). Each cell is described by $(k+1)$ integers, where k is the number of inputs in the logic MV cell. The first k numbers describe the cell inputs and the final integer defines the functional type of the cell. If this final integer is positive then the cell is assumed to be a MUX gate, otherwise it refers to one of the gate types listed in Table 1. The gate types listed in Table 1 have the following definitions (for an r valued 2-input function). Note that the over-bar in all the expressions in Table 1 refers to the complement (or inversion) operator defined below:

$MAX(x_1 , x_2)$ is the maximum of inputs x_1 and x_2 and in the binary case becomes identical to the inclusive-OR operation. $MIN(x_1 , x_2)$ is the minimum of inputs x_1 and x_2 and in the binary case becomes identical to the AND operation. $TSUM(x_1 , x_2)$ is referred to as the truncated sum operator and is defined as $MIN(x_1+x_2 , r-1)$, this again reduces to the inclusive-OR operation for the binary case. The truncated product operator $TPRODUCT(x_1 , x_2)$ is defined as $MAX(x_1+ x_2-(r-1) , 0)$, which in the binary case becomes the AND operation. $MODSUM(x_1, x_2)$ and $MODPRODUCT(x_1, x_2)$ are defined as (x_1+x_2) and $x_1 \cdot x_2$ in modulo r arithmetic. The former becoming the exclusive-OR operation in the binary case, while the latter reduces to the AND function. Finally the complement of a r-valued symbol x, indicated with an over-bar is defined as $(r-1)-x$ and represents the NOR operation in binary logic.

The user can choose any subset of this list of gates and later we examine the relative performances of various sub-sets. Note that the list of MV gates presented in Table 1 is just a small subset of possible gates proposed in the literature [5, 6].

In the discussion which follows we will use the following terminology. Associated with each rectangular array of logic cells are three geometrical constants: the number of rows, N_{rows}, the number of columns, $N_{columns}$, and the connectivity of the circuit which we refer to as the levels-back parameter, l.

The levels-back parameter l for a cell j ($j = 1, ..., N_{rows}$) in column i ($i = 1, ..., N_{columns}$) defines how many columns of cells to the left of column i can have their outputs connected to the inputs of the cell, this also applies to the final circuit outputs, and if $(i-l) <= l$, then any of the primary inputs can be connected to the cell in question. Also the primary outputs can be connected to the cell outputs according to the levels-back parameter. For example if the levels-back parameter is 2 for the circuit geometry shown in Figure 1, then the cells numbered from 17 to 21 can be additionally defined as outputs of this circuit.

Table 1. Cell gate functionality according to negative gene value in chromosome

Gene value	Gate function
-1	$MAX(x_1,x_2)=x_1 \vee x_2$
-2	$\overline{MAX(x_1,x_2)} = x_1 \vee x_2$
-3	$TSUM(x_1, x_2)=x_1 \oplus x_2$
-4	$\overline{TSUM(x_1,x_2)} = x_1 \oplus x_2$
-5	$TPRODUCT(x_1, x_2)=x_1 \otimes x_2$
-6	$\overline{TPRODUCT(x_1,x_2)} = x_1 \otimes x_2$
-7	$MIN(x_1, x_2)=x_1 \wedge x_2$
-8	$\overline{MIN(x_1,x_2)} = x_1 \wedge x_2$
-9	$MODSUM(x_1, x_2) = x_1 + x_2$
-10	$\overline{MODSUM(x_1,x_2)} = x_1 + x_2$
-11	$MODPRODUCT(x_1, x_2) = x_1 \cdot x_2$
-12	$\overline{MODPRODUCT(x_1,x_2)} = x_1 \cdot x_2$
-13	$\overline{x}$

A *chromosome* defines the connections in the network between the MV logic cells. In a circuit of 2-input, 3-valued logic gates each gate is represented by triple of integer numbers $<c^1 c^2 c^3>$ which define the connectivity between gates and the type of the examined gate.

The i-th cell is represented in the circuit by the integers $c_i=<c_i^1 c_i^2 c_i^3>$. Gene c_i^3 defines the type of MV gate. If $c_i^3<0$, then the i-th cell is two-input one-output logic gate, else when $c_i^3 \geq 0$ the i-th cell a 3-1 MUX gate with single control input c_i^3. In the case of two-input one-output logic gate, the genes c_i^1 and c_i^2 represent the first and second input of logic gate respectively. If c_i^1, $c_i^2 < (2n+r)$, then the inputs of the i-th gate are connected to the primary inputs of circuit, otherwise these inputs are linked to the outputs of the c_i^1-th and c_i^2-th gates respectively. The gene $c_i^3<0$ defines the type of logic gates which are coded according to Table 1. For example the cells numbered 13 and 20 shown in Fig

1 are 2-input 3-valued gates. The cell 13 describes function $f_{13} = \overline{MODPRODUCT(2, x_1)}$ because gene $c_{13}{}^1=2$ defines logical constant, $c_{13}{}^2=3$ corresponds to the primary variable x_1 and the gene $c_{13}{}^3=-12$ describes *MODPRODUCT* gate with inverted output. The cell 20 describes logic function $f_{20} = TPRODUCT(\overline{x_3}, f_{13})$, where function f_{13} denotes behavior of cell 13. The gene $c_{20}{}^1=8$ determines the inverted variable $\overline{x_3}$. The gene $c_{20}{}^2=13$ shows the connection between 13 and 20 cells and is considered as second variable of *TPRODUCT* function. The third gene $c_{20}{}^3=-5$ encodes *TPRODUCT* type for the cell 20. So, the output of cell 20 can be described as $f_{20} = TPRODUCT(\overline{x_3}, \overline{MODPRODUCT(2, x_1)})$. In the case of MUX gate the gene $c_i{}^1$ describes the first input of MUX gate, the next inputs of MUX gate are connected to the cells numbered $(c_i{}^1+1)$, $(c_i{}^1+2)$, ..., $(c_i{}^1+r-1)$. Note that the gene $c_i{}^2$ is not used in this case. For example the cell 12 shown in Fig 1 is T-gate (MUX gate) with control input 3 ($c_{12}{}^3=3=x_1$) and with inputs 4, 5 and 6 respectively (the first input is $c_{12}{}^1=4=x_2$), the second input is $5=x_3$ and the third input is $6= \overline{x_1}$. Note that the gene $c_{12}{}^2=7$ is not used. The last genes of chromosome describe the m outputs of implemented function and define the outputs of cell, which should be considered as circuit outputs. For instance the output of cell 20 defines the logic function y_1 described by the expression: $y_1 = f_{20} = TPRODUCT(\overline{x_3}, \overline{MODPRODUCT(2, x_1)})$.

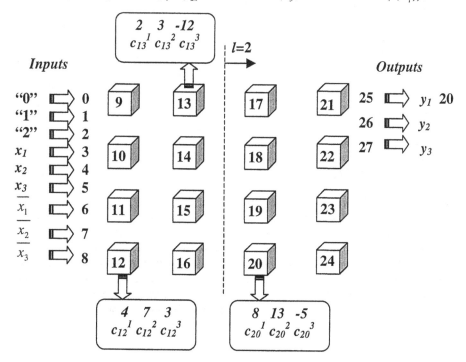

Fig 1. A 4 x 4 geometry of uncommitted MV logic cells with netlist numbering

The *fitness function* is defined as the percentage of the correct output digits for every input combination of the implemented MV functions. The *uniform crossover* and *mutation operators* are built in the traditional way for integer representation chromosome. The number of genes mutated in current population is defined by the *mutation rate*. The *crossover rate* shows how many chromosomes breed in current population. *Elitism* is used to promote the best chromosome obtained from one population to the next. The *selection operation* used is a variation on standard tournament selection (of size two) in which the winner of the tournament (the chromosome with the greater fitness) between two chromosomes chosen at random is accepted with certain probability (otherwise the loser of the tournament is chosen). This probability is called the *tournament discriminator*. If this probability is set to unity, then the tournament becomes the standard tournament sized two selection mechanisms. If this probability is less than unity, then the selection pressure on the population is reduced. Note that often the chromosome contains cells that are not actually connected to any of the outputs. The process of removing these redundant cells is carried out for chromosomes with 100% functionality after the GA has completed.

3   Relationship Between the Levels-Back Parameter, Circuit Geometry and Ga Performance

Table 2. Initial data

Population size	30
Crossover rate	100
Mutation rate	15
Number of generations	200
Number of GA runs	100
PLA file processing	Add32.pla
Radix of logic	3
Tournament Discriminator	70

The experiments in this paper were aimed at correctly evolving the functionality of a one-digit 3-valued adder with output carry. This is a circuit with 2 inputs and 2 outputs and requiring 9 input and output conditions for full specification (see Appendix 1.). The main purpose of these experiments was to investigate how the levels-back parameter and circuit geometry affect the performance of the GA. We investigate how the percentage of the 100% functionality circuit evolved by the GA and mean fitness values depend on the levels-back parameter and the number of rows and columns. In this series of experiments the functional basis contains all the MV gates shown in Table 1. The *functional basis* is the set of MV operators used to synthesize the circuit. The logical constants as well as the primary and inverted input variables can only be connected in accordance with the levels-back parameter. This prevents cells becoming wires in the circuit. For example cell MAX with one of the input 0 passes the signal of another input to next cell without change. The GA parameters used here are shown in Table 1.

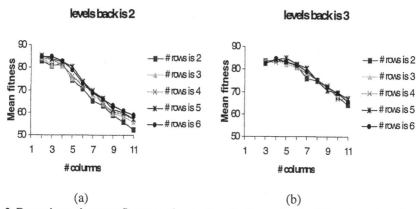

Fig. 2. Dependence the mean fitness on the number of columns for the different number of rows

Figures 2 and 3(a) show that the GA performance is not influenced strongly by the number of rows. However it is clear that the number of columns in the chosen geometry has a significant effect.

We can see that the number of rows used in the geometry has some effect since with $l=1$ the number of rows must be at least as large as the number of outputs.

Associated with every logic function having a minimum number of logic cells is a minimum depth (number of columns) which we denote, d. If the number of columns is less than d then clearly we cannot evolve 100% solutions.

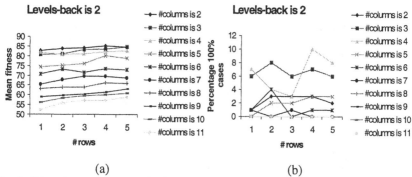

Fig. 3. Dependence of percentage 100% cases and the mean fitness on the number of rows for the different number of columns

Examining Figure 4 (a)-(d) (note that the graphs have been plotted as continuous functions for ease of viewing) we see that as l increases the minimum geometry at which we can obtain 100% solutions has to increase. This is understandable since the probability that we would connect the inputs to the first column of cells and the outputs to the last column of cells decreases quickly. Thus the lowest cut-off point for circuit geometry (which indicates the minimum number of columns at which the 100% functionality circuit are not evolved) must increase. The marked cut-off of the highest point, b, appears to indicate that only circuits within a certain size range are likely to be evolved so that when the geometry becomes too large (in terms of columns) it becomes no longer possible to connect up even the largest circuit.

There are certainly a number of interesting features of these graphs (i.e. the marked modality) and it looks as if there are species of circuits each with a characteristic size which become more or less likely to be evolved as the number of columns of cells and the levels-back parameter are altered.

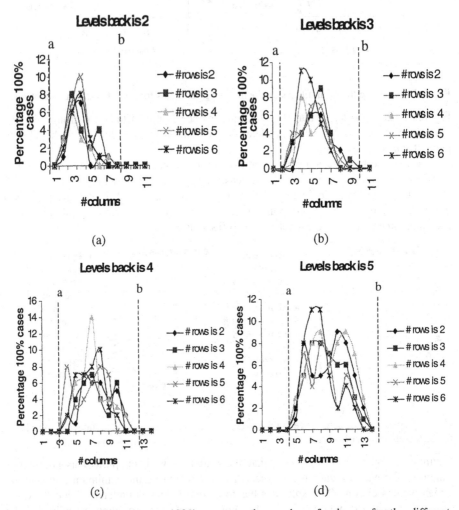

Fig. 4. Dependence percentage 100% cases on the number of columns for the different number of rows (levels-back = 2, 3, 4 and 5)

The dependence of the mean fitness with the number of columns is shown in Figure 5 for levels-back parameters equal to 2 and 3 respectively. Points a' and a'' show the minimum fitness at which 100% functionality for the one-digit 3-valued adder with carry can be achieved. At these points the number of columns in the circuit geometry are 8 and 9 respectively. It is clear that if the mean fitness is less than 60% then the probability of obtaining a circuit with 100% functionality drops to zero. If the mean fitness is 85% or more for the optimal number of columns (points *b′* and *b″*) circuits with 100% functionality can be evolved with the highest probability.

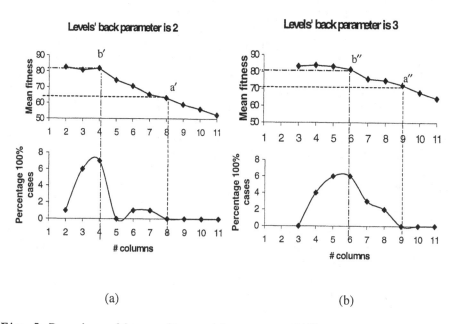

(a)   (b)

Fig. 5. Dependence of the mean fitness and the percentage of 100% cases on the number of columns.

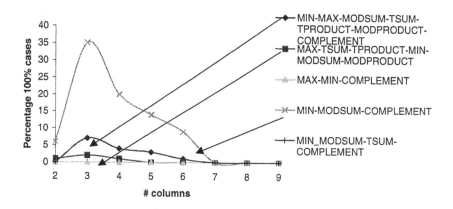

Fig.6. Dependence the GA performance on the set of MV gates

4 Influence of the MV Gate Set on the GA Performance

In this section we will discuss how GA performance depends on the set of MV gates chosen for circuit design.

Figure 6 shows how GA performance depends on the set of MV gates which are used in the circuit design. It was found that of the gate sets used the best set was MIN-MODSUM-COMPLEMENT. This basis is startlingly better than some of the other bases. This set allows a five fold improvement in the GA performance in comparison with the MIN- MAX- TSUM- TPRODUCT- MODSUM- MODPRODUCT- COMPLEMENT set and a ten fold improvement over the MIN-MAX-TSUM-TPRODUCT-MODSUM-MODPRODUCT set. Note that the attempts to evolve circuits using only MIN-MODSUM-TSUM gates or MAX-MIN-COMPLEMENT gates didn't give any 100% circuits. These experiments were carried out using the same initial parameters as before but with a fixed number of generations equal to 500. For each MV basis the GA runs 100 times. The vertical axis of Fig.6 shows the percentage of MV logic circuits evolved with 100% functionality in 100 runs of the GA.

It is interesting to note that we expected to receive the highest percentage of 100% functionality for the case when we use *all* well-known MV gates because it gives a bigger choice of any MV gate for evolved circuit than any particular functionally complete basis with has a much more restricted number of MV gates. However experimental results show that the correct choice of MV logic gate set allows us significantly increase the GA performance without changing any of the GA parameters. This indicates the importance of choosing the architecture on which to conduct the evolutionary process, and indeed, such a choice has a much more significant impact on the success of the evolution than fine tuning the GA parameters.

5 An Evolved One-Digit 3-Valued Adder with Output Carry Circuit

The arithmetic MV circuits synthesized using the proposed method have an unusual structure. It is for this reason that the obtained results are very interesting and allow us to consider arithmetic MV circuits from a new viewpoint.

Now we present some of the arithmetical circuits that we have so far been able to evolve using the above method. This circuit implements the 3-valued one-digit adder with output carry.

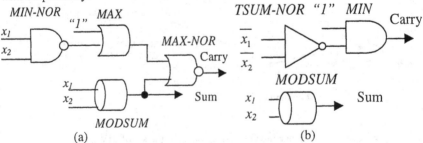

Fig. 7. Evolved solution for the one-digit 3-valued adder with output carry using (a) MIN-MAX-MODSUM (b) TSUM-MIN-MODSUM gates (note small circles represent inversion)

Using only MIN-MAX-MODSUM allowed the circuit shown in Figure 7(a) to be evolved (with a 3 column , 2 row geometry). The sum component in this design is implemented in the optimum way as it uses only one MODSUM gate. The analytical description of MV logic gates is given in Table 1. This circuit can be described analytically as follows:

$$y_{sum} = x_1 + x_2$$

$$y_{carry} = \overline{((x_1 \wedge x_2) \vee 1) \vee (x_1 + x_2)}$$

When a set of basic gates had been changed to (with the 2x2 geometry) TSUM, MAX, MIN and MODSUM the design shown in Figure 7 (b) was evolved. The sum component in this circuit is the familiar sum-digit 3-valued circuit of conventional adders. Note that the sum and carry components in this design are implemented separately and can be expressed as follows:

$$y_{sum} = x_1 + x_2$$

$$y_{carry} = \overline{\overline{x_1} \oplus x_2} \wedge 1.$$

When the geometry was constrained to 2x2 and the set of basis gates chosen contained only MODSUM, TSUM and MAX gates with direct and inverted inputs and outputs the optimum designs shown in Figure 8 was obtained. This was a gratifying result to obtain as it is clear that these designs are an optimum solution. These circuits were previously unknown. The analytical representation of these designs are shown in Fig. 8 also.

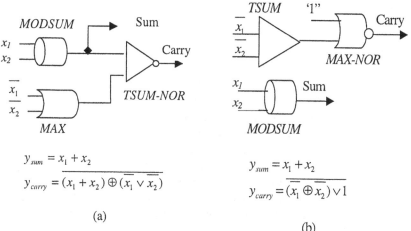

$$y_{sum} = x_1 + x_2$$
$$y_{carry} = (x_1 + x_2) \oplus \overline{(x_1 \vee x_2)}$$

(a)

$$y_{sum} = x_1 + x_2$$
$$y_{carry} = \overline{(\overline{x_1} \oplus x_2) \vee 1}$$

(b)

Fig. 8. Evolved optimum solution for the one-digit 3-valued full adder using MODSUM, TSUM and MAX

Evolving the one-digit 3-valued adder was easier to do with a larger geometry but resulted in a less efficient circuit. For instance, the circuits shown in Figure 7 have also been obtained with much larger geometries (3x4, 4x3, 4x4, 5x5, 4x6, 4x8, 4x10). Choosing too small a geometry ran the risk that no 100% solutions could be found because it was physically impossible to build the required functionality with few gates.

While using too large a geometry simply gave the GA too many possibilities to work with and it struggled to find the fully functional solutions.

6 Summary

In this paper it has been shown that by evolving a linear chromosome of cell functionalities and connectivities based on a rectangular array of logic cells it is possible to evolve both traditional and novel designs for arithmetical MV circuits. The method uses a genetic algorithm to evolve both a netlist structure and functionality for MV logic cells. The fitness function tests the functionality of the circuit. This approach for the first time allows us to synthesize combinational MV circuits in different functionally complete basis or a combination of these. The method allows the synthesis of very novel circuit structures, which have never been seen before. We have evolved the 3-valued 2-digit adder with carry as an example. The results of some of the experiments showed that it is possible to improve the GA performance considerably by choosing carefully the number of columns of cells used and the levels-back parameter, but that the number of rows in the geometry was less important.

There are still many avenues for further work. Other ways of representing rectangular arrays of logic cells may be devised and also, the relationship between cell connectivity and the evolvability of designs has still to be explored, this would involve examining a suitable concept of cell-neighborhood. It is a feature of the current technique that one has to specify the functionality of the target circuit using a complete truth table, however this is impractical for circuits with large numbers of inputs. Further investigations are under way to see if these findings are carried over to other MV benchmarks.

References

1. Fogarty T. C., Miller J. F., Thomson P.:Evolving Digital Logic Circuits on Xilinx 6000 Family FPGAs. The 2nd Online Conference on Soft Computing (1997), now pulished in Soft Computing in Engineering Design and Manufacturing, Roy R., and Pant R. K. (eds), Springer-Verlag, London, (1998) 299 - 305

2. Goeke M., Sipper M., Mange D., Stauffer S., Sanchez E., Tomassini M.: Online autonomous evolware, in Proceedings of The First International Conference on Evolvable Systems: From Biology to Hardware (ICES96), now published in Lecture Notes in Computer Science Vol. 1259, Springer-Verlag, Heidelberg, (1997) 96-106.

3. Hurst S.L.: MV Logic: Its Status and its Future, IEEE Transactions on Computers. Vol. C-33, (1984) 1160 - 1179

4. Iba I., Iwata M., Higuchi T: Machine Learning Approach to Gate-Level Evolvable Hardware. Proc. Of the 1st Int. Conference on evolvable Systems: From Biology to Hardware (ICES'96), now published in Lecture Notes in Computer Science, Springler-Verlag, Heidelberg, (1996) 118 - 135

5. Kalganova T., Miller J. F.: Evolutionary Approach to Design MV Combinational Circuits. Proc. of the 4th Int. conference on Applications of Computer Systems, ACS'97. Szczecin, Poland (1997) 333 – 339

6. Kalganova T., J.F. Miller and N. Lipnitskaya Multiple-Valued Combinational Circuits Sythesised using Evolvable Hardware Approach, Proc. of the 7th Workshop on Post-Binary Ultra Large Scale Integration Systems (ULSI'98) in association with ISMVL'98, Fukuoka, Japan, May 27-29, 1998.

7. Koza J. R.: Genetic Programming, The MIT Press, Cambridge, Massachusetts (1992)

8. Koza J.R., Andre D., Bennet III F.H., Keane M.A: Design of a High-Gain Operational Amplifier and Others Circuits by Means of Genetic Programming. Proc. Of the 6th Int. Conference on Evolutionary Programming, now published in Lecture Notes in Computer Science, Vol. 1213, (1997) 125 - 135

9. Mariani R., R. Roncella, R. Saletti, P. Terrini A Useful Application of CMOS Ternary Logic to the Realization of Asynchronous Circuits. Proc. of the 27th IEEE Int. Symposium on MV Logic (1997) 203 - 208

10. Miller J. F., Thomson P., Fogarty T. C.: Designing Electronic Circuits Using Evolutionary Algorithms. Arithmetic Circuits: A Case Study, chapter 6, in Genetic Algorithms and Evolution Strategies in Engineering and Computer Science: Recent Advancements and Industrial Applications. Editors: D. Quagliarella, J. Periaux, C. Poloni and G. Winter, published by Wiley, (1997)

11. Thompson A.: An evolved Circuit, Intrisic in Silicon, Entwined with Physics. Proc. Of the 1st Int. Conference on Evolvable Systems: From Biology to Hardware (ICES'96), now published in Lecture Notes in Computer Science, Springler-Verlag, Heidelberg, (1996) 390 - 405

12. Zebulum R., M. Vellasco, M. Pacheco: Evolvable Hardware Systems: Taxonomy, Survey and Applications. Proc. Of the 1st Int. Conference on Evolvable Systems: From Biology to Hardware (ICES'96), now published in Lecture Notes in Computer Science, Springler-Verlag, Heidelberg, (1996) 344 - 358

APPENDIX 1.

Truth table for 3-valued 2-input adder with output carry (Add32.pla)

x_1	x_2	carry	sum
0	0	0	0
0	1	0	1
0	2	0	2
1	0	0	1
1	1	0	2
1	2	1	0
2	0	0	2
2	1	1	0
2	2	1	1

Adaptation in Co-evolving Non-uniform Cellular Automata

Vesselin Vassilev and Terence Fogarty

Department of Computing, Napier University, Edinburgh EH14 1DJ,
v.vassilev@dcs.napier.ac.uk and t.fogarty@dcs.napier.ac.uk

Abstract. *Cellular Programming* is a model of co-evolving dissipative systems which *absorb* information and *dissipate* the useless information about the fitness landscapes on which co-evolution is performed. The information convection in the population determines a *dissipative structure* of the co-evolving non-uniform cellular automata which depends on how far from equilibrium the system is. We show that Cellular Programming is capable of demonstrating an adaptive behaviour somewhere between perfect order and complete disorder where the dissipative structure of co-evolving non-uniform cellular automata is complex.

1   Introduction

Non-uniform cellular automata (CAs) [23, 24, 20, 19] are discrete self-organising systems of simple locally connected interacting elements which may contain different transition rules. The massive parallelism and heterogeneity of the non-uniform CAs, as well as the CAs capabilities to exhibit emergent computation [25, 13, 14] means that the non-uniform CAs have the potential to perform complex computations [17], and therefore, to be used in the implementation of complex hardware and artificial models [8, 18].

Recently, genetic algorithms [9, 1] have been used to evolve CAs to perform a particular computational task [13, 14, 5, 6]. The goal has been to find a single CA rule that applied to all elements of the CA lattice to perform density or synchronisation tasks. However, it has been shown that the tasks can be arduous for uniform CAs [10], and thus, a *Cellular Programming* (CP) approach has been proposed where non-uniform CAs are evolved [17].

Cellular Programming is a model of an adaptive machine for building emergent computation by co-evolution [17]. It is a system in which self-organisation is implemented by a population of cells which co-evolve to attain a global coordination in solving a particular computational task. Each cell has a fitness value, given by a fitness function that evaluates the cell's contribution to the solution of the task. Thus the problem is transformed to a fitness landscape [26] in which each point is represented by a cell and its temporal fitness value, and the cells of two neighbouring points are different from one another by one locus. The evolution driven by *mutation*, *recombination*, and *selection* is performed on fitness landscapes which *fluctuate* since the fitness value of each cell depends on the efficiency of co-operation in the population [4].

In this paper, we are interested in how adaptation appears in CP and especially how to build an adaptive cellular machine. We investigate CP for density and synchronisation tasks in order to show that CP is capable of exhibiting an adaptive behaviour when the complexity of the dissipative structure of the population or the co-evolved non-uniform CA is high. By moving the co-evolving system from perfect order to complete disorder, we reveal that when the average complexity of the population structure is maximal, CP performs an adaptive evolutionary search. We show that the adaptive search is a sequence of epochs in which the population is in punctuated equilibrium.

The next section introduces Cellular Programming and the computational tasks by which CP are investigated. Section three offers the underlying idea of the paper that a relation between the evolutionary search and the structure of the co-evolved non-uniform CA exists. In section four we propose a brief qualitative and quantitative overview of CP in order to reveal how the complexity of population structure is related to the evolutionary dynamics. Finally, discussion is made and conclusions are derived.

2 Cellular Programming for Density and Synchronisation Tasks

Initially, emergent computation in two-state, one-dimensional uniform CAs has been studied [5, 6, 13, 14] where a CA rule has been searched that is capable of performing some non-trivial computation as density or synchronisation tasks. In short, the density classification task is to locate CA which can converge to a configuration of 1's when the number of 1's in the initial configuration of the lattice of N cells exceeds $N/2$, and to 0's otherwise. Respectively, the global synchronisation task is to determine CA which can subside to a simple oscillation between 0's and 1's configurations of the lattice (see Figure 1). The tasks are non-trivial because a particular global behaviour of the CA is expected to be attained by local interactions among the cells.

In order the performance of the model to be increased non-uniform CAs for non-trivial computational tasks have been studied [19]. Thus, Cellular Programming has been introduced in which the elements of the population co-evolve to attain an ensemble of CA rules or one non-uniform CA for the computational task that has to be solved. The algorithm has been defined as follows:

1. Initialise the population with CA rules at random, uniformly distributed among different λ values [11].
2. Generate $2 * (N + 1)$ initial configurations at random with uniformly distributed density of 1's over the interval $[0, 1]$, N is the size of the population.
3. Run the CA $N + 1$ steps for each configuration and attach a fitness value to each cell by determining the number of correct states of the cell in the final configurations.
4. Label each cell with a fitness score that specifies the number of fitter neighbouring cells and update the population as follows - when the fitness score of

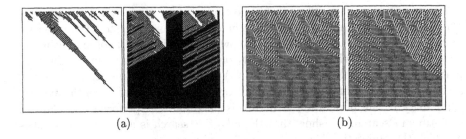

Fig. 1. The (a) density ($\rho(0) \approx 0.483$ (left) and $\rho(0) \approx 0.51$ (right)) and (b) synchronisation tasks ($\rho(0) \approx 0.497$ (left) and $\rho(0) \approx 0.503$ (right)). The results are obtained by co-evolving non-uniform CAs with grid size 149 and rule radius 2. The average fitness and computation accuracy of the non-uniform CAs are (a) 0.93877 and 0.8433, and (b) 0.9432 and 0.8633, respectively.

the cell is: 0, leave the rule unchanged; 1, replace the rule with the mutated fitter neighbour; 2 or higher, replace the rule by recombination of two fitter neighbouring rules, followed by mutation.

5. If not finished, go to step 2.

Here, in order to estimate how successful an evolutionary run is, we examine the average fitness of the population, or the average of the fitness values of all cells within the population, and the computation accuracy which is the percentage of the initial grids, used in the estimation of the average fitness, on which successful computations of the non-uniform CA are performed. Figure 1 represents the results of two successful evolutionary runs of the cellular programming algorithm for density and synchronisation tasks.

3 Cellular Programming: Behaviour and Patterns

As a model of systems of strongly interacting elements Cellular Programming inherits the features of self-organisation phenomena studied in the physics of inanimate systems [15, 16]. CP algorithms are nonlinear dissipative systems in which *bulk* motions under the effect of information inhomogeneties within the ensemble of interacting cells can be observed [21, 22]. The dissipation begets self-organisation which leads to the emergence of the structural organisation of the entities [2, 3]. The information convection within the ensemble determines a *dissipative structure* which is dependent on how far from equilibrium the system is. The analogy to the dissipative inanimate systems is straightforward, where for instance in fluids, by changing the temperature inhomogeneties the structural organisation of the systems can be altered from a simple linear termal conduction flows through the cells of Bénard to complex turbulent patterns even further to simple highly entropic structures [16]. In CP, the source of randomness in the population is the probability for mutation or the probability to alter each gene of the cell. If we move the system from order to disorder by varying the mutation probability, we can observe different spatio-temporal patterns within the population whose quantity characteristics are strongly related to the dynamical behaviour of the system. For instance, for low values of the mutation probability the population dynamics vanish and the population co-evolves to a homogeneous ensemble of CA rules. On the contrary, when the mutation probability is too high, the population becomes a strongly heterogeneous ensemble of cells which represent a heterogeneous set of landscape optima, and the behaviour is specified as completely disordered.

The adaptivity is somewhere between perfect order and complete disorder. Adaptation is related to the ability of the population to investigate effectively the fitness landscape. An adaptive search can be observed far enough from equilibrium and disequilibrium to avoid becoming frozen or aimless, respectively, where the population dynamics are well fitted to the structure of the landscape. It leads to appearance of variety of coherent structures within the population or spatio-temporal *patterns* with complex structural organisation. In the context of the co-evolutionary search the phenomenon *adaptaion* can be described as a population flow on the landscape which is able to avoid trapping in local optima. Initially, the cells are randomly scattered on the landscape and the information represented within the population is *perturbed*. The system dissipates the initial information in an attempt to structure the information and to extract a promising fitness landscape area in which the corresponding non-uniform CA is capable of emergent computation. The process of dissipation of the initial information subsides, and the population is clustered around the most attractive CA. The population is capable of escaping the located optimum by a perturbation, caused by the landscape fluctuations, which is followed by shrinking around another attractive local optimum. The process leads to an adaptive population flow on the landscape which can be described as a sequence of epochs in which the population is in punctuated equilibrium.

4 Complexity and Adaptation

Various concepts for measuring complexity of systems of interacting elements exist [2,3]. In [21] we proposed a measure of complexity of evolving systems which quantifies the self-dissimilarity of the population structural organisation. However, our measure is inapplicable to the co-evolving systems. A major impediment is that the fitness landscapes fluctuate and the evaluation of the population structure in the landscape scale is computationally hard. The complexity of the dissipative structure of a population of cells can be quantified by measuring the population randomness [12, 7]. Consider a non-uniform CA, P_t, obtained on an evolutionary step t. Then the complexity of the population structure is

$$\mathcal{C}_\mu(P_t) = \mathcal{H}_\mu(P_t)\mathcal{D}_\mu(P_t), \tag{1}$$

where $\mathcal{H}_\mu(P_t)$ is the information entropy of the population, $\mathcal{D}_\mu(P_t)$ is the population relative information, and μ is the source of randomness. The complexity,

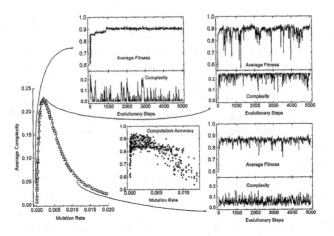

Fig. 2. The density task: Average complexity of the population structure versus mutation probability (each point is a mean of the average complexities of 5 evolutionary runs). The selected plots represent the computation accuracy of the best ensemble of CA rules for each evolutionary run, and the average fitness and complexity of three evolutionary runs with different average complexities. The results are obtained by co-evolving non-uniform CAs with grid size 75 and rule radius 2.

$\mathcal{C}_{\mu}(P_t)$, estimates the balance between the forces by which CP absorbs and dissipates, respectively, information about the landscapes on which co-evolution is performed.

Now, having the measure of complexity, we show that adaptation in CP can be achieved when the complexity of the dissipative structure of the population of CAs is high. We performed two series of experiments - density and synchronisation tasks - in which by varying the mutation probability, we explored how the behaviour of CP is related to the complexity of the population structure. The plots depicted in Figure 2 represent the results obtained for the density classification task. Each point of the average complexity plot is a mean of the average complexities, $\sum_t \mathcal{C}_{\mu}(P_t)$, of 5 evolutionary runs. The other plots represent the computation accuracy (the percentage of initial grids on which computation is successfully performed) of each evolutionary run, and the average fitness and complexity of three evolutionary runs with surmised deterministic, adaptive and chaotic behaviours.

Figure 2 reveals that highest average fitness and highest computation accuracy are attained when the average complexity of the dissipative structure of the co-evolved non-uniform CA is high. The plots illustrate that for low values of the probability for mutation the population dynamics are low and CP performs simple hillclimbing to frozen lowly entropic ensembles of CA rules where the co-evolution subsides. On the other side, when the mutation probability is too high, the average fitness plot is typical, demonstrating no preference to any landscape optimum, and the population is co-evolved to highly entropic struc-

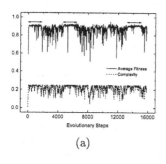
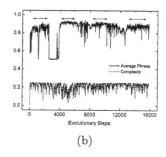

(a)	(b)

Fig. 3. The average fitness and complexity of an adaptive search performed by cellular programming for (a) density and (b) synchronisation tasks. The results are obtained by co-evolving non-uniform CAs with grid size 149 and rule radius 2. The CAs with highest computation capabilities are shown in Figure 1.

turless non-uniform CA. Note that the both examples are characterised with low complexity of the population structure.

Now looking at the plots of average complexity and computation accuracy, and the average fitness plot, for which complexity is maximal (Figure 2), we can see how the performance of CP and the population dynamics are related to the complexity of the structural organisation of the population. The adaptive evolutionary search appears to be a sequence of epochs in which the population is in punctuated equilibrium. The epochs are outlined by higher amplitudes of the average fitness plot's fluctuations. The phenomenon of metastability appears as a consequence of the population efforts to preserve the located local optimum. Since the co-evolution is performed on fluctuating fitness landscapes each epoch may end and a transition to another local optimum may begin.

Metastability can also be observed in Figure 3 which represents the results of co-evolving non-uniform CAs with grid size 149 and rule radius 2 for density and synchronisation tasks. It can be seen that the average fitness increases when an epoch is commenced, and decreases when a transition to another epoch must be performed. Again, the plots in Figure 3 suggest that adaptive evolutionary search can be attained when the dissipative structure of the population is complex, and it can be presented as a sequence of epochs in which the adaptive population investigates the most attractive local optima. The difference between the average fitness plots for density and synchronisation tasks suggests that adaptation is also related to the structure of the fitness landscapes.

5   Discussion and Conclusions

In this paper, Cellular Programming for density and synchronisation tasks were investigated. The underlying idea was that during the evolution the population of co-evolving CAs adopts a dissipative structure which depends on how far from equilibrium the system of interacting cells is. Thus, in order to explore when CP can be successful we proposed to measure the complexity of the structural organisation of the co-evolved non-uniform CA. It can be suspected that when the average complexity of the population structure is high, CP will only look for non-uniform CAs with strongly heterogeneous structures. However, we saw that during the search, the complexity is fluctuating, and therefore, non-uniform CAs with less complex structure can be located. We performed a lot of experiments for CAs with rule radius 1, where the attained optima (average fitness approximately 0.933) were populations with very simple structure (see [19]). Consequently, the average complexity is related to efficiency of the evolutionary search, or the adaptive capabilities of the population of co-evolving cells. We showed that adaptation can be achieved between order and disorder where the average complexity of the population structure is high.

By measuring complexity, we observed that the adaptive evolutionary search is a sequence of epochs in which the population is in punctuated equilibrium. Since the fitness landscapes are fluctuating each epoch may end, the equilibrium attained by co-evolution is infringed and a transition to another epoch can be observed. We showed that the transitions between the epochs can be different and that they depend on the structure of the CAs landscapes which fluctuate. In this context, reasonable questions are how the adaptive capabilities of the cellular programming algorithms are dependent on the landscape fluctuations and how the non-uniform CAs landscapes fluctuate in different stages of the evolutionary search? These questions are open for future research.

Acknowledgements

We are grateful to Dominic Job and Vanio Slavov for stimulating discussions and constructive criticism.

References

1. Th. Bäck, D. B. Fogel, and Z. Michalewicz, eds., *Handbook of Evolutionary Computation* (Oxford University Press, NY, 1997).
2. C. H. Bennett, Dissipation, Information, Computational Complexity and the Definition of Organization, in: D. Pines, ed., *Emerging Syntheses in Science* (Addison-Wesley, 1985) 215–233.
3. C. H. Bennett, On the Nature and Origin of Complexity in Discrete, Homogeneous, Locally-Interacting Systems, *Foundations of Physics* **16**(6) (1986) 585–592.
4. T. Cheetham, Adaptation on Fluctuating Fitness Landscapes: Speciation and the Persistence of Lineages, *J. Theor. Biol.* **161** (1993) 287–297.
5. R. Das, M. Mitchell, and J. P. Crutchfield, A Genetic Algorithm Discovers Particle-based Computation in Cellular Automata, in: Y. Davidor, H.-P. Schwefel, and R. Männer, eds., *Parallel Problem Solving from Nature III*, Lecture Notes in Computer Science 866 (Springer-Verlag, 1994) 344–353.
6. R. Das, J. P. Crutchfield, M. Mitchell, and J. E. Hanson, Evolving Globally Synchronised Cellular Automata, in: L. J. Eshelman, ed., *Proc. 6th Int. Conf. on Genetic Algorithms* (Morgan Kaufmann, San Francisco, CA, 1995) 336–343.

7. D. P. Feldman, and J. P. Crutchfield, Measures of Statistical Complexity: Why?, *Physics Letters A* **238**(4-5) (1998) 244–252.

8. F. A. Gers, and H. de Garis, CAM-Brain: A New Model for ATR's Cellular Automata Based Artificial Brain Project, in: T. Higuchi, M. Iwata, and W. Liu, eds., *Proc. 1st Int. Conf. on Evolvable Systems*, Lecture Notes in Computer Science 1259 (Springer-Verlag, 1996) 437–452.

9. J. Holland, *Adaptation in Natural and Artificial Systems* (MIT Press, Cambridge, MA, 1992).

10. M. Land, and R. Belew, No Perfect Two-State Cellular Automata for Density Classification Exists, *Physical Review Letters* **74**(25) (1995) 5148–5150.

11. C. G. Langton, Computation at the Edge of Chaos: Phase Transitions and Emergent Computation, *Physica D* **42** (1990) 12–37.

12. R. López-Ruiz, H. L. Mancini, and X. Calbet, *Phys. Lett. A* **209** (1995) 321–326.

13. M. Mitchell, P. T. Hraber, and J. P. Crutchfield, Revisiting the Edge of Chaos: Evolving Cellular Automata to Perform Computations, *Complex Systems* **7** (1993) 89–130.

14. M. Mitchell, J. P. Crutchfield, and P. T. Hraber, Evolving Cellular Automata to Perform Computations: Mechanisms and Impediments, *Physica D* **75** (1994) 361–391.

15. G. Nicolis, and I. Prigogine, *Self-Organization in Nonequilibrium Systems*, (Wiley Interscience, NY, 1977).

16. G. Nicolis, Physics of Far-From Equilibrium Systems and Self-Organisation, *The New Physics* (1988) 316–347.

17. M. Sipper, Co-evolving Non-Uniform Cellular Automata to Perform Computations, *Physica D* **92** (1996) 193–208.

18. M. Sipper, The Evolution of Parallel Cellular Machines: Toward Evolware, *BioSystems* **42** (1997) 29–43.

19. M. Sipper, *Evolution of Parallel Cellular Machines: The Cellular Programming Approach*, Lecture Notes in Computer Science 1194 (Springer-Verlag, 1997).

20. T. Toffoli, and N. Margolus, *Cellular Automata Machines* (MIT Press, Cambridge, MA, 1987).

21. V. Vassilev, T. Fogarty, and V. Slavov, Complexity and Adaptation in Evolutionary Algorithms, Technical Report (submitted for publication).

22. V. Vassilev, and T. Fogarty, Co-evolving Non-uniform Cellular Automata on the Edge of Chaos, Technical Report (will be submitted).

23. J. von Neumann, *Theory of Self-Reproducing Automata*, (University of Illinois, 1966).

24. S. Wolfram, Cellular Automata as Models of Complexity, *Nature* **311** (1984) 419–424.

25. S. Wolfram, Universality and Complexity in Cellular Automata, *Physica D* **10** (1984) 1–35.

26. S. Wright, The Roles of Mutation, Inbreeding, Crossbreeding and Selection in Evolution, in: D. F. Jones, ed., *Proc. 6th Int. Conf. on Genetics* **1** (1932) 356–366.

Synthesis of Synchronous Sequential Logic Circuits from Partial Input/Output Sequences

Chaiyasit Manovit, Chatchawit Aporntewan and Prabhas Chongstitvatana

Department of Computer Engineering, Faculty of Engineering, Chulalongkorn University
Bangkok 10330 Thailand
prabhas@chula.ac.th

Abstract. This work takes a different approach to synthesize a synchronous sequential logic circuit. The input of the synthesizer is a partial input/output sequence. This type of specification is not suitable for conventional synthesis methods. Genetic Algorithm (GA) was applied to synthesize the desired circuit that performs according to the input/output sequences. GA searches for circuits that represent the desired state transition function. Additional combination circuits that map states to the corresponding outputs are synthesized by conventional methods. The target of our synthesis is a type of registered Programmable Array Logic which is commercially available as GAL. We are able to synthesize various types of synchronous sequential logic circuit such as counter, serial adder, frequency divider, modulo-5 detector and parity checker.

1 Introduction

The conventional method to synthesize a sequential logic circuit requires knowledge of the circuit's behavior in the form of a state diagram. A sequential network is a common starting point for sequential synthesis system such as Berkeley Synthesis System SIS [1]. We aim to realize an evolvable hardware that can "mimic" other sequential circuit by *observing its partial input/output sequences*. In this case, circuit specifications are in the form of partial input/output sequences which are not suitable to synthesize a circuit by the conventional method. Genetic Algorithm (GA) [2,3] is used to search for circuits that represent the desired state transition function. Additional combination circuits that map states to the corresponding outputs are synthesized by conventional methods. The simulated evolution has been used to synthesize a finite state machine (FSM) in [4,5] where the resulting FSM can predict the output symbol based on the sequence of input symbols observed. In contrast to representing circuits as FSMs [6] proposes the automated hardware design at the Hardware Description Language level using GA. [7] describes the evolution of hardware at function-level based on reconfigurable logic devices. [8,9] evolved circuits at the lowest level, in the actual logic devices, using real-time input/output. Our work is similar to [2,3] in the use of FSM but we use FSM as the *model* of the desired circuit behavior. We aim to evolve the circuit at the logic device level similar to [9]. The following sections describe our synthesis method, the experiment in synthesizing various simple sequential circuits and the analysis of the result. Notably the analysis about the length of the input/output sequence which has implication on the efficiency of the synthesizer.

2 The Synthesizer

We use Programmable Logic Device (PLD) as our target structure. For simplicity, we used GAL structure which is composed of rows of two-level sum-of-product Programmable Array Logic (PAL) connected to D flip-flops. The implemented model has four logic terms per one flip-flop, and has a total of four flip-flops. A circuit is specified by a linear bit-string representing all connection points, which is entirely 256 bits in length. We applied Genetic Algorithm to synthesize simple circuits, such as counter, serial adder, frequency divider, modulo-5 detector and parity checker. The specification of a desired circuit is in the form of a partial input/output sequence. The circuit acquired from the evolution process will realize only the state transition part. The outline of the synthesizer's steps is as follows:

1 sample a partial input/output sequence from the target circuit
2 use the sequence as inputs of the synthesizer program
3 verify the resulting circuit if the run yields a solution within 50,000 generations (because GA is a probabilistic algorithm, not all runs are successful in yielding a solution by the specific generation)

3.1 Genetic Operations

Each individual is represented by a 256-bit bit-string. We defined genetic operators as follows:

1 Reproduction: Ten new offsprings survive in the next generation by selecting the first 10 fittest individuals ordered by combined rank method [10] (calculated from each individual's fitness rank and its diversity rank).
2 Crossover: More individuals being added to the population are produced by uniform crossover [11]. All possible pairs among 10 already selected individuals are used to produce new offsprings. That is, we will have 90 new individuals.
3 Mutation: Last 10 individuals being added are mutated version of the first 10 selected individuals. The mutation process is controlled so that it changes exactly 5 bits of each individual.

We also add some more constraints. First, we avoid creating a product term which always be "0". Second, connection points to unused input signals are left unwired. This reduces the search space of the problem and lets the program concentrate on the connection points that do affect the function of the circuit.

3.2 Fitness Evaluation

An individual is evaluated by the following steps:
1. feed one input to the circuit and clock the circuit
2. next state of the circuit would be mapped with the corresponding output, record the number of times the state has been mapped to output "0" and "1", independently
3. repeat steps 1 and 2 until the end of the sequence

After the sequence is completed, the fitness value of the individual, F, is computed by:

$$F = \sum_{i=0}^{i=S-1} f_i \quad \text{where} \quad f_i = \begin{cases} \max(p_i, q_i) & ; p_i = 0 \text{ or } q_i = 0 \\ -\min(p_i, q_i) & ; \text{otherwise} \end{cases} \tag{1}$$

where
f_i is the fitness value of state i
p_i is the number of times in which state i has to be mapped with output "0"
q_i is the number of times in which state i has to be mapped with output "1"
S is the number of states, equal to 16 for the GAL structure

Based on Moore's model, a state of an FSM must be mapped to only one output value. Therefore, any state that is mapped to both "0" and "1" will cause a penalty in the fitness value as shown in the formula. For Mealy's model, the evaluation is similar. The difference is that output values are mapped to transition paths instead of states.

3.3 Size of Input/Output Sequences

A partial input/output sequence, which is used as a circuit's specification, is attained by generating a sequence of inputs, feeding each one to the target machine and then recording the corresponding output that the machine gives. To be a general approach and yield a simple analysis, the input sequence is created at random with uniform distribution (i.e. at any time, the probability that the input bit be "0" and "1" are equal).

The input sequence should be long enough to exercise all aspects of the circuit's function. In other words, it should be able to test all paths of the state diagram of the circuit. As mentioned earlier in this paper, we use the GAL structure which can be programmed as a 16-state state machine, larger than the desired circuit's need. Therefore, the input sequence should also be long enough to exercise all paths of any 16-state circuit. If the sequence is too short, it may cause an ambiguity in describing the desired circuit. In this paper, we will call the length of the input sequence which is long enough to describe (exercise) only the desired circuit as *lowerbound length* and call the length of the input sequence which is long enough to exercise any 16-state circuit as *upperbound length*.

We can find the proper size of the input sequence by making an analogy to a dice rolling problem. Consider rolling a dice, how many times do we have to roll it, until all of its faces appear? This problem is called *waiting times in sampling* [12], in which we can find the *expected value* of the number of times by the following formula:

$$E(n) = n\left\{\frac{1}{1} + \frac{1}{2} + \cdots + \frac{1}{n}\right\} \quad ; \text{n is the number of faces}. \tag{2}$$

Consider a circuit that has i bit inputs, giving possible $I = 2^i$ input patterns, and has S states. We assume that at any time, the probability that the circuit be in any state is equal. So, from the starting state, we expect the number of state transitions to be $E(S)$ to traverse through all states of the state diagram. And at any state, we expect the number of inputs to be $E(I)$ to traverse through all paths from that state. Therefore, to traverse through all paths of the state diagram, we would expect the number of inputs, the length of the input sequence, to be:

$$L = E(S) \times E(I). \tag{3}$$

The lowerbound length is computed by using S equal to the number of states of the circuit. While the upperbound length is computed by using S equal to the maximum number of states which the GAL structure can represent, in this case, 16.

Table 1. Description of circuits in the experiment

Circuit	# of input bits	# of states		lower bound length		upper bound length
		Moore	Mealy	Moore	Mealy	
Frequency Divider	0	8	8	22	22	55
Odd Parity Detector	1	2	2	9	9	163
Modulo-5 Detector	1	6	5	45	35	163
Serial Adder	2	4	2	70	25	451

Note
Frequency Divider   give square wave outputs of clock's frequency divided by 8
Odd Parity Detector   give an output "1" when the number of "1" in inputs is odd
Modulo-5 Detector   give an output "1" when the current input is the $5n^{th}$ "1" ; $n \in I^+$
Serial Adder   give the sum of 2 inputs, inputs are feeding from LSB to MSB

The experiment was done with many input/output sequences of different lengths for each problem. The selected lengths are 10, 100, 1000, lowerbound and upperbound. For each length, at least 3 different random sequences of input/output are used. And the synthesizer was repeatedly run 50 times on each sequence. Thus, we had at least 3×50, equal to 150, independent runs for each length of the sequence for one problem. Table 1. shows the details of desired circuits and the calculated lengths of each problem.

3.4 Circuit Verification

In this experiment, the state diagrams of desired circuits are known. The state transition part of the state diagram of the resulting circuit can be derived by fixing the current state and feeding in all patterns of inputs and recording state transitions. We then compare to check if it is equivalent to the known one. They can be considered to be equivalent if and only if each state of one diagram can be matched with one state of the other.

In a general case, when we do not know the state diagrams of desired circuits, we propose one way to verify it. The concept is similar to the determination of the size of input/output sequences. An upperbound size of input/output sequence should be used to test the circuit. Only one sequence, however, might not suffice to test the correctness, especially of the very first states. It is possible that the input sequence leads the circuit to pass one state and never go back to that state again. In this case, we cannot be sure that other out paths from that state lead to the correct next states. Consequently, we should use several test sequences to be more confident.

4 Effort

The experiment is done on a 200 MHz workstation. In the worst case, one run uses 40 minutes and terminates at generation 50,000 with no solution. We will calculate and compare empirical computational effort using a method which is slightly adapted from Koza [13] as follows:

i generation number

M population size

$P(M, i)$ cumulative probability of yielding a correct solution by generation i

$W(M, i)$ cumulative probability of yielding a wrong solution by generation i

$X(M, i)$ instantaneous probability of yielding a wrong solution by generation i

$I(M, i, z)$ individuals that must be processed by generation i with probability z

$R(z)$ number of independent runs required to yield at least one successful run with probability z

z probability of satisfying the success predicated by generation i at least once in R independent runs $= 0.99$

$$z = 1 - \left[1 - P(M,i)\right]^{R} = 0.99 \tag{4}$$

$$R = R(M,i,z) = \left\lceil \frac{\log(1-z)}{\log(1-P(M,i))} \right\rceil \tag{5}$$

$$I(M,i,z) = M \times (i+1) \times R(z) . \tag{6}$$

We pay attention only to the correct solutions. We could not count the run yielding a wrong solution as a normal unsuccessful run. If we do count it, the calculated effort will be over-estimated. We have to subtract $I(M,i,z)$ by the number of individuals which are not produced after the run stopped. Empirically, there are $X(M, j) \times R$ runs that yield wrong solutions at generation j. Therefore, if we let the program run till generation i, $i > j$, the number of individuals which are not produced is equal to

$$M\sum_{j=0}^{j=i-1}(i-j)X(M,j)R = M\left\{\sum_{j=0}^{j=i-1}iX(M,j)R - \sum_{j=0}^{j=i-1}jX(M,j)R\right\} \tag{7}$$

$$M\sum_{j=0}^{j=i-1}(i-j)X(M,j)R = M\left\{R\times i\times W(M,i) - R\sum_{j=0}^{j=i-1}jX(M,j)\right\}. \tag{8}$$

Consequently,

$$I(M,i,z) = M\times(i+1)\times R - M\left\{R\times i\times W(M,i) - R\sum_{j=0}^{j=i-1}jX(M,j)\right\}. \tag{9}$$

The computational effort value, E, is the minimum $I(M,i,z)$, varying value of i. E is the minimum number of individuals needed to produced to yield a correct solution with a satisfactorily high probability ($z = 0.99$). However, the time in evaluating a circuit also depends on the length of input/output sequences. Thus, it is noted that the length of input/output sequences should be considered along with the computational effort.

5 Experiment and the Results

Fig. 1 shows the evolution of a serial adder. The circuit that performs accordingly to the input/output sequence appears in the 53rd generation. We know it is a correct serial adder after we verified it. We could observe many redundant states in the resulting circuit because the given space (16-state machine) is larger than the solution.

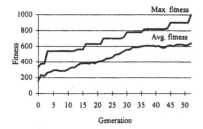
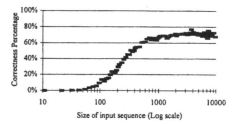

Fig. 1. Evolution of a Serial Adder (Mealy, size of input sequence = 1000)

Fig. 2. Correctness and size of sequence (Serial Adder, Mealy's model)

We define the correctness percentage as

$$\text{Correctness Percentage} = \frac{\text{number of runs yielding correct solutions}}{\text{number of runs yielding solutions}}. \tag{10}$$

Fig. 2 shows the relation of the size of input sequence to the percentage of correct results of synthesis from the sequence of that size. We used five random input sequences for each point and run each input sequence for 100 times to make the average correctness. The longer input sequence increases the correctness. However, the correctness percentage becomes saturated at the large size of input sequence.

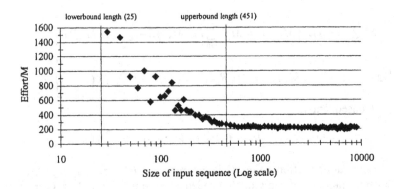

Fig. 3. Effort and size of input sequence (Serial adder, Mealy's model, M=110)

Fig. 3 shows the effort of evolving a serial adder using different sizes of sequences. This result confirms our lowerbound and upperbound length hypotheses. Any input sequence that is shorter than the lowerbound length can not yield any correct solution, consequently, the effort can not be computed. The input sequence longer than the upperbound length yields saturated effort value because of the approximately constant correctness percentage. It is possible to select the suitable size of input sequence to minimize effort and maximize correctness. The size of input sequence should be small to minimize the running time

Table 2. summarizes the computational effort of each problem using input/output sequences of upperbound length.

Table 2. The summary of computational effort

Circuit	Effort	
	Moore	Mealy
Frequency Divider	770	440
Odd Parity Detector	1,210	1,760
Modulo-5 Detector	87,967,440	7,018,000
Serial Adder	3,035,120	26,730

6 Conclusion

This paper described synthesis of synchronous sequential logic circuit from a partial input/output sequence. GA was applied to synthesize a circuit that based on Moore and Mealy's model on the GAL structure. We can approximate the suitable size of the input sequence to yield high correctness and low effort. This work can be extended in many ways. One major aspect is the enhancement of the evolutionary process in both effort and time. An implementation of this work in real hardware to realize an on-line evolware [14,15] is one interesting approach.

References

1. Sentovich, E., Singh, K., Moon, C., Savoj, H., Brayton, R. and Sangiovanni-Vincentelli, A.: Sequential circuit design using synthesis and optimization. Proc. of Int. Conf. on Computer Design (1992)
2. Holland, J.: Adaptation in natural and artificial systems. MIT Press (1992)
3. Goldberg, D.: Genetic Algorithm in search, optimization and machine learning. Addison-Wesley (1989)
4. Fogel, L.: Autonomous Automata. Industrial Research 4 (1962) 14-19
5. Angeline, P., Fogel, D., Fogel, L. : A comparison of self-adaptation methods for finite state machine in dynamic environment. Evolution Programming V. L. Fogel, P. Angeline, T. Back (eds). MIT Press (1996) 441-449
6. Mizoguchi, J., Hemmi, H., Shimohara K.: Production Genetic Algorithms for automated hardware design through an evolutionary process. Proc. of the first IEEE Int. Conf. on Evolutionary Computation (1994) 661-664
7. Higuchi, T., Murakawa, M., Iwata, M., Kajitani, I., Liu, E., Salami, M.: Evolvable hardware at function level. Proc. of IEEE Int. Conf. on Evolutionary Computation. (1997) 187-192
8. Thompson A., Harvey, I., Husbands, P.: The natural way to evolve hardware. Proc. of IEEE Int. Conf. on Evolutionary Computation (1996) 35-40
9. Thompson, A.: An evolved circuit, intrinsic in silicon, entwined with physics. Proc. of the First Int. Conf. on Evolvable Systems: From Biology to Hardware (ICES96). Lecture Notes in Computer Sciences. Springer-Verlag (1997)
10. Winston, P.: Artificial Intelligence. Addison-Wesley (1992) 505-528
11. Syswerda, G.: Uniform crossover in Genetic Algorithms. Proc. of the Third Int. Conf. on Genetic Algorithms, J. D. Schaffer (ed). Morgan Kauffman (1989) 2-9
12. Feller, W.: An introduction to probability theory and its applications vol. I. Wiley (1968) 224-225
13. Koza, J.: Genetic Programming. MIT Press (1992)
14. Tomassini, M.: Evolutionary algorithms. Towards Evolvable Hardware. E. Sanchez and M. Tomassini (eds). vol. 1062 of Lecture Notes in Computer Science, Springer-Verlag, Heidelberg. (1996) 19-47
15. Sipper, M., Goeke, M., Mange, D., Stauffer, A., Sanchez, E., Tomassini, M.: The Firefly machine: Online evolware. Proc. of IEEE Int. Conf. on Evolutionary Computation. (1997) 181-186

Data Compression for Digital Color Electrophotographic Printer with Evolvable Hardware

Masaharu Tanaka[1], Hidenori Sakanashi[2], Mehrdad Salami[2]
Masaya Iwata[2], Takio Kurita[2], Tetsuya Higuchi[2]

[1]Mitsubishi Heavy Industries Ltd., Iwatsuka-Cho, Nakamura-ku, Nagoya 453, Japan
tanaka@etd.nmw.mhi.co.jp
[2]Electrotechnical Laboratory, 1-1-4 Umezono, Tsukuba 305-8568, Japan
{h_sakana, miwata, higuchi}@etl.go.jp

Abstract. This paper describes a data compression system using Evolvable Hardware (EHW) for digital color electrophotographic (EP) printers. EP printing is an important technology within digital printing, which is currently having a significant impact on the printing and publishing industry. Although, it requires data-compression to reduce the cost for transferring and storing large EP images, traditional techniques can not handle this data-compression well. This paper explains how EHW can be used as a compression system. EHW can change the compression method according to the characteristics of the image. The proposed EHW-based compression system can compress approximately twice as much data as JBIG, the current international standard.

1   Introduction

Digital printing is the latest generation system in the publishing and printing industry, and electrophotographic (EP) printing is one of its key technologies. However, currently available commercial EP printers are very large and expensive, because they require large storage devices and wide data-buses. These are necessary to process image data at speeds of giga-bytes per minute. To overcome these problems, image data must be compressed as much as possible and must be restored to its original form very quickly. Unfortunately, traditional data compression methods can not satisfy these requirements, because EP images have different characteristics from usual bi-level images.

This paper applies Evolvable Hardware (EHW) to a data compression system for EP printer to meet these requirements. EHW can quickly reconfigure its hardware structure to improve the method of coding according to the characteristics of the image. The EHW-based system can compress input images more efficiently at higher speeds than with the JBIG, the current international standard. EHW can also meet the speed requirements of the printer because the data-decompression is done by the hardware.

In the rest of this paper, Section 2 explains data compression for EP printing. An overview of EHW is given in Section 3. In section 4, the EHW-based data compression system is described in detail, and the results of computational simulations are presented which show that EHW outperforms other methods. Section 5 details the architecture of the proposed system, before a summary in Section 6.

2   Data Compression of Electrophotographic Image

This section explains about the requirements of EP printers, and discusses the reasons why traditional data-compression methods can not satisfy these needs.

2.1   Digital Printing

The concept of digital printing is to provide the *required number* of color copies of the *photo-quality* printed matter *from any printer*, *at anytime*, and *at low costs*. This short sentence contains the following major features of digital printing; (a) support of small number of copies, (b) photo-quality, (c) distributed printing, (d) on-demand, (e) low-cost.

For printing and publishing, we already have two technologies, namely, offset printing and color copying. Offset printers are suitable for mass-production and superior to copy machines with respect to features (b) and (e). However, they can not provide features (a), (c) and (d), because it is very time-consuming and expensive to make the special masks called *plates*. Also, the large sizes of these machines limits the usage of offset printing to personal use in normal offices. Conversely, although, copy machines are compact and very flexible, they are not suitable for digital printing because they can not produce high quality or high-speed printing, and the cost per page is too expensive, so they can not satisfy features (b), (c) and (e).

EP printing represents the latest technology in digital printing. It can print photo-quality images very fast without requiring plates, and hence the initial costs are avoided. Moreover, using the digital electronic data, we can obtain the required number of copies of the photo-quality printed matter from any printer, at anytime, and at low costs. The next section explains EP printing in more detail.

2.2 Electrophotographic Printing System

Fig.1 illustrates the basic structure of EP printers. The original image, represented in Postscript format, is created using a desktop publishing (DTP) system. This is transformed into four EP images by a raster image processor (RIP). The four images are halftone dot images, which are bi-level images, corresponding to the colors: cyan, magenta, yellow and black (CMYK). They are sent separately to the hard-disk arrays of the printing unit, and transcribed onto the paper by 4 drums settled in tandem.

In this printing process, severe bottlenecks occur at data-transfer, because the printing unit uses electrophotography, which provides high imaging quality. In other words, electrophotography is very fast (1800 Mbytes per minute), and can theoretically satisfy the speed requirements. However, usually hard disks can only process 300 Mbytes per minute, and it is not sufficient for hundreds pages of A4-size EP images of 1200 dpi (about 70 Mbytes). As a result, the printer must have (1) wide data-buses to transfer the image data faster than the printing unit, and (2) large hard-disk arrays to store large amounts of EP image data.

One of the major approaches to solve this problem in data-transfer is compression. Smaller data sizes allows for faster transfer and requires less storage. However, there are no commercial data-compression chips which can solve this problem, because none have sufficient performance levels to satisfy the following three important criteria:

Lossless data compression: The compressed data must be reconstructed completely to retain the quality of the images, otherwise the printed image quality deteriorates, such as blurred letters and fine distortions in the images. Therefore, the EP images must be compressed in *lossless* ways, where the reconstructed data is identical to the original.

High compression ratio: The compression ratio required by EP printing systems is equal to $\left(S \times (N/60)/4\right)/V$, where S is the data size, N is printing speed (page/min), and V is the data transfer rate of a hard-disk drive (about 5 Mbyte/sec). If we wish to print an A4-size EP image of 1200 dpi (about 70 Mbytes) at the speed of 100 page/min, then the compression ratio must be more than 6.0.

Quick decompression: The decompression speed also influences the cost of data-

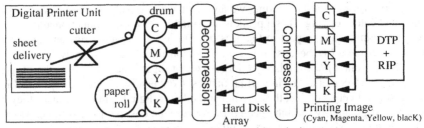

Fig.1: Structure of electrophotographic printing system.

transfer. The required speed for decompressing the compressed image is calculated as $S \times (N/60) \times 8/4$ (Mbps: Mega bit per second). The required decompression speed must be more than 240 Mbps in the above case.

These criteria are very severe, and there is no method of satisfying all of them, including the international standard method, explained in the next subsection.

2.3 Compression of Electrophotographic Images

Generally, the value of each pixel in an image tends to tightly correlate with those of neighboring pixels, and the pixel value can be predicted using the state of its neighborhood. If the prediction is correct, we do not have to represent the value of the predicted pixel explicitly, and the data size can become smaller. This is the basic principle behind JBIG (Joint Bi-level Image coding experts Group), the international standard for bi-level images [1]. The accuracy of the prediction strongly influences the compression performance of JBIG. It is also effected by the pattern of the pixels used in the prediction (called the *template*) and the prediction mechanism. In JBIG, the QM-Coder has the functionality of effective prediction using *context*, which is the pattern of the values of pixels in the template [2].

JBIG has a template which consists of ten reference pixels (Fig.2). In the figure, the black rectangle represents the position of a pixel being coded and the hatched rectangles indicate the positions of reference pixels. JBIG has a reference pixel called the *adaptive template*, which can move in the area indicated by the rectangles with crosses whilst compressing one image according to simple statistical calculations. However, with EP images, each pixel correlates with neighboring pixels a little bit further away than the adjacent ones. This means that JBIG can not compress EP images very well, because (1) the template configuration is not suitable, and (2) the simple mechanism to move the adaptive template can not work well [3,4]. Therefore, to solve these two problems, we tried to apply Evolvable Hardware to data compression in a EP printing system. The architecture proposed in this paper will compress EP images much better than other methods, without reductions in the speeds of compression or decompression.

3 Evolvable Hardware

Evolvable Hardware (EHW) is a relatively new paradigm in Evolutionary Computation, combining Genetic Algorithms (GAs) and the software-reconfigurable devices [5, 6].

GAs were proposed to model the adaptation of natural and artificial systems through evolution [7], and are well known as powerful search procedures, which exhibited superior performances in many types of problem [8,9]. However, few industrial applications or commercial products of GAs have been reported so far, because of their computational costs. EHW is a promising approach to overcome such problem, and makes use of the adaptive capability of GA by reducing its computational costs.

EHW is hardware device which is built on software-reconfigurable device, e.g. PLD (Programmable Logic Device) and FPGA (Field Programmable Gate Array). Its structure can be determined by downloading binary bit strings called the *architecture bit*. Generally, the architecture bit will be produced in the EHW by GA (Fig.3). The archecture bits are treated as chromosomes in the population by the GA, and can be downloaded to the reconfigurable device resulting in changes to the hardware structure. The changed functionality of the device can then be evaluated and the fitness of the chromosome is calculated. The performance of the device is improved as the population is evolved by GA according to fitness.

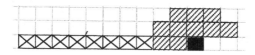

Fig.2: Configuration of reference pixels of JBIG.

Table 1: Two parts of the proposed system and their two function-modes.

		Learning Mode	Compression Mode
Template Generator	Statistical Calculation Genetic Algorithm	Search for best template	---
	Template Memory	Store discovered templates	Output stored templates
Data Compressor		Calculate compression ratio	Output compressed data

EHW includes two important features: fast computation of hardware devices and the adaptability of GAs. EHW has three advantages over traditional hardware and software systems: First, EHW can autonomously improve its performance by changing its hardware configuration according to the GA. Second, it processes information much faster than software systems. Third, the reconfigurable devices can change their functionality in an on-line fashion during execution. EHW can therefore be applied to new areas of application, where more inflexible traditional hardware systems are not efficient [10,11]. In particular, because it can reconfigure its hardware structure to adapt to the changes, EHW exhibits superior performance in dynamically adapting to changing environments. Image compression is just such an environment.

4 Data Compression System with EHW

The proposed system in this paper consists of two parts, a *template generator* and a *data compressor* (Fig.4), which can function in two modes: a *learning mode* and a *compression mode* (Table 1). In both modes, the image data is divided into some *stripes*, composed of L lines of the image, with the stripes being processed sequentially.

4.1 Data Compressor

The data compressor returns the size of the data compressed by using the template received from the template generator [12]. It roughly consists of 2 components, the *context selector* and *QM-Coder*. The context selector determines the context for each pixel during encoding, by collecting the values of 10 reference pixels specified by the template. In Fig.5, the locations of 10 reference pixels are pointed by x_k ($k = 0, ..., 9$), and a black rectangle represents the currently coded pixel. In this paper, each location is represented by 8 bits, and a template with 10 reference pixels can be represented by 80 bits. The 8 bits specifying one location is called a *slot*. A reference pixel can occupy one location over possible $2^8 = 255$ locations. The determined context is a vector of the binary values of the pixels pointed by x_k, and is sent to the QM-Coder with the value of the currently coded pixel. The QM-Coder attempts to predict the pixel values using the contexts, and compress the sequence of the received pixels.

The most advantageous feature of the data compressor is the flexible context selector, which can change the locations of all reference pixels at once, whereas JBIG can only change one. Furthermore, the pixels can be selected over a much wider area

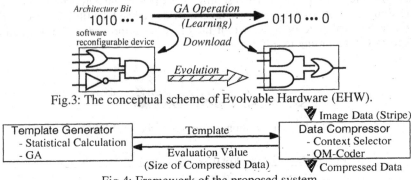

Fig.3: The conceptual scheme of Evolvable Hardware (EHW).

| Template Generator
- Statistical Calculation
- GA | Template

Evaluation Value
(Size of Compressed Data) | Image Data (Stripe)
Data Compressor
- Context Selector
- QM-Coder
Compressed Data |

Fig.4: Framework of the proposed system.

than with JBIG. These two enhancements improve the performance of our system for EP images, and can also efficiently compress the facsimile images. However, because of these enhancements, the template generator must effectively change the locations of all the reference pixels, in order to optimize the template for every stripe. Therefore, the template generator explained in the next section adopts a genetic algorithm to search for the best template.

4.2 Template Generator

The template generator consists of three components to produce templates (Table 1). A *template memory* stores the best templates, which minimizes the size of the corresponding stripe, discovered by *GA* and the *statistical calculation* in the learning mode, and automatically sends them to the data compressor one by one in compression mode. All of the templates stored in the template memory are sent to the decoder (receiver) after completing the compression.

The statistical calculation is used to initialize the population for the GA. Using the first stripe, it determines the *initial template* by the procedure given in the Fig.6. The initial template is used to compress the first stripe. It is also copied as one of the chromosomes in the population, and its mutated copies are set to the others. The procedure is very simple, but very efficient at increasing the search speed of GAs compared to starting from a random initialization.

After initializing the population, the GA runs from the second stripe. Each chromosome is sent to the context selector as a template, and receives the size of the compressed stripe from the data compressor as of its fitness. The number of generations for one stripe is represented by the parameter G. From the viewpoint of GA, the data compression is a nonstationary problem because the target stripe for compression changes with every G generations. It is well known that nonstationary problems are difficult for GAs [13]. To resolve these difficulties, we need to enhance the search performance of the GA, in order to cope with the dynamically changing environment.

4.3 Enhancement in Genetic Algorithm

The basic idea for the enhancement to the GA is to make the population keep the previously good templates, even if the environment changes. The characteristics of the adjacent stripes tend to be similar. Optimal templates for adjacent stripes are not completely different, and hence they have some common reference pixels. It is therefore advantageous for the chromosomes in the population to keep the locations of the pre-

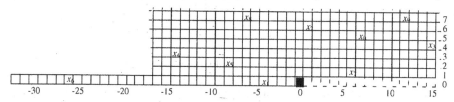

Fig.5: An example of configuration of reference pixels of the proposed system.

1. The default template of JBIG is set as the temporary template.
2. The temporary template is sent to the data compression part and receives the compression ratio.
3. From 10 slots in the temporary template of JBIG, the most useless slots are determined. The most useless slots specifies the position of the reference pixel without which the template gives the largest compression ratio.
4. From all possible locations in the 32x8 area, the most efficient location is selected. The most efficient location yields the largest compression ratio when it is assigned to the useless slot.
5. The new template is created by assigning the effective location to the useless slot.
6. If the compression ratio using the new template is better than the previous template's one, the previous temporary template is exchanged by the new one and go to step 3.
7. The temporary template is set as the initial template, and the procedure is finished.

Fig.6: Procedure of the statistical calculation for initializing the population of GA.

Table 2: Parameter setting in computational simulations.

Population size	20	Slot-copy ratio	0.1
Tournament size	3	Slot-swap ratio	0.1
Crossover ratio	0.8	Number of lines in one stripe (L)	40
Mutation ratio	0.03	Number of generation in one stripe (G)	100
Number of slots	20		

viously good reference pixels even after the compressed stripes are changed.

The enhanced representations have $M \geq 10$ slots, with only the active slots being allowed to compose a template (Fig.7). The activation of slots is determined by M additional bits, called *activation bits*, like those observed in structured GA [14]. The inactive slots can keep the previously good reference pixels, while the active slots are changed to search for better locations for the reference pixels. Crossover and mutation are not allowed to change the activation bits or the inactive slots. Two more genetic operations are introduced to handle the inactive slots, *slot copy* and *slot swap*. Using slot copy, the contents of one randomly selected active slot is copied to one of the inactive slots. The slot swap operation chooses one active slot and one inactive slot at random, and exchanges their activation bits. These two additional operations of slot copy and slot swap correspond to memorization and remembrance. The probabilies that a chromosome is changed by these operators are given by the parameters, the *slot-copy ratio* and *slot-swap ratio*.

4.4 Performance

This section shows the results of some computational simulations. For the simulations, we prepared a printer image (JIS/ISO standard color EP image N2), and eight facsimile images (CCITT standard images). The printer image was divided into eight subimages that were compressed separately. To compare the performance of the proposed system, the following 4 methods were also applied to the same images: (1) JBIG, (2) "compress" command of Unix, (3) the statistical method, and (4) the limited exhaustive search. Method 3 executes the procedure shown in Fig.6 using the first stripe, and the discovered template is applied to the remaining stripes. In method 4, to prevent combinatorial explosion, only one reference pixel in a template is allowed to be changed for one stripe by the exhaustive search, and the other pixels are changed for other stripes.

Tables 2 and 3 give the parameter settings and the results of the simulations. In Table 3, two values for each method are given, (a) the average compression ratio for the eight divided images of printer image, and (b) the average compression ratio for the eight facsimile images. The compression ratios in Table 3 include the discovered templates. These results show that, although the proposed method could not outperform method 4 very much, it represents the best compression method for both types of images. Besides, the proposed method can obtain the better compression ratio by carefully setting its parameters, whereas the other methods can never improve their performances.

From these results we can make the following points: First, the simple statistical calculation shows the best performance for compression of both images among methods 1, 2 and 3, with their major differences being context selection. The proposed context selector was the most suitable for modeling the general image data. Second, the effect GA's adaptability during compression of an image is demonstrated by the fact that the proposed method exhibited a better compression ratio than achieved using method 3. Third, the proposed method exhibited a better performance than that using

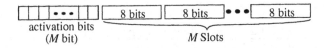

Fig.7: Enhanced representation of chromosome.

Table 3: Compression ratios of various method.

	Printer Image	Fax Image
(1) Lempel-Ziv	3.34	8.41
(2) JBIG	3.35	14.67
(3) Statistical calculation	6.23	19.49
(4) Exhaustive search for one reference pixel	6.61	19.73
Proposed method (standard)	6.66	19.86
(with random initialization of population)	6.51	19.82
(without enhanced representation)	6.45	19.74

random initialization of the population for GAs. Fourth, the enhanced representation gave higher compression ratios compared with the proposed method.

5   Architecture of Data Compression EHW

We develop a compression/decompression system with EHW for an EP printer, as briefly shown in Fig.8. The compressor consists of a *GA processor*, a *template memory*, an *image data buffer & sampling unit*, a *context modeling unit* and *QM-Coder*.

The GA processor executes learning to find the optimal template for each stripe by GA. The template memory maintains the best templates discovered by the GA processor. The image data buffer with sampling unit caches the images, and extracts the reference pixels and the current coded pixel to be transferred to the next step. The context modeling unit, which consists of a reconfigurable device, generates the context from the values of the reference pixels using the template given by the template memory. The QM-Coder encodes the current coding pixel using the corresponding context. It also counts the code length for the fitness calculation in GA in the training mode.

Currently, we are developing the EHW chip of the compressor, which has a processor for running GA without any host computers during compression [15]. Additionally, we are starting to design a new chip which can execute compression and decompression. This will be adopted in commercial EP printers.

6 Conclusion and Discussion

This paper has described a data compression system for high-precision electrophotographic (EP) printers using Evolvable Hardware (EHW). The EP printer is the latest generation technology for the digital printing and publishing industry. Data compression is required to reduce the costs of transferring and storing a large number of EP images at high speed. Unfortunately, traditional data compression methods are unable to compress images at speeds required, and to treat EP image with completely different characteristics from ordinary bi-level images. Therefore, this paper applied EHW to a data compression system in order to adaptively change coding methods, according to the local characteristics of the image, and to satisfy the severely limited criteria of the

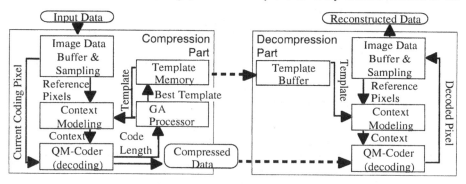

Fig.8: Basic architecture of compression/decompression system with EHW.

printer. Because of the fast computation of EHW, the proposed data compression system can execute more complex calculations to compress the image data. Moreover, the adaptability of EHW enables changes to be made to the compression configuration. The proposed system can compress EP images as well as ordinal bi-level images.

From the view point of GA in EHW, data compression is an unstable problem, because the optimal compression configuration varies in local areas even for one image. Therefore, a simple statistical calculation was used to prepare an initial population for the GAs to accelerate their search speed, and an enhanced representation of the chromosomes in the population was proposed to improve the search performance of the GA. The results of simulations showed that these enhancements in the data compression system and GA made the proposed system exhibit even better compression ratios in both of the printer and the facsimile images compared with traditional methods.

This paper has also explained the basic architecture of the data compression chip which we are developing. The severe requirements encumbered by EP printers can be satisfied with this chip. GA in the EHW can be executed for learning the compression method without a host computer.

Acknowledgements: This work is supported by MITI Real World Computing Project (RWCP). We thank Dr. Otsu and Dr. Ohmaki in Electrotechnical Laboratory.

Reference

[1] International Telegraph and Telephone Consultative Committee (CCITT): *Progressive Bi-level Image Compression*, Recommendation T.82 (1993).

[2] Pennebaker, W. B., et al.: *An overview of the basic principles of the Q-coder*, IBM Journal of Research & Development 32 (6), pp.717-726 (1988).

[3] Forchhammer, S. and Jansen, K. S.: *Data Compression of Scanned Halftone Images*, IEEE Transacctrion on Communications, vol.42, no.2, pp.1881-1893 (1994).

[4] Forchhammer, S.: *Adaptive Context for JBIG Compression of Bi-Level Halftone Images*, Proc. of the 1993 Data Compression Conference (DCC'93), pp.431, IEEE Computer Society Press (1993).

[5] Higuchi, T., et al.: *Evolvable Hardware with Genetic Learning*, Proceedings of Simulated Adaptive Behavior, The MIT Press (1992).

[6] Yao, X. and Higuchi, T.: *Promises and Challenges of Evolvable Hardware*, Evolvable Systems: From Biology to Hardware, pp.55-78, Springer (1996).

[7] Holland, J. H.: *Adaptation in Natural and Artificial Systems*, p.183, University of Michigan Press (1989).

[8] Goldberg, D. E.: *Genetic Algorithms in Search, Optimization and Machine Learning*, p.412, Addison-Wesley (1989).

[9] Davis, L. (ed.): *Handbook of Genetic Algorithms*, p.385: Nostrand (1991).

[10] Bennett III, F. H., et al.: *Evolution of a 60 Decibel Op Amp Using Genetic Programming*, Evolvable Systems: From Biology to Hardware, pp.455-469, Springer (1996).

[11] Murakawa, M., et al.: *On-line Adaptation of Neural Networks with Evolvable Hardware*, Proc. of ICGA7, pp.792-799, Morgan Kaufmann (1997).

[12] Salami, M. et al.: *On-Line Compression of High Precision Printer Images by Evolvable Hardware*, Proc. of the 1998 Data Compression Conference (DCC'98), IEEE Computer Society Press.

[13] Grefenstette, J. J.: *Genetic Algorithms for changing environment*, Parallel Problem Solving from Nature, 2, pp.137-144, Elsevier (1992).

[14] Dasgupta, D. and McGregor, D. R.: *Nonstationary Function Optimization using the Structured Genetic Algorithm*, Parallel Problem Solving from Nature, 2, pp.145-154. Elsevier (1992).

[15] Sakanashi, H., et al.: *Data Compression for Digital Color Electrophotographic Printer with Evolvable Hardware*, Proc. of the 15th National Conf. on Artificial Intelligence (AAAI-98), AAAI Press (1998).

Comparison of Evolutionary Methods for Smoother Evolution

Tomofumi Hikage, Hitoshi Hemmi, and Katsunori Shimohara

NTT Communication Science Laboratories
2-4 Hikaridai, Seika-cho
Soraku-gun, Kyoto 619-0237, JAPAN
E-mail:{hikage, hemmi, katsu}@cslab.kecl.ntt.co.jp

Abstract. Hardware evolution methodologies come into their own in the construction of real-time adaptive systems. The technological requirements for such systems are not only high-speed evolution, but also steady and smooth evolution. This paper shows that the Progressive Evolution Model (PEM) and Diploid chromosomes contribute toward satisfying these requirements in the hardware evolutionary system AdAM (Adaptive Architecture Methodology). Simulations of an artificial ant problem using four combinations of two wets of variables — PEM vs. non-PEM, and Diploid AdAM vs. Haploid AdAM — show that the Diploid-PEM combination overwhelms the others.

1 Introduction

The ultimate goal of evolutionary system research is to create a system able to provide the flexibility, adaptability, and robustness of living organisms with the speed necessary for engineering application. Evolutionary methods, like Genetic Algorithms and Genetic Programming, have already been applied in various fields like optimization and pattern recognition. These methods, however, require a large amount of computations. Solving this problem will provide a large number of new applications for evolutionary systems.

Simulated evolution using electronic circuits, or in general hardware evolution, represents a significant breakthrough in speeding up the computations. However, simply speeding up individual operations will not necessarily solve the computation problem.

Yao et al. [6] discussed "the danger of relying too much on hardware speed". They described the possibility that the time required in evolution is exponential in terms of the target circuit scale. However, if we could navigate such evolution aptly, it would perhaps be possible to substantially reduce the problem space, i.e., the space in which the evolutionary process actually performs exploration, and to avoid, or at least to reduce, the difficulty. In this paper, we show that the Progressive Evolution Model (PEM) along with Diploid (DL) chromosomes can be employed for this purpose.

In our earlier work on PEM [3], evolution took place in a stepwise manner to match the stepwise environmental changes. This is an efficient way to lower the problem "hurdle". An essential part of PEM involves environmental changes, which often destroy well-adapted individuals. Individuals specialized to a previous environment may be unable to cope with environmental changes. Although this is a necessary cost for PEM, techniques for easing the damage are strongly desired.

On the other hand, our diploid-chromosomes-based model broadens the population diversity [2], and this makes it easier for the population to find the optimal solution. Combining these two techniques provides an efficient method for smoother evolution, as will be shown.

2 Diploid AdAM and Progressive Evolution Model

2.1 Hardware Evolutionary System: AdAM

In experiments, we used our HDL (Hardware Description Language) - based hardware evolutionary system named AdAM [4].

An initial chromosome population is first created at random according to a set of production rules, which are of the Backus-Naur form of the HDL named SFL (Structured Function description Language). Each chromosome then generates an individual SFL program through interpretation. Next, the individual SFL programs, i.e., individual hardware behaviors, are simulated and evaluated by a certain fitness criterion. Finally, the selection and a set of genetic operations are performed on the chromosome population to produce the next generation.

To expresses SFL efficiently, the chromosomes of this system are in a tree structure (see Fig. 2). The tree is a parse tree. By assuming the above production rules to be rewriting rules, information on these rewriting rules is stored in each node of the tree structure chromosomes. Two or more rewriting rules exist in one non-terminal symbol. The stored information on rewriting rules include rewritten non-terminal symbols and information of which rewriting rule is to be used, or terminal symbols.

2.2 Biological Background of Diploid-chromosomes-based Model

Most multi-cellular organisms are diploid, having a set of diploid chromosomes, and exhibit dominant and recessive heredity; a haploid, by comparison, has only one set of chromosomes. Since a model in which one individual has one "chromosome" is usually employed, we postulate that a diploid has a pair of chromosomes and a haploid has a single chromosome to simplify the following discussion.

In a haploid, a change in its chromosome immediately appears as a change in the pheno-type experiencing selection pressure. In general, most changes in a chromosome are harmful in the sense that the probability of destroying a gene that generates an effective function has been predicted to be higher than that of gaining a new gene with a superior function. For a haploid to acquire a new function effectively, a gene should be duplicated before it is mutated. This would allow multiple copies of a gene for an existing function, so the chromosome could acquire a new gene by mutation while maintaining the original gene. That is, gene duplication must precede mutation; as a result, haploid evolution would generally take a long time [1].

A diploid has a recessive gene, and a change in one of the chromosomes does not appear in the pheno-type if the change occurs in the recessive gene. Therefore, it is very likely that harmful changes in a chromosome would survive and be stored up in a selected lineage until a new surpassing combination is eventually constructed. From this viewpoint, the genetic diversity of a diploid is larger than that of a haploid.

The process of meiosis, i.e., a cell division process into a germ cell, is as follows. First, each chromosome in a diploid cell is replicated. Next, mother chromosomes and father chromosomes become involved with each other. From this, crossover occurs in this stage. Then, at division I of meiosis, the involved four chromosomes are divided into two pairs of chromosomes. Finally, at division II of meiosis, each pair of chromosomes is divided into two germ cells which have only one chromosome each.

When a germ cell meets a germ cell of another individual, the two cells unite into one cell which has two chromosomes (i.e., fertilization, Fig. 1).

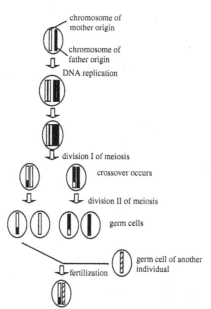

Fig. 1. Simplified illustration of the biological process of meiosis and fertilization

2.3 Chromosome of Diploid AdAM

A diploidy with a bit string chromosome is treated in [5]. Here, we introduce a tree structured chromosome with dominant-recessive heredity (see Fig. 2). We simplify the model of dominant-recessive heredity by assigning two sub-trees to one node corresponding to dominant-recessive alleles, which are initially defined randomly as dominant or recessive. A node having dominant-recessive alleles is initially selected at random.

2.4 Genetic Operations

Along with introducing dominant-recessive heredity into a chromosome, we employ the following genetic operations.
  - Mutation: This changes the rewriting rule in a node, or changes the terminal symbol (constant, variable name, etc.) in a node.
  - Duplication/Deletion: These duplicate or delete a node.
  - HL (Haploid)-type crossover: This exchanges the rewriting rules of two nodes, each of which belongs to a different chromosome.
  - DL (Diploid)-type crossover: This newly introduced operation exchanges the rewriting rules of two nodes, each of which belongs to either of two sub-trees corresponding to alleles in the same chromosome (see Fig. 3).

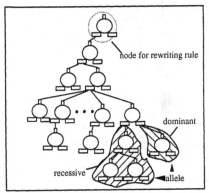

Fig. 2. A chromosome with dominant-recessive heredity

  - Meiosis/Fertilization: These newly introduced operations are the operations most characteristic of dominant-recessive heredity. They exchange the sub-trees of two nodes that have the same non-terminal symbol, but each of which belongs to a different chromosome.

Fertilization fuses two "germ cells", i.e., the germ cells of parent A and parent B are fused to yield child C and child D. The dominant or recessive tags of the parent sub-trees are already determined before the fertilization. In Fig. 3, if we assume that sub-trees 'm' and 'o' are dominant over sub-tree 'n', then sub-tree 'n' can not appear in the pheno-type in the generation of the parents. However, in the generation of the children, child D has two copies of the same recessive sub-tree in its two sub-trees, that is, both sub-trees are recessive and a recessive sub-tree appears in the pheno-type.

2.5 Progressive Evolution Model

Basic idea Organisms evolve while acquiring new functions to match environmental changes. We therefore believe that environmental changes drive evolution, so our idea is to use environmental changes actively to accelerate evolution.

In the progressive evolution model, evolution occurs in environments that change in a stepwise manner toward the final target environment. The purpose of the model is to divide a large "hurdle" into a series of small steps that the evolutionary process can easily handle.

Because this model depends on the applied problem and is implemented, this model is concretely explained after the problem is described in the following section.

3 Experiments

Our target is to evolve a hardware system to control the autonomous behaviors of a robot. In order to test our evolutionary method, a software simulation of the ant problem explained above is used.

3.1 The Ant Problem

In our artificial ant problem, artificial ants adapt to a certain environment so that they can gather food faster and more effectively; the best artificial ant is an ant that collects the most food in the fewest number of steps. The environment is usually an arrangement of food in a toroidal space of m × n cells. There are q pieces of food, indicated by black squares. In our case, each artificial ant takes five inputs showing the existence of food and performs one of three actions (Fig. 4).

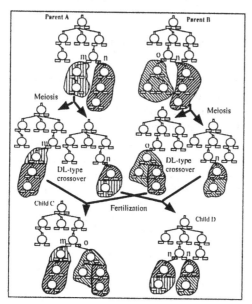

Fig. 3. Example of Meiosis/Fertilization and DL-type crossover with dominant and recessive heredity.

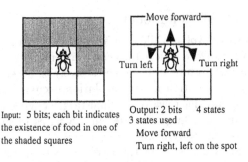

Input: 5 bits; each bit indicates the existence of food in one of the shaded squares

Output: 2 bits   4 states
3 states used
Move forward
Turn right, left on the spot

Fig. 4. Artificial Ant

3.2 Measuring the Environmental Complexity

In PEM, a measure of the environmental complexity is needed to divide the environment into a series of easier environments. As shown in Fig. 4, each ant has five inputs and two outputs. In the ant's inner mechanism, each input needs to be taken into account, and the arbitration among them is indispensable; an ant may encounter two different food items.

It is thought that, from the artificial ant's standpoint, the count of input-action pairs defines the environmental complexity. In this case, there are 96 ($= 2^5 \times 3$) combinations of inputs and actions. We call each combination an action primitive. These action primitives express the difficulty of the environment; the number of action primitives an ant needs to successfully collect all food from the environment is the complexity factor of the environment.

We write action primitive as follows. {Input form left, left front, front, right front, right: ACTION}, where ACTION means "move forward", "turn left", and "turn right". For instance, {0 0 1 0 0: "move forward"}, {0 0 1 0 1: "turn right"} means a complexity factor of 2.

3.3 Progressively Changing Environments

We designed a series of environments for artificial ants. Figure 5(e) shows the final target environment, and Figs. 5(a – d) show intermediate environments leading to the target environment. These environments are designed so that action primitives are obtained incrementally.

The first intermediate environment [Fig. 5(a)] requires four action primitives, {0 0 1 0 0: "move forward"}, {0 0 0 1 0: "move forward"}, {0 0 0 0 0: "move forward"}, {0 0 0 0 1: "turn right"}, so the complexity is 4. The second intermediate environment [Fig. 5(b)] requires three action primitives, {0 1 0 0 0: "move forward"}, {0 1 1 0 0: "move forward"}, {1 0 0 0 0: "turn left"}, in addition to those of the first, so the complexity of this environment is 7. In the same way, the third intermediate environment, [Fig. 5(c)] requires two more action primitives, {0 0 1 1 0: "move forward"}, {0 1 0 0 1: "turn right"}, so the complexity is 9. The fourth environment [Fig. 5(d)] adds three more action primitives, {1 1 1 0 0: "move forward"}, {0 0 1 1 1: "turn right"}, {1 1 0 0 0: "turn left"}, for a complexity of 12. Finally, the target environment requires two more action primitives in addition to the previous ones, {0 1 1 1 0: "move forward"}, {1 0 0 0 1: "turn right"}, and therefore the complexity measure is 14. The timing to exchange the environments is set at five generations after the best individual appears in each environment. In addition, due to the elite strategy, the first successful individuals spread through the population and most of the circuits adapt to the environment during the five generations.

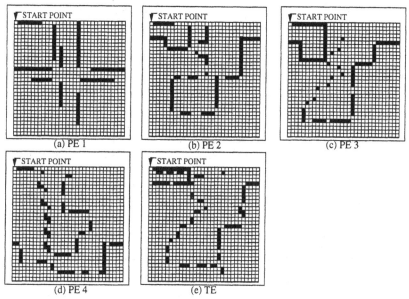

Fig. 5. Progressive Environments and Target Environment (TE)

3.4 Simulation Details

We carried out an experiment in which the controller circuit of an artificial ant was generated evolutionarily. We compared PEM to nonPEM by examining their performance in terms of the number of generations until convergence over all environments used. We also compared Diploid AdAM to Haploid AdAM. There were four types of simulations. The experimental conditions were as follows.

- Food: 89 types of food/environment
- Population size: 256 individuals
- Selection Methods: roulette model and elite strategy
- Fitness function
  - Fitness = Score + Limit - Step + 1
    Score: units of food collected; Limit: Maximum step number allowed in this system (350 steps); Step: move from one square to another

4 Results

The results for Diploid AdAM with PEM and Haploid AdAM with nonPEM are shown in Figs. 6(a – c) and 6(d), respectively. The experimental results obtained in early generations with PEM are shown in Fig. 6(a). The best individuals in the first, second, and third intermediate environments appear in the 1st, 11th, and 54th generation, respectively. Figure 6(b) shows the result for the last stage of PEM. From these figures, one can see that evolution takes place in a stepwise manner to match the environmental changes. Figure 6(c) shows the fitness transition for all generations. It takes 3,331 generations to produce the best individual in the final target environment.

Figure 6(d) shows the result obtained for evolution in the final target environment with an ordinary evolution model. The best individual appears in 18,427 generations, indicating that a progressive model is more efficient. Table 1 shows results for all types. Only Diploid AdAM with PEM could obtain the best ant within 10,000 generations. Accordingly, PEM is effective for Diploid AdAM, which is probably because of the ability of Diploid AdAM to treat dominant and recessive heredity works well for environmental changes. Conversely, it seems that in a stable environment, there is hardly any difference between Haploid AdAM and Diploid AdAM, as indicated by Haploid AdAM with nonPEM and Diploid AdAM with nonPEM. We think that the effect of PEM is lost for Haploid AdAM with PEM , in which the environment is changed in Haploid AdAM, because Haploid AdAM is unable to follow the changing environment.

Table 1. Number of generations to produce the best individual in the final target environment

	Haploid AdAM	Diploid AdAM
PEM	13532	3331
nonPEM	18427	21698

5 Conclusion

Smooth evolution is a must for real-time evolutionary systems. The combination of the progressive evolution model and diploid-chromosome-based model was proposed to achieve smooth evolution. The effect of using this method was shown by comparing four types of evolutionary methods for an ant problem.

Although the experiments were limited to one application, the comparison clearly shows that the combination of the two methods is effective. It remains to be seen whether this is the case with other applications as well. The result, however, are so promising that we plan to establish the generality of the method in a wide range of other potential applications like job scheduling. The measurement of the environmental complexity requires further research, especially with respect to how this measurement can be generalized.

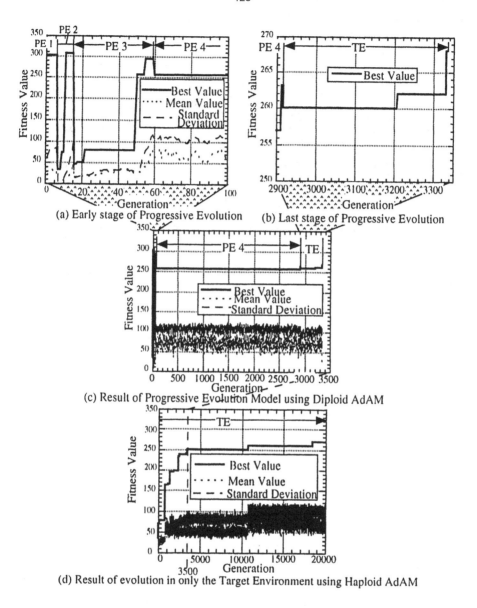

(a) Early stage of Progressive Evolution

(b) Last stage of Progressive Evolution

(c) Result of Progressive Evolution Model using Diploid AdAM

(d) Result of evolution in only the Target Environment using Haploid AdAM

Fig. 6. Results of simulations

References

1. Bruce Albers et al. *MOLECULAR BIOLOGY OF THE CELL SECOND EDITION*, chapter 15. Garland Publishing, Inc., 1989.
2. Tomofumi Hikage, Hitoshi Hemmi, and Katsunori Shimohara. Hardware evolution system introducing dominant and recessive heredity. In *Evolvable Systems: From Biology to Hardware (LNCS 1259)*, pages 423–436. Springer, 1996.
3. Tomofumi Hikage, Hitoshi Hemmi, and Katsunori Shimohara. Progressive evolution model using a hardware evolution system. In *Artificial Life and Robotics*, pages 18–21, 1997.
4. Jun'ichi Mizoguchi, Hitoshi Hemmi, and Katsunori Shimohara. Production genetic algorithms for automated hardware design through an evolutionary process. In *IEEE Conference on Evolutionary Computation*, 1994.
5. R. E. Smith and D. E. Goldberg. Diploidy and dominance in artificial genetic search. *Complex Systems*, 6(3):251–285, 1992.
6. Xin Yao and Tetsuya Higuchi. Promises and challenges of evolvable hardware. In *Evolvable Systems: From Biology to Hardware (LNCS 1259)*, pages 55–78. Springer, 1996.

Automated Analog Circuit Synthesis Using a Linear Representation

Jason D. Lohn[1] and Silvano P. Colombano[2]

[1] Caelum Research Corporation, NASA Ames Research Center,
Mail Stop 269-1, Moffett Field, CA 94035-1000, USA
email: jlohn@ptolemy.arc.nasa.gov
[2] Computational Sciences Division, NASA Ames Research Center,
Mail Stop 269-1, Moffett Field, CA 94035-1000, USA
email: scolombano@mail.arc.nasa.gov

Abstract. We present a method of evolving analog electronic circuits using a linear representation and a simple unfolding technique. While this representation excludes a large number of circuit topologies, it is capable of constructing many of the useful topologies seen in hand-designed circuits. Our system allows circuit size, circuit topology, and device values to be evolved. Using a parallel genetic algorithm we present initial results of our system as applied to two analog filter design problems. The modest computational requirements of our system suggest that the ability to evolve complex analog circuit representations in software is becoming more approachable on a single engineering workstation.

1   Introduction

Analog circuits are of great importance in electronic system design since the world is fundamentally analog in nature. While the amount of digital design activity far outpaces that of analog design, most digital systems require analog modules for interfacing to the external world. It was recently estimated that approximately 60% of CMOS-based application-specific integrated circuit (ASIC) designs incorporated analog circuits [1]. With challenging analog circuit design problems and fewer analog design engineers, there are economic reasons for automating the analog design process, especially time-to-market considerations.

Techniques for analog circuit design automation began appearing about two decades ago. These methods incorporated heuristics [13], knowledge-bases [4], and simulated annealing [11]. Efforts using techniques from evolutionary computation have appeared over the last few years. These include the use of genetic algorithms (GAs) [5] to select filter component sizes [6], to select filter topologies [3], and to design operational amplifiers using a small set of topologies [10]. The research of Koza and collaborators [8] on analog circuit synthesis by means of genetic programming (GP) is likely the most successful approach to date. Unlike previous systems, the component values, number of components, and the circuit topologies are evolved. The genetic programming system begins with minimal knowledge of analog circuit design and creates circuits based on a novel circuit-encoding technique. Various analog filter design problems have been solved using genetic programming (e.g., [9]), and an overview of these techniques, including eight analog circuit synthesis problems, is found in [8]. A comparison of genetic-based techniques applied to filter design appears in [14] and work on evolving CMOS transistors for function approximation [12] has also recently appeared.

The system we present here was motivated by the genetic programming system described above. Our investigation centers on whether a linear representation and simple unfolding technique, coupled with modest computer resources, could be effective for evolving analog circuits. In the GP system, a hierarchical representation is manipulated by evolution, and a biologically-inspired encoding scheme is used to construct circuits. In our system we use a linear genome representation and a simple unfolding process to construct circuits. As mentioned, our current system is topology-constrained, yet such constraints were deemed reasonable since a vast number of circuit topologies are attainable. Our technique presented below differs from the previous GA techniques in that we allow both topology and component sizes to be evolved. In [14], a GA approach is presented in which topologies and component values are evolved for circuits containing up to 15 components. Here we use dynamically-sized representations in the GA so that circuits containing up to 100 components can be evolved. Using a cluster of six engineering workstations (1996 Sun Ultra), we present evolved circuit solutions to two filter design problems.

2 Linear Representation

Circuits are represented in the genetic algorithm as a list of bytecodes which are interpreted during a simple unfolding process. A fixed number of bytecodes represent each component as follows: the first is the opcode, and the next three represent the component value. Component value encoding is discussed first.

Using three bytes allows the component values to take on one of 256^3 values, a sufficiently fine-grained resolution. The raw numerical value of these bytes was then scaled into a reasonable range, depending on the type of component. Resistor values were scaled sigmoidally between 1 and 100K ohms using $1/(1 + \exp(-1.4(10x - 8)))$ so that roughly 75% of the resistor values were biased to be less than 10K ohms. Capacitor values were scaled between approximately 10 pF and 200 μF and inductors between roughly 0.1 mH and 1.5 H.

The opcode is an instruction to execute during circuit construction. In the current design of our system, we use only "component placement" opcodes which accomplish placement of resistors, capacitors, and inductors. The five basic opcode types are: x-move-to-new, x-cast-to-previous, x-cast-to-ground, x-cast-input, x-cast-to-output, where x can be replaced by R (resistor), C (capacitor), or L (inductor). In a circuit design problem involving only inductors and capacitors (an LC circuit), ten opcodes would be available to construct circuits (five for capacitors and five for inductors).

The circuit is constructed between fixed input and output terminals as shown in Fig. 1. An ideal AC input voltage source v_s is connected to ground and to a source resistor R_s. The circuit's output voltage taken across a load resistor R_l.

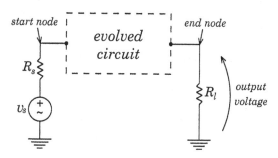

Fig. 1. Artificially evolved circuit is located between fixed input and output terminals (v_s is an ideal ac voltage source, R_s is the source resistance, R_l is the load resistance).

To construct the circuit, a "current node" register (abbreviated CN; with "current" used in the sense of present, not electrical current) is used and initialized to the circuit's input node. The unfolding process then proceeds to interpret each opcode and associated component values, updating the CN register if necessary. The x-move-to-new opcode places one end of component x at the current node (specified by the CN register) and the other at a newly-created node. The CN register is then assigned the value of the newly-created node. The "x-cast-to-" opcodes place one end of component x at the current node and the other at either the ground, input, output, or previously-created node. After executing these opcodes, the CN register remains unchanged. The meanings of each opcode are summarized in Table 1. All five opcode types place components into the circuit, although they could be designed to do other actions as well, e.g., move without placement.

Opcode	Destination Node	CN Register
x-move-to-new	newly-created node	assigned the newly-created node
x-cast-to-previous	previous node	unchanged
x-cast-to-ground	ground node	unchanged
x-cast-to-input	input node	unchanged
x-cast-to-output	output node	unchanged

Table 1. Summary of opcode types used in current system. x denotes a resistor, capacitor, or inductor.

The list of bytecodes is a variable-length list (the length is evolved by the GA). Thus, circuits of various sizes are constructed. When the decoding process reaches the last component to place in the circuit, we arbitrarily chose to have the last node (value in CN) connected to the output terminal by a wire. By doing so, we eliminate unconnected branches.

We had two goals in designing the above encoding scheme. First, we wanted to see if a very simple set of primitives encoded in a linear fashion could indeed be used to successfully evolve circuits. Second, we wanted to minimize computer time during the genetic algorithm run. By keeping the decoding process minimal, the total time for fitness evaluations is thus reduced. Along the same lines, we wanted to keep circuit "repair" operations (e.g., removal of unconnected nodes) to a minimum since these also slow the system down.

The most significant restriction of our technique is that it cannot support all possible circuit topologies: circuit branches off of the main "constructing thread" cannot, in general, contain more than one node (there are some exceptions to this). The constructing thread is the sequence of components that are created by the x-move-to-new opcode. The constructing thread itself can be of varying lengths and can contain both series and parallel configurations. In spite of these limitations, our system allows creation of circuits with a large variety of topologies, especially topologies seen in hand-designed circuits (e.g., ladder constructs). We have lessened the topology restrictions somewhat by allowing "move-to" opcodes and will report on these efforts in the future.

3 Genetic Algorithm

The genetic algorithm operates on a population of dynamically-sized bytecode arrays. In practice we imposed a maximum size of about 400 bytes (100 circuit components) in order to accommodate population sizes of up to 18,000 individuals in our GA runs. The crossover and mutation (per locus) rates were set at 0.8 and 0.2 respectively. An overview of the evaluation process is depicted in Fig. 2. As in the GP system mentioned above, we used the Berkeley SPICE circuit simulation program to simulate our circuits. The array of bytecodes was interpreted in the manner previously described, and resulted in a SPICE netlist representation. The netlist is processed by SPICE and the output is then used to compute fitness for the individual. Fitness was calculated as the absolute value of the difference of the individual's output and the target output. These error values were summed across evaluation points, with error being the distance between the target and the value the individual produced.

The parallel genetic algorithm implemented uses master/slave style parallelism [2] over a network of UNIX-based computers. A controlling host computer performs GA functions and distributes a population of bytecoded-individuals to specified number of worker nodes using socket connections. The worker nodes decode the individuals into SPICE netlists which are then fed into SPICE via FIFO pipes to minimize disk activity. Fitness is calculated using SPICE's output, and then sent back to the host. Hundreds of individuals (and fitness scores) are packaged into a single message so that external network congestion delays are minimized. The SPICE program itself required little modification since it runs

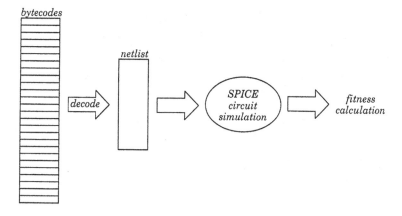

Fig. 2. Overview of circuit evaluation process starting with bytecoded representation and ending with fitness score.

as a separate process. Written in the C programming language, the system currently runs on Sun workstations and is portable to other UNIX systems (e.g., we have ported the software to PCs running UNIX). This allows the system to run on UNIX-based clusters comprised of computers from different manufacturers.

4 Experimental Results

We attempted to evolve two analog filter circuits. The choice of using passive analog filters was inspired by the previous studies and is a good choice for testing the effectiveness of our system for three reasons. First, all components have two-terminals, the minimum number possible. If the proposed system could not evolve useful circuits using two-terminaled devices, then attempting to evolve circuits using more complex components (e.g., transistors) would likely prove ineffective. Second, there are no energy sources required within the circuit which further reduces the complexity. Lastly, filter design is a well-understood discipline within circuit design. Its "design space" has been greatly explored [7] which allows us to compare our evolved designs to well-known designs.

The problems we present below are both low-pass filters. A low-pass filter is a circuit the allows low frequencies to pass through it, but stops high frequencies from doing so. In other words, it "filters out" frequencies above a specified frequency. The unshaded area in Fig. 3 depicts the region of operation for low-pass filters. Below the frequency f_p the input signal is passed to the output, potentially reduced (attenuated) by K_p decibels (dB). This region is known as the passband. Above the frequency f_s, the input signal is markedly decreased by K_s decibels. As labeled, this region is called the stopband. Between the passband and stopband the frequency response curve transitions from low to high attenuation. The parameter located in this region, f_c, is known as the cutoff frequency.

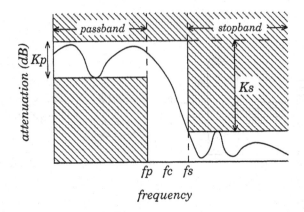

Fig. 3. Low-pass filter terminology and specifications. The crosshatched regions represent out-of-specification areas. An example frequency response curve that meets specifications is shown.

4.1 Electronic Stethoscope Circuit

The first circuit we attempted to evolve is one that is suitable for use in an electronic stethoscope. In this application, it is desired to filter out the extraneous high-frequency sounds picked up by a microphone which make it difficult to listen to (low-frequency) bodily sounds (e.g., a heart beating). As such, the frequency response specifications do not need to be extremely accurate since we are dealing with audible frequencies and the human ear cannot discern frequencies that are close together. The target frequency response data was taken from an actual electronic stethoscope, which was built with a cutoff frequency of 796 Hz corresponding to an output voltage of approximately 1 volt. This circuit is relatively easy to design and so we chose it as our first problem to solve.

The GA was allowed to use resistors and capacitors during evolution, resulting in an RC low-pass filter. The evolved circuit is shown in Fig. 4 and its frequency response, which matches almost exactly the target is shown in Fig. 5.

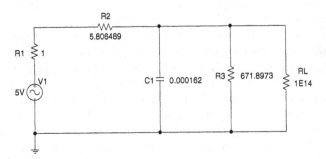

Fig. 4. Evolved low-pass filter for use in an electronic stethoscope (units are ohms and farads).

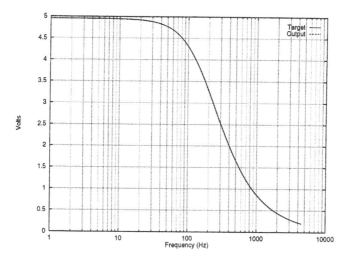

Fig. 5. Nearly identical frequency response curves for evolved and actual electronic stethoscope circuit. The frequency axis is scaled logarithmically.

4.2 Butterworth Low-Pass Filter

The second low-pass filter we evolved was more difficult. We chose a circuit that can be built using a 3rd-order Butterworth filter [7]. The specifications are as follows:

$$f_p = 925 \text{ Hz} \qquad K_p = 3.0103 \text{ dB}$$
$$f_s = 3200 \text{ Hz} \qquad K_s = 22 \text{ dB}$$

Such a filter design can be derived using a ladder structure and component values found in published tables. The GA was allowed to use capacitors and inductors during evolution, resulting in an LC low-pass filter. The evolved circuit that meets these specifications is shown in Fig. 6 and its frequency response is shown in Fig. 7. It was found in generation 22 of a GA run that lasted approximately four hours using six Sun Ultra workstations working in parallel.

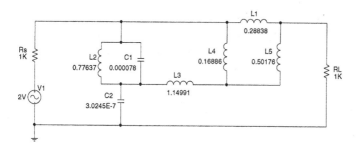

Fig. 6. Evolved 3rd-order Butterworth low-pass filter (units are ohms, farads, and henries).

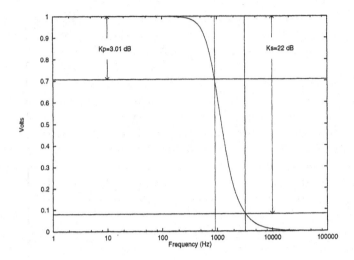

Fig. 7. Frequency response curve for evolved 3rd-order Butterworth low-pass filter. Attenuation specifications are also shown. The frequency axis is a scaled logarithmically.

5 Discussion

We have shown that a genetic algorithm using a simple linear circuit representation is capable of evolving two circuits of low to medium difficulty. The circuit construction method devised uses a very simple set of primitives encoded in a linear fashion. Such a method helps to minimize the computer time required to evolve circuits by keeping the decoding and repairing processes shorter. Although this technique is topology-limited, the ability of our system to produce useful circuits was demonstrated. It is likely that these topological space restrictions are favorable to many filter designs, especially filters that are known to have less complex branching patterns (e.g., ladder structures). We intend to build upon this technique to allow for greater topologies and three-terminal devices such as transistors. With the previous successes in evolving analog circuits, and the encouraging early results of our system, we are optimistic that a subset of analog circuit design tasks may be routinely accomplished by means of evolutionary computation in the future.

6 Acknowledgments

The authors would like to thank M. Lohn, D. Stassinopoulos, G. Haith, and the anonymous reviewers for their helpful suggestions and comments.

References

1. G. Gielen, W. Sansen, *Symbolic Analysis for Automated Design of Analog Integrated Circuits*, Boston, MA: Kluwer, 1991.
2. D.E. Goldberg, *Genetic Algorithms in Search, Optimization, and Machine Learning*, Addison-Wesley, Reading, Mass, 1989.
3. J.B. Grimbleby, "Automatic Analogue Network Synthesis using Genetic Algorithms," *Proc. First Int. Conf. Genetic Algorithms in Engineering Systems: Innovations and Applications (GALESIA)*, 1995, pp. 53-58.
4. R. Harjani, R.A. Rutenbar, L.R. Carey, "A Prototype Framework for Knowledge-Based Analog Circuit Synthesis," *Proc. 24th Design Automation Conf.*, 1987.
5. J.H. Holland, *Adaptation in Natural and Artificial Systems*, Univ. of Michigan Press, Ann Arbor, 1975.
6. D.H. Horrocks, Y.M.A. Khalifa, "Genetically Derived Filters using Preferred Value Components," *Proc. IEE Colloq. on Linear Analogue Circuits and Systems*, Oxford, UK, 1994.
7. L.P. Huelsman, *Active and Passive Analog Filter Design*, New York: McGraw-Hill, 1993.
8. J.R. Koza, F.H. Bennett, D. Andre, M.A. Keane, F. Dunlap, "Automated Synthesis of Analog Electrical Circuits by Means of Genetic Programming," *IEEE Trans. on Evolutionary Computation*, vol. 1, no. 2, July, 1997, pp. 109–128.
9. J.R. Koza, F.H. Bennett, J.D. Lohn, F. Dunlap, M.A. Keane, D. Andre, "Use of Architecture-Altering Operations to Dynamically Adapt a Three-Way Analog Source Identification Circuit to Accommodate a New Source," in *Genetic Programming 1997 Conference*, J.R. Koza, K.Deb, M.Dorigo, D.B. Fogel, M. Garzon, H. Iba, and R.L. Riolo, (eds), Morgan Kaufmann, 1997, pp. 213–221.
10. M.W. Kruiskamp, *Analog Design Automation using Genetic Algorithms and Polytopes*, Ph.D. Thesis, Dept. of Elect. Engr., Eindhoven University of Technology, Eindhoven, The Netherlands, 1996.
11. E.S. Ochotta, R.A. Rutenbar, L.R. Carley, "Synthesis of High-Performance Analog Circuits in ASTRX/OBLX," *IEEE Trans. Computer-Aided Design*, vol. 15, pp. 273–294, 1996.
12. A. Stoica, "On Hardware Evolvability and Levels of Granularity," *Proc. 1997 Int. Conf. Intell. Systems and Semiotics*, 1997, pp. 244-247.
13. G.J. Sussman, R.M. Stallman, "Heuristic Techniques in Computer-Aided Circuit Analysis," *IEEE Trans. Circuits and Systems*, vol. 22, 1975.
14. R.S. Zebulum, M.A. Pacheco, M. Vellasco, "Comparison of Different Evolutionary Methodologies Applied to Electronic Filter Design," *1998 IEEE Int. Conf. on Evolutionary Computation*, Piscataway, NJ: IEEE Press, 1998, pp. 434–439.

Analogue EHW Chip for Intermediate Frequency Filters

Masahiro Murakawa[1] Shuji Yoshizawa[1]
Toshio Adachi[2] Shiro Suzuki[2] Kaoru Takasuka[2]
Masaya Iwata[3] Tetsuya Higuchi[3]

[1] University of Tokyo, 7-3-1 Hongo, Bunkyo, Tokyo, Japan
[2] Asahi Kasei Microsystems, 3050 Okada, Atsugi, Kanagawa, Japan
[3] Electrotechnical Laboratory, 1-1-4 Umezono, Tsukuba, Ibaraki, Japan

Abstract. This paper describes an analogue EHW (Evolvable Hardware) chip for Intermediate Frequency (IF) filters, which are widely used in cellular phones. When analogue Integrated Circuits (ICs) and Large-Scale Integrated Circuits (LSIs) are manufactured, the values of the analogue circuit components, such as resistors or capacitors, often vary from the precise design specifications. Analogue LSIs with such defective components can not perform at required levels and thus have to be discarded. However, the chip proposed in this paper can correct these discrepancies in the values of analogue circuit components by genetic algorithms (GAs). Simulations have shown that 95% of the chips can be adjusted to satisfy the specifications of the IF filters. Using this analogue EHW chip has two advantages, namely, (1) improved yield rates and (2) smaller circuits, which can lead to cost reductions and efficient implementation of LSIs. The chip is scheduled to appear in the fourth quarter of 1998.

1 Introduction

Due to remarkable advances in recent CPUs and DSPs, applications with analogue circuits are rapidly being replaced with digital computing. However, there are still many applications that require high-speed analogue circuits. Communication is one such application.

An inherent problem in implementing analogue circuits is that the values of the manufactured analogue circuit components, such as resistors and capacitors, will often differ from the precise design specifications. Such discrepancies cause serious problems for high-end analogue circuit applications. For example, in intermediate frequency (IF) filters, which are widely used in cellular phones, even a 1% discrepancy from the center frequency is unacceptable. It is therefore necessary to carefully examine the analogue circuits, and to discard any which do not meet the specifications.

In this paper, we propose an analogue EHW chip for IF filters, which can correct these variations in the analogue circuits values by genetic algorithms (GAs). Using this chip provides us with two advantages.

1. Improved Yield Rates

   When analogue EHW chips are shipped, which do not satisfy specifications, then GAs can be executed to alter the defective analogue circuit components in line with specifications.

2. Smaller Circuits

One way to increase the precision of component values in analogue LSIs has been to use large valued analogue components. However, this involves larger circuits, and accordingly higher manufacturing costs and greater power consumption. With the EHW chip, however, the size of the analogue circuits can be made smaller.

This approach could be applied to a wide variety of analogue circuits.

In the IF filter analogue EHW chip, there are 39 G_m components (transconductance amplifiers) whose values can be set genetically. Each G_m element value may differ from the target value up to a maximum of 20%. Initial simulations have shown that 95% of the chips can be corrected to satisfy the IF filters specifications. In contrast to digital EHW [1], where the acquisition of improved and new hardware functionality is of primary importance, the analogue EHW chip reconfigures its hardware structure in order to conform to the permissible specification range of the original design.

This paper is organized as follows: In section 2, the concept of analogue EHW is briefly described. Section 3 introduces the analogue EHW chip for IF filters, and Section 4 describes the results of a performance simulation. In section 5, the advantages of the analogue EHW are discussed, as well as related issues, before a summary section.

2 Proposal of Analogue EHW

2.1 Variations in Analogue Circuits

The characteristics of analogue circuits can vary widely due to environmental influences. In particular, component values (e.g. capacity and resistance) may vary by several tens of percent, due to changes in temperature. When an analogue circuit is implemented as an integrated circuit, there are large variations in the values for components, caused by the manufacturing process or interference from other components. These variations impair the performance of the circuits, which can be especially serious when the specifications of the circuit require greater precision.

2.2 Analogue EHW

In order to compensate for variations in analogue circuits, we propose analogue EHW. Analogue EHW is hardware that is built on software-reconfigurable analogue circuits, and whose architecture can be reconfigured through genetic learning. As a first application of analogue EHW, we have designed an analogue EHW chip for IF filters. IF filters are band path filters, which are widely used in cellular phones. If IF filters are implemented with CMOS technology, it is difficult to conform to specifications because of variations in component values.

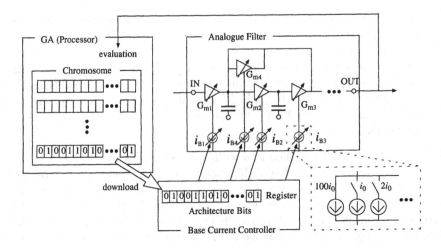

Fig. 1. Analogue EHW Chip for Intermediate Frequency Filters

Fig. 1 illustrates the analogue EHW chip for IF filters. The chip includes 39 G_m amplifiers, whose transconductance can be varied. This is done by subtly varying the base currents of the CMOS, which constitute the G_m amplifier. The configurations can be changed by downloading a binary string to the control register in the chip. This string is called the architecture bits. In genetic learning, the architecture bits are regarded as GA chromosomes. GAs are robust search algorithms which use multiple chromosomes (usually represented as binary strings) and apply natural selection-like operations to them in seeking improved solutions [2]. The GA identify the optimal architecture bits for the analogue EHW chip. The goal of this evolution is to satisfy the specification of the filter.

Every chromosome is downloaded into the control register. A fitness value is calculated by giving inputs to the filter and observing the outputs. Transconductance values are varied to improve the total performance of the filter.

3 Analogue EHW Chip for Intermediate Frequency Filters

This section introduces the analogue EHW chip for intermediate frequency filters in detail.

The ideal frequency response [3] and typical specifications for an IF filter are shown in Fig. 2 and Fig. 3. The filter has a pass bandwidth of 21.0kHz centered at 455kHz, and stop bands specified at attenuations of 48dB and 72dB. Filter gain should be within the doted lines in the figures. The −3dB points should be within $455 - 10.5 \pm 1$kHz and $455 + 10.5 \pm 1$kHz. These specifications are very hard to meet because the −3dB points become outside these limits if the center frequency is shifted even by 1 percent.

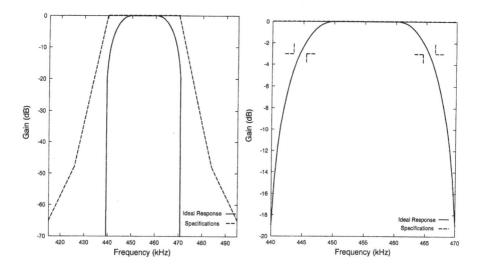

Fig. 2. Ideal Frequency Response and Specifications of an IF filter

Fig. 3. Magnification of the Fig. 2

The analogue EHW chip for IF filters can conform to these specifications by compensating for variations in component values. A detailed description of the IF filter and how to compensate for such variations using genetic algorithms is given below.

3.1 Filter Architecture

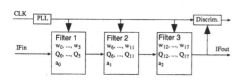

Fig. 4. IF Filter Architecture

Tab. 1. Ideal Parameter Values

$w_0, \cdots, w_{17}$	$2\pi f_0$
$Q_0, \cdots, Q_{17}$	$2\pi f_0/22$
$a_0, \cdots, a_2$	$2\pi f_0/11$
	$(f_0 = 455 \times 10^3)$

The IF filter which we have adopted is an 18th order filter [4]. The architecture for the filter is depicted in Fig. 4. The filter consists of three 6th order G_m-C leapfrog filters. A block diagram of the 6th order filters is shown in Fig. 5.

This IF filter has 39 parameters in total. 16 of these are related to the center frequency $(w_0, \cdots, w_{15})$, 16 for band width $(Q_0, \cdots, Q_{15})$, and 3 for filter gain $(a_0, \cdots, a_2)$. Table 1 shows the target values for these parameters.

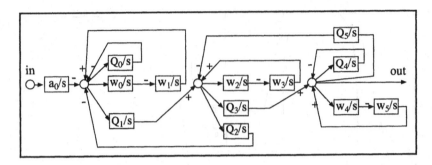

Fig. 5. 6th Order G_m-C Filter

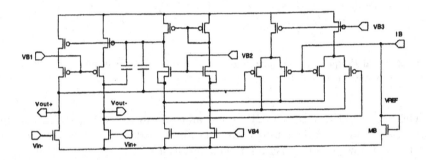

Fig. 6. G_m/common-mode feedback amplifier

In the integrated circuit of the filter, these parameters correspond to the transconductance of the G_m amplifiers. A schema of a G_m amplifier is shown in Fig. 6. Transconductance varies greatly from the target values, by up to 20 percent. Of this 20 percent variation, 5 to 10 percent is due to temperature. The remaining 10 to 15 percent is caused by the manufacturing process. A control circuit with a PLL (Phase-Locked Loop) is usually used to correct this kind of variation. This circuit can correct variations if all the transconductance values differ equally from the target values. Thus, variations due to temperature can be corrected by this circuit. However, because the relative differences between transconductance values can be between 20 to 30 percent, it is difficult to satisfy the specifications with the control circuit. However, the analogue EHW chip can compensate for such variations by using GAs.

3.2 EHW Chip

In the EHW chip, transconductance, which corresponds to the filter's parameters, can be subtly varied as shown in Fig. 1. The GA determines the optimal configuration of the transconductance values.

A chromosome for the GA consists of 39 genes which correspond to the filter parameters (i.e., $a_0, \cdots, a_2$, $w_0, \cdots, w_{15}$ and $Q_0 \cdots, Q_{15}$). Each gene has N bits that determine the transconductance. For example, if $N = 2$, there are 4 transconductance values for selection by the GAs. The genes 00, 01, 10 and 11 mean that the parameter is multiplied by $1.0 - 2 \times D, 1.0 - D, 1.0 + D$ and $1.0 + 2 \times D$ respectively, where D is a constant value.

Multiple chromosomes are prepared as a population for the GAs. In the initial population, no individual can satisfy the specifications, because of variations in transconductance. By repetitive applications of GA operations to the population, a configuration that satisfies the specifications emerges.

3.3 Genetic Operation

The fitness function is defined as follows:

$$\text{fitness} = \sum_{i=1}^{n} w_i |S(f_i) - O(f_i)| \tag{1}$$

Fitness is the weighted sum of the deviations between the ideal gain $S(f_i)$ and the gain obtained $O(f_i)$ by the EHW chip at frequency f_i.

Chromosomes with lower values from equation 1 are reproduced according to the tournament selection rule. The tournament size is 2. By using an elitist strategy, the chromosome with the best fitness value is always reproduced. After reproduction, child chromosomes are generated using single-point cross-over with a probability of P_c, and mutations are applied to each bit in the chromosome with a probability of P_m.

4 Simulation Results

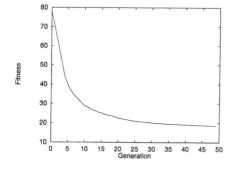

Fig. 7. Fitness versus Generation

Tab. 2. Effect of Evaluation Weights for −3dB points

Weights for −3dB points	Successful Runs
1.0	26
3.0	79
5.0	95
7.0	86
9.0	4

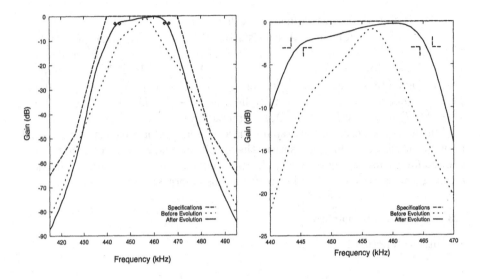

Fig. 8. Frequency Response of the Analogue EHW Chip

Fig. 9. Magnification of the Fig. 8

In this section, we present the results of simulations of the analogue EHW chip's performance. In the simulations, each transconductance value in the circuit was assumed to vary from the target value by Gaussian distribution of $\sigma = 5\%$. The parameters N, D were set to 2, and 0.025 respectively. This meant that each transconductance value could vary by –5%, –2.5%, +2.5% or +5.0% from the actual values in the circuit.

The fitness was defined using the gain obtained by the EHW chip at frequencies 440.0, 444.5, 449.75, 455.0, 460.25, 465.5 and 470.0 (kHz) (i.e. $n = 7$). The weights for the –3dB points (i.e. 455 ± 10.5kHz) were 5.0. The other weights were set to 1.0.

We used a population of 50 individuals, of which chromosome length was 78. The crossover rate P_c was 0.5 and the mutation rate P_m was 0.013. A run terminated after the 40th generation.

Fig. 7 shows the fitness values of the best-of-generation individual. This curve was obtained by averaging the results of 100 independent runs. As can been seen, the fitness value decreases as the generation increases. The GAs gradually found the optimal configuration for the EHW chip. Figs. 8 and 9 show the frequency responses for the best individual in a run. After evolution, the best individual could satisfy the specifications, which the initial population was unable to meet.

Out of 100 runs, 95% of the EHW chips conformed to specifications. For comparison, we conducted simulations with the hill climbing method instead of the GA. All the parameters for the filter were the same as for the above simulations. A run terminated after fitness was evaluated 1000 times. The result of this was that only 75% could meet the specifications. This result shows the effectiveness of the GA in avoiding the local minimums of the fitness function.

In Table 2, the effects of evaluation weights for −3dB points are given. As can be seen, 5.0 is the best of the 5 values. In the simulations, when the weights were less than 5.0, many EHW chips were unable to conform to the specification at the −3dB points. On the other hand, when the weights were more than 5.0, many of the EHW chips could change to within the specifications at −3dB points, but not at the stop bands.

5 Discussion

This section discusses the advantages of analogue EHW and required reconfiguration frequency, beyond the considerations of the proposed EHW chip. Related works of analogue EHW are also discussed.

5.1 Advantages of Analogue EHW

Analogue EHW is hardware which is built on software-reconfigurable analogue circuits, and whose architecture can be reconfigured using genetic learning. The basic idea behind analogue EHW is to regard the architecture bits of the reconfigurable analogue circuits as chromosomes for GAs and to identify optimal hardware structures by running these GAs.

Analogue EHW has two advantages. Firstly, analogue EHW can modify its performance to fit within an acceptable range of specifications set out in the original designs. This advantage is illustrated by the analogue EHW chip for the IF filter. The chip can satisfy the specifications by compensating for variations in the integrated circuit.

Secondly, if the functional requirements for a piece of analogue EHW are drastically changed, the analogue EHW can respond by alter its own hardware structure quickly and thus accommodate changes in real-time. Such requirements are of primary importance for digital EHW chips (e.g. EHW chips for adaptive equalization in digital communication [5] and chips for printer image compression [6]). Such applications for analogue EHW will be presented in the near future.

To fully realize the second advantage, it may be necessary to reconfigure the topology of analogue circuits. For example, when the function required of an analogue EHW is changed from a low-pass filter to a band-pass filter, then the topology would have to be quickly reconfigured using GAs.

5.2 Reconfiguration Frequency

Analogue EHW chips can be categorized into two types, static analogue EHW chips and dynamic analogue EHW chips, according to the frequency with which their analogue circuits are reconfigured. Dynamic analogue EHW chips can be reconfigured on-line at regular intervals by observing the outputs from the chip. These chips are for use in environments that vary over time. In such cases, GA hardware [7] can be incorporated into the analogue EHW, especially for analogue

LSIs in industrial applications that ideally should be compact for embedded systems. The GA hardware can handle self-reconfiguration without host-machine control. In contrast, static analogue EHW chips are reconfigured only once or only at longer intervals. Reconfiguration in these cases is usually at the chip manufacturers or by the chip user. For these chips, the GA program can be executed by the external host-machine. The EHW chip for IF filters is of this static type.

5.3 Related Works

This subsection discusses some related works for analogue EHW. To compensate for the variations in analogue circuits, self-correcting circuits and dummy elements are used generally. However, the design process of these circuits requires experience, intuition and a thorough knowledge of the process characteristics. There is accordingly a shortage of design engineers.

To solve this problem, Koza, et al. have proposed the automatic synthesis of analog circuits using genetic programming [8]. Circuit topologies and component values are coded into a tree, which is evolved by the Genetic Programming. A number of circuits, such as low-pass filters, amplifiers using transistors, and voltage reference source circuits have been successfully synthesized. However, this study did not seek to correct variations in the actual circuits, only to automatically synthesize the circuits without the need for human knowledge and intelligence. However, the problem of variations still arises when circuits synthesized by GP are implemented as integrated circuits.

It is therefore important to establish methods to compensate for variations in actual circuits, without trial and error at the design stage. The use of analogue EHW is a promising method of achieving this. A fitness evaluation is a measurement of performance of the actual circuit on a reconfigurable device, and so can be used to detect variations among circuits. Thompson has also emphasized the importance of using real hardware [9]. He has evolved a frequency discriminator on a Xilinx FPGA XC6216. Regarding the circuits more as dynamic systems than as static ones, he was able to utilize the device more efficiently for natural and dynamic physical behavior of the medium. Our research differs from this in that our aim is not to evolve improved hardware functionality, but is to correct discrepancies in analogue IC so that they conform to specifications.

6 Conclusion

In this paper, we have proposed an analogue EHW chip for IF filters, which can correct discrepancies in analogue IC/LSI implementations. Even if some analogue circuit elements have different values from the design specifications, which prevent the analogue LSI from performing at required levels, the LSI can change the values of variant analogue circuit elements by genetic algorithms. Using analogue EHW chips has two advantages, namely; (1) improved yield rates, and (2) smaller circuits. These advantages can lead to cost reductions and to the realization of more effective analogue LSIs. Under secured industrial cooperation, the EHW chip described in this paper is currently being manufactured in order to evaluate mass production costs, and is scheduled to appear in the fourth quarter of 1998.

References

1. T. Higuchi, T. Niwa, T. Tanaka, H. Iba, H. Garis, and T. Furuya. Evolvable hardware with genetic learning. In *Proceedings of Simulation of Adaptive Behavior*. MIT Press, 1992.
2. J. H. Holland. *Adaptation in Natural and Artificial Systems*. The University of Michigan Press, 1975.
3. J. Proakis. *Digital Communications*. Prentice Hall Inc., 1988.
4. T. Adachi, A. Ishikawa, K. Tomioka, S. Hara, K. Takasuka, H. Hisajima, and A. Barlow. A low noise integrated AMPS IF filter. In *Proceedings of the IEEE 1994 Custom Integrated Circuits Conference*, pages 159–162, 1994.
5. M. Murakawa, S. Yoshizawa, I. Kajitani, and T. Higuchi. Evolvable hardware for generalized neural networks. In *Proceedings of the Fifteenth International Conference on Artificial Intelligence*, pages 1146–1151. Morgan Kaufmann, 1997.
6. H. Sakanashi, M. Salami, M.Iwata, S. Nakaya, T. Yamauchi, T. Inuo, N. Kajihara, and T.Higuchi. Evolvable hardware chip for high precision printer image compression. In *Proceedings of the Fifteenth National Conference on Artificial Intelligence (AAAI98)*, 1998.
7. I. Kajitani, T. Hoshino, D. Nishikawa, H. Yokoi, S. Nakaya, T. Yamauchi, T. Inuo, N. Kajihara, and T. Higuchi. A gate-level EHW chip; implementing GA operations and reconfigurable hardware on a single LSI. In *Proceedings of the Second International Conference on Evolvable Systems*, 1998.
8. J. R. Koza, F. H. Bennett III, D. Andre, M. A. Keane, and F. Dunlap. Automated synthesis of analog electrical circuits by means of genetic programming. *IEEE Transactions on Evolutionary Computation*, 1(2):109–128, 1997.
9. A. Thompson. An evolved circuit, intrinsic in silicon, entwined with physics. In *Proceedings of the First International Conference on Evolvable Systems*, pages 390–405. Springer Verlag, 1996.

Intrinsic Circuit Evolution Using Programmable Analogue Arrays

Stuart J Flockton and Kevin Sheehan

Royal Holloway, University of London,
Egham Hill, Egham, Surrey, TW20 0EX, UK
{S.Flockton, Kevin.Sheehan}@rhbnc.ac.uk

Abstract. The basic properties of programmable analogue arrays are described and the problem of quantifying the fitness of an analogue circuit is discussed. A set of blocks appropriate for use in an evolutionary algorithm is described and results presented showing how an evolutionary algorithm using these blocks can learn to produce a given input-output characteristic. Finally an example is presented showing how the evolutionary algorithm can exploit any looseness in the specification of the desired characteristic.

1 Introduction

The majority of work published so far on evolvable hardware has been carried out with digital logic systems. However the world is inherently analogue and the applications that many devices are put to involve analogue inputs and outputs. Some work has been done to exploit the inherently analogue nature of 'digital' circuits, using properties of the circuit that were not envisaged by the designer. Whilst this enables devices to be used for tasks well outside the range for which they were originally designed, the results can be highly irreproducible from one instance of a device to another [1], [2]. Circuits for analogue applications are normally designed so that their properties are relatively independent of the detailed behaviour of most of the circuit elements and depend critically on just a few components, the stability of which can be considered particularly carefully by the designer. It would seem better, therefore, to use building blocks explicitly designed for their analogue properties when trying to develop evolutionary analogue design methods, as these building blocks are more likely to exhibit the desired stable behaviour.

One way of exploring the use of evolution for design of analogue systems is 'extrinsic' evolution in which an evolutionary algorithm is used to suggest circuits and a software simulator to test these candidates, such as in [3]. In our work we are looking at evolution as a method for obtaining the set of instructions to apply to programmable analogue devices. These are devices which have been designed to process analogue signals, but the nature of the analogue operations that are performed on the signals are specified by the contents of digital registers on the device. Hence candidate circuits can be implemented immediately and tested on real signals, providing 'intrinsic' evolution.

Using evolutionary methods to programme these new analogue devices has the advantage not only of providing a method of automating the design process but also allows candidate designs to be drawn from a wider repertoire of combinations than would be practical for a human designer. This may allow a given performance to be obtained from a system in a more efficient way than would otherwise have been discovered.

2   Programmable Analogue Arrays

Programmable analogue arrays [4], [5], [6] are similar to their digital cousins, FPGAs, in that the functions that they implement are chosen from the available set by programming particular binary patterns into them, but different in that the available set of functions are defined and thought of in terms of their analogue, rather than digital, performance. The devices are usually based on a grid of cells which can have their properties and interconnects altered digitally. The cells are, for the most part, composed of identical analogue circuitry throughout the device, except that the outside cells have Input/Output pins. Typically each individual cell is based around an operational amplifier surrounded by digitally controlled switches to alter the parameters and interconnects. These switches are usually programmed by writing to a shift register or static RAM. This can be done from an EPROM or a micro-processor, and can usually be accomplished in well under a second. This ability to change the performance of the device simply by re-writing the contents of digital registers brings some of the advantages of digital systems to the analogue domain, as it enables the same piece of hardware, the cost of which can be made small as a result of producing it in high volume, to be used for many different purposes.

This new flexibility in reconfiguration of an analogue circuit brings a new challenge: how to use it efficiently. Most existing design methods are predicated on the assumption that implementation and testing of a particular design is comparatively expensive, so it is worthwhile expending a large amount of effort in the design process before testing a prototype. Rapidly reprogrammable devices turn these assumptions on their head. Implementation of a particular configuration is cheap and easy, so many different combinations can be tested without any physical change to the board being necessary. They offer, therefore, an attractive route into using intrinsic evolution for analogue system design.

3   Use of Evolutionary Design Methods

One of the problems of extrinsic evolutionary design methods is that, since analogue circuits have a much richer set of available dynamics than digital circuits, they are more difficult to simulate and require much more computing power to evaluate them.

When using an evolutionary design method the behaviour of many candidate solutions must be inspected, so evolving an analogue circuit extrinsically is generally a lengthy process. A further problem is that the result is still only a *simulation* of the real circuit. The behaviour of an analogue system is difficult to predict precisely, and therefore the differences in behaviour between the design and its implementation may be significant.

These difficulties do not arise in intrinsic evolution as simulation is not required. The system that is being tested is the actual final system and not just a design for it, so there is no room for discrepancies to arise, and the fitness of the circuit can be tested on real analogue signals, so in principle the time required is limited only by the time necessary to exercise the system over the range of inputs given in its specification.

3.1 Evaluating the Fitness of an Analogue Circuit

A major problem that has to be overcome when using evolution to design an analogue circuit is to decide how to evaluate the fitness of a candidate solution. With a digital circuit a truth table can be used to provide a measure of a circuit's fitness, whereas with an analogue circuit specifying every possible input and output would be intractable. Consequently, we need to develop new ways of calculating a fitness score.

Fig. 1 shows the type of system that is appropriate for testing and comparing candidates. It shows the same inputs being fed to both the candidate and a realisation of the desired circuit, and the two outputs being captured for comparison. Although the diagram shows two actual circuits the target can be replaced, if appropriate, by a paper (or computational) specification.

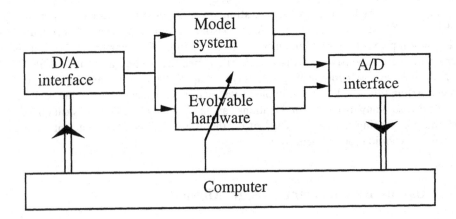

Fig. 1. Block diagram of a system suitable for evolving a hardware implementation when a model system is available

4 Some Examples of Analogue Evolution

In this section we show some results obtained using a system of Zetex TRAC (Totally Reconfigurable Analogue Hardware) devices, which consist of twenty operational amplifiers laid out as two parallel sets of ten interconnectable amplifiers. Each amplifier has various components around it that may be switched in and out of circuit by programming a shift register to give one of eight distinct operations. These are: off, pass, invert, exponentiate, take logarithm, rectify (these last three functions involve the use of internal diodes in one portion or the other of the amplifier feedback circuit), sum two inputs, and finally a state where user-supplied external components provide the feedback. Four chips were installed on a test board connected to the parallel port of a PC, enabling the control registers to be programmed externally. The PC also had an analogue interface installed to enable the generation and detection of analogue signals. In the results presented here we were looking at the ability of the system to learn a steady-state input-output characteristic. This is simpler than looking for a frequency-dependent characteristic as the timing of the measurements from which the fitness is calculated is not critical.

4.1 Building Blocks

When using an automatic design system one must ensure that: (i) the system is provided with a sufficient set of building blocks to be capable of producing an

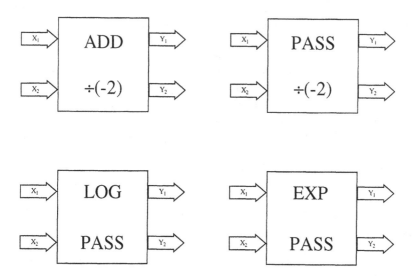

Fig. 2. The four different building blocks used in the experiments

adequate design; (ii) that damaging configurations either cannot be generated or can be identified before being loaded into the device for testing; (iii) the probability of each possible configuration providing some useful behaviour distinct from all or most other configurations is maximised. Obviously it is important during early stages of experimentation to have a clear idea of at least one specific achievable configuration by which the desired performance may be obtained; only when one has acquired sufficient experience in well-understood situations will it be worthwhile to branch out into problems where the level of prior understanding is less. One way in which all these criteria can be met is to choose a well-understood problem and constrain the set of building blocks available to the system so that they are sufficient to achieve the known solution, but also can be interconnected in any achievable way without damage.

The Zetex TRAC device does not have any on-chip programmable resistors or capacitors, so it needs to be provided with external components in order to make useful circuits. Since we wished to create new circuits purely by programming the control registers of the chips we needed to install an appropriate configuration of off-chip external components. The criteria that these components had to meet was that they should enable the generation of several different building blocks, depending on the contents of the control register, but that the circuit they produced should not be damaging when the bits in the control shift register were being moved in, as the analogue section of the chip is *not* disabled during this process.

It was these considerations that made us choose to use the set of two-input/two-output blocks shown in Fig. 2 for the experiments reported here. A set of fixed resistors were installed on the test board in such an arrangement that, depending on the block type selected by the control register, they either provided the appropriate gain control for the adjacent op-amps or had the same voltage at each end and therefore did not affect the circuit. The output of the upper part of the ADD/divide-by-minus-2 block (called the ADD block subsequently) is the negative of the sum of the two inputs, while the output of the lower part is the negative of half the input of the lower part. The lower part of the PASS/divide-by-minus-2 block (subsequently referred to as the PASS block) is identical to that of the ADD block, while the upper part simply negates the upper input. [Note: the reason for the negating is to maximise the processing power of the available circuitry. This is somewhat analogous to the way that using NAND instead of AND in digital logic widens the functions that can be realised with blocks of just one type.] In the LOG/PASS (subsequently referred to LOG) block the upper part performs the operation

$$Y_1 = \frac{1}{k_2} \ln\left(\frac{X_1}{k_1} + 1\right)$$

where k_1 and k_2 are constants, while the lower part is non-inverting pass function. [Note: the fact that this is non-inverting, while the ADD and PASS blocks both change the sign of the lower part makes the sign of the output to depend on the number of blocks of each type cascaded together.] The EXP/PASS (subsequently

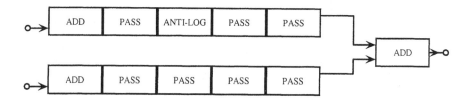

Fig. 3. Block configuration used as a prototype.

referred to as EXP) block performs the inverse operation to the LOG block, that is to say the upper part performs the operation

$$Y_1 = k_1 \left[\exp(-k_2 X_1) - 1 \right].$$

while the lower part is again a non-inverting pass function.

Various considerations were important in making the choice of choice of blocks to be used; these included the following. (i) To have a small number of blocks but to be able to achieve "interesting" dynamics. The blocks shown allow inversion, addition and a form of multiplication (achieved by cascading a LOG block with an ADD block and an EXP block). (ii) To ensure that all voltage swings inside a cascaded system of the blocks were visible (albeit attenuated) from the outputs. The reason for this, related to the concept of observability in control theory, was to ensure that there would be no completely unobservable dynamics in the system as they might lead to unpredictable behaviour. The objective was achieved by ensuring that all blocks gave non-zero outputs from non-zero inputs.

4.2 Modelling a Configuration

We set up a the block configuration shown in Fig. 3. and measured its input/output characteristic; this is shown in Fig. 4. The curve is antisymmetric about the origin and has sufficient structure to make it a worthwhile modelling task. The evolutionary algorithm was set up to use the same basic configuration of two parallel banks of five blocks each with the outputs added together, but was given no explicit information about either the type or position of the blocks. The number of possible ways of arranging four types of block in this general arrangement is about half a million, so it is certainly not feasible to find the desired result by exhaustive search.

We used a population based evolutionary algorithm for our search. This was essentially a genetic algorithm of the type proposed by Holland [7] in that each member of the population was represented as a bit string, but the bit string was structured and the genetic operators used to produce new candidate solutions operated on the bit string in a structured way. The quality of each candidate solution was evaluated by finding the sum of the absolute difference between the output of the

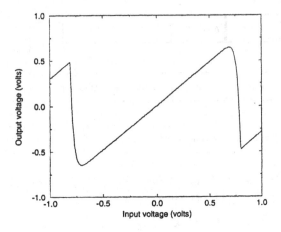

Fig. 4. Input-output voltage characteristic of the target system

candidate and the target for five different positive input voltages, chosen to delineate the key features of the target characteristic. We used only positive values as we knew that all the operators had antisymmetric characteristics, so could (we thought) give only antisymmetric outputs.

The operators used to generate new candidate solutions were a swap operator which exchanged two randomly-chosen blocks with each other and a mutation operator that acted on single randomly-chosen blocks. The mutation operator behaved differently depending on which type of block had been chosen for mutation. The blocks can be divided into two classes, the linear blocks (PASS and ADD) and the non-linear blocks (LOG and EXP). Changing a block of one class to the other block of the same class will generally make much less difference to the output than changing to a block of the other class. Hence the probabilities of the various mutations were set to make within-class mutation much more likely than trans-class mutation. This allows wide exploration to take place on a less frequent basis alongside more frequent local search.

The success of the strategy is demonstrated in the results presented here. Fig. 5 shows an example of the error of the best candidate as a function of generation number. It can be seen that a very good solution has been found within less than 25 generations. Typically the evolved solutions matched the original block structure other than that a few ADDs and PASSes in low signal parts of the circuit had been exchanged with each other. It is interesting to see how the error distribution changes with time. Fig. 6 shows histograms of the error at the start of the run (for the randomly generated initial population), then at stages through the rest of the run. The general creep towards lower error produced by the less drastic mutation and swapping can clearly be seen alongside a gradually smaller number of higher error candidates

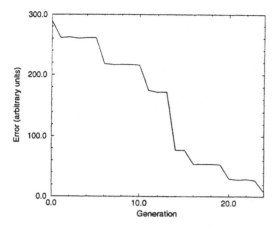

Fig. 5. Time history of the error for the system programmed to mimic the characteristic shown in Fig. 4

being created in each generation by the more drastic operators. The algorithm therefore seems successfully to balance exploration and exploitation.

As the population size was 25, the number of fitness evaluations required to find a good solution was around 600, a small fraction of the approximately half million members of the search space. However in order to judge whether our method was indeed providing good performance we needed to know whether many members of the search space might have similar characteristics to the chosen target, as this would mean that our result could be being achieved purely by random sampling of this rather undifferentiated search space. We therefore investigated the variability of fitness within the search space by looking at the fitness distribution of 3000 (i.e. five times the number of candidates used in a typical run of the evolutionary algorithm) randomly generated candidates. The histogram of these results showed a wide variation of fitness, and none of the randomly-generated patterns had errors anywhere near as small as those found by the algorithm. We therefore deduced that our algorithm was indeed investigating the search space in a much better than random manner.

4.3 A Cautionary Example

It was pointed out earlier that, as we knew that the characteristics of each block were antisymmetric, we used only one side of the characteristic when testing candidate solutions. The folly of this is shown in Fig. 7 which compares the characteristics of the target and the solution to which the algorithm converged on one particular run. It can be seen that the two characteristics match well in the region where the error was evaluated, but far less well in the negative input region. Further investigation showed

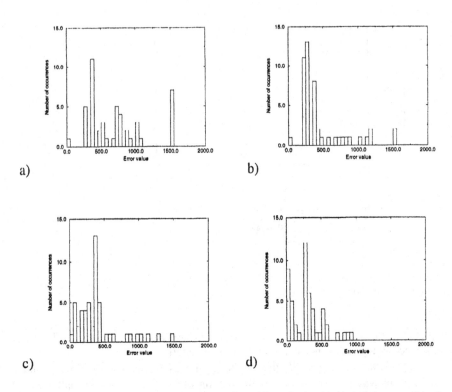

Fig. 6. Histograms of the error of the population: (a) at the beginning of the run; (b) after 8 generations; (c) after 16 generations; (d) after 23 generations

that the solution contained two pairs of LOG and EXP blocks cascaded with each other. If the circuitry were exactly matched this would have no effect, but clearly the characteristics of the diodes in the two circuits differed sufficiently to produce the asymmetry seen. Once again one is reminded of the ability of evolutionary algorithms to exploit any laxity in specification to produce solutions which satisfy the explicit criteria tested, but not necessarily any untested assumed criteria

5   Conclusions

We have demonstrated the feasibility of using an evolutionary algorithm and a programmable analogue array to perform intrinsic evolution of an analogue circuit to approximate a given steady-state input-output characteristic. In order to do this we have designed a variety of useful building blocks which can be connected together within the array in any order without damage. The system has been shown to be capable of finding good solutions to a given problem within comparatively short spaces of time and encourages us to move on to the more difficult task of evolving circuits having particular frequency-dependent behaviour. Finally the crucial importance of having a sufficiently rich test suite to detect and remove candidates which superficially have the right properties but are not in fact appropriate has again been demonstrated.

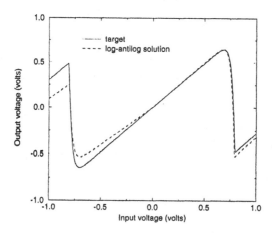

Fig. 7. Characteristics of the target and the false asymmetric solution

References

1. Thompson, A.: An evolved circuit, intrinsic in silicon, entwined with physics. In: Proceedings of The First International Conference on Evolvable Systems: from Biology to Hardware (ICES96). Higuchi, T. and Iwata, M. (eds.), Springer Verlag LNCS 1259, (1997) 390-405.
2. Harvey, I., Thompson, A.: Through the Labyrinth Evolution Finds a Way: A Silicon Ridge. In: Proceedings of The First International Conference on Evolvable Systems: from Biology to Hardware (ICES96). Higuchi, T. and Iwata, M. (eds.), Lecture Notes in Computer Science, Vol. 1259. Springer Verlag, (1997) 406-422.
3. Koza, John R., Bennett III, Forrest H, Andre, David, Keane, Martin A, and Dunlap, Frank. Automated synthesis of analog electrical circuits by means of genetic programming. IEEE Transactions on Evolutionary Computation. 1(2) (1997) 109-128.
4. University of Toronto Field Programmable Analogue Array Web site: http://www.eecg.toronto.edu/~vgaudet/fpaa.html
5. Motorola FPAA Web site: http://www.mot-sps.com/fpaa/index.html
6. TRAC Web site: http://www.fas.co.uk
7. Holland, John H.: Adaptation in Natural and Artificial Systems, University of Michigan Press, Ann Arbor, (1975).

Analog Circuits Evolution in Extrinsic and Intrinsic Modes

Ricardo S. Zebulum[1,2] Marco Aurélio Pacheco[2,3] Marley Vellasco[2,3]

[1] CCNR, University of Sussex, Brighton, BN1 9SB UK,
e-mail:ricardoz@cogs.susx.ac.uk
[2] ICA - Pontificia Universidade Catolica do Rio de Janeiro - Brasil
[3] Depto de Engenharia de Sistemas e Computação, UERJ -RJ, Brasil

Abstract. Our work focuses on the use of artificial evolution in Computer Aided Design (CAD) of electronic circuits. Artificial evolution promises to be an important tool for analog CAD development, due to the nature of this task, which has been proven to be much less amenable for standard tools than its digital counterparts. Analog design relies more on the designer's experience than on systematic rules or procedures. The recent appearance of Field Programmable Analog Arrays (FPAAs) allows evolution to be performed in real silicon, which opens new possibilities to the field. Our work addresses the evolution of amplifiers and oscillators, through the use of a standard simulator and a programmable analog circuit respectively. Furthermore, the issue of the implementability of the circuits evolved in simulation is also examined.

1 Introduction

This work applies artificial evolution as a tool for automatic synthesis of analog circuits. Some recent works [1] [5] have demonstrated that evolution based tools have the potential to be used in analog circuit design, showing promising results. The nature of this kind of design makes it less amenable for automation than its digital counterparts [13]: it is not as systematic as digital design, and good designs still depend much on the designer's intuition [6]. These features indicate that biological inspired operators, such as selection, recombination and mutation, may be successfully applied to the area and produce new automatic design tools, more powerful than the ones available at the present moment.

Our work addresses the question of how to evolve analog circuits, i. e., how to establish the means whereby analog circuits will be represented and evaluated. In terms of representation, there are basically two approaches being currently used: *circuit constructing trees* [7] and *string representation*[5]. In terms of evaluation, there are two modes of performing analog circuit evolution: *intrinsic and extrinsic* [2]. In the former, evolution is performed in real silicon through reconfigurable VLSI chips; the latter is performed by the use of standard simulators.

Our work reports applications using both the extrinsic and intrinsic evaluation modes. The extrinsic experiments have been carried on using the standard simulator SMASH [14] and constraints had to be incorporated into the simulator in order to evolve amplifiers whose behaviours could be reproduced in reality; the intrinsic evolutionary experiments were carried on through the use of a new field programmable analog array VLSI chip based on switching capacitors[10].

This article is composed of five additional sections: section 2 examines the different forms of analog circuit representation for evolutionary applications; section 3 discusses the issue of evaluating circuits in intrinsic and extrinsic modes; sections 4 and 5 describe, respectively, the experiments performed using extrinsic and intrinsic evolution. Finally, section 6 concludes this work.

2 Analog Circuits Representation

The representation of analog circuits concerns the way they are encoded into the genotypes. According to the kind of evolutionary system employed, different representations are used. The authors have identified basically two explicit encoding schemes employed so far: circuit constructing trees [7] and linear chain of genes or string representation [5] .

The *circuit constructing trees* representation is a way devised by Koza et. al. to encode the cyclic graphs which characterise electronic circuits into genetic programming trees. The population will now be composed of this type of tree, which will determine the development of a particular initial circuit, called embryonic. These constructing trees can contain *component creating functions* and *connection-modifying functions*. The former inserts particular components, such as capacitors, resistors and transistors, into developing circuits; the latter modifies the developing circuit topology, creating, for instance, series or parallel compositions of a particular component. Using this representation, passive filters [7] and high gain amplifiers [8] have been synthesised.

Whenever using evolutionary systems based on a string representation, a possible way of encoding analog circuits is through the association of each component to a gene of the string, as proposed in [5]. The number of genes of the overall genotype will then be equal to the number of components of the analog circuit expressed by it. Each gene will determine the following information about its encoded component: connecting points, nature of the component (i. e., resistor , capacitor, inductor or transistor) and its value, so that both topology and circuit sizing are encoded in the genotype. Figure 1 illustrates this representation.

The main advantages of this representation are that:

1. It leads to a compact representation of each component, which may facilitate schema processing [4];
2. it is a very straightforward representation, in the sense that the genotype decoding, which generates a ".cir", file to the simulator, is direct.

The latter representation has been used by the authors throughout the extrinsic experiments.

When using reconfigurable devices, there is not an explicit representation of the circuit, since the genotype is now a binary string that is downloaded into

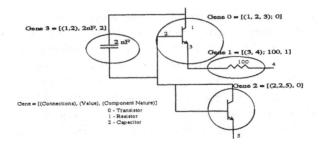

Fig. 1. Circuit encoding based on string representation

the chip memory. This genotype, though meaningless to the user, is interpreted in a particular way by the reconfigurable board.

3 Analog Circuits Evaluation

There are two standard approaches to assessing circuits along the evolutionary process: extrinsic and intrinsic [2].

The extrinsic evaluation mode requires the use of simulators, such as SPICE and SMASH. Each individual will be assessed by the execution of a standard circuit analysis, such as small signal, DC transfer or transient analysis. The main problem with the use of simulators is that they usually consider devices as mathematical entities, ignoring some physical limitations they may present. It has been verified, for instance, that no warning is given if overcurrent or overvoltage conditions are achieved. This problem stems from the fact that some simulators assume a previous knowledge of the user, i.e., components limitations, about the circuit being analysed.

Our particular work has focused mainly on bipolar transistor based circuits. In the particular case of bipolar transistors, the following precautions should be taken when using simulators:

1. Checking that the base-emitter voltages do not surpass the nominal value, which is usually around 0.7V;
2. checking that the collector current does not surpass the maximal possible value, usually around 100mA for low-power transistors [12];
3. adjusting the transistor parameters, such as I_s and β [9], to the values of the transistor to be used in the real implementation.

Summarising, it is necessary to match the simulation conditions to the devices' features. Within the evolutionary system conceived by the authors, after each circuit simulation, violating conditions are checked and individuals presenting these conditions are heavily penalised.

Whenever performing *intrinsic experiments*, the user does not need to constrain evolution as shown in the previous case (except to avoid hardware damage), since there is not difference between design and implementation anymore

[16]. The main advantage of using intrinsic evolution is that the silicon properties may now be fully explored and, consequently, novel and smaller circuits (comparing to conventional ones) are synthesised [15]. It is interesting to observe that the exploration of the physics of silicon goes beyond the scope of its electric properties, as performed by conventional designs; electromagnetic and thermal properties are also used [15].

Field programmable gate arrays, FPGAs, have recently been used in intrinsic evolutionary experiments, such as the evolution of frequency discriminators [15] and oscillators [16] , with successful results. Our work focuses on the intrinsic evolution of analog circuits by the use of FPAAs, which are the analog counterparts of the FPGAs. FPAAs have just recently started being manufactured, and the authors use one, which is based on switching-capacitors technology [6], in the intrinsic experiment reported in this work.

4 Extrinsic Evolution

The issue of operational amplifiers design, which has already been tackled in a pioneering work [1] through the genetic programming methodology, is analysed in this section. The design of operational amplifiers for practical applications is a complex multi-objective task [9], in which around 20 factors have to be taken into account, such as gain, bandwidth, slew-rate, power consumption, area, noise, CMRR and PSRR. As proposed in [1], we adopt the approach of evolving circuits with transfer characteristics similar to the one observed in OpAmps (Figure 2), which is a first step towards the evolution of commercial operational amplifiers.

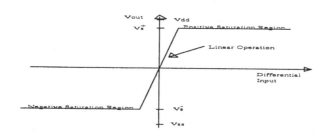

Fig. 2. Typical DC response of an OpAmp

The curve of Figure 2 shows that the range of the OpAmp differential input for linear operation is very low, due to the large differential gain [9]. The operational amplifier saturates for large positive and negative voltages, and, generally, these operating points are not symmetrical, i.e. $v_s^+ \neq v_s^-$. v_s^+ and v_s^- stand for the saturation voltages for positive and negative differential inputs respectively. If $v_s^+ < 0$ and $v_s^- > 0$, we have an inverting amplifier.

Therefore, our goal has been to use the DC transfer analysis of our simulator to achieve a transfer characteristic similar to the one shown in Figure 2. We have used the representation shown in Figure 1 and the evolutionary system is allowed to use npn and pnp transistors as well as resistors to try to meet the requirements; the genotypes have consisted of a collection of 15 genes. As we are not interested in phase response nor in bandwidth in this particular experiment, we did not use compensation capacitors, whose effect is to add a dominant pole to the circuit [11]. Each component encoding gene may use up to eight different connecting points to place the correspondent component: three of them represent a 12 V power supply, a -12 V power supply, and the input ac signal, which represents the differential input. Therefore, we do not impose the particular points in which the input signal or the power supplies will be applied. Additionally, the simulation constraints described previously have been used in these experiments.

The fitness evaluation function rewards the magnitude of the difference between the two saturation values and also their symmetry , which is a requirement of many operational amplifier's applications. The following equation expresses our fitness evaluation function, that is based on the one used in [1]:

$$Fitness = |v_s^+ - v_s^-| - A.v_s^+.v_s^- \tag{1}$$

The first term of the above equation accounts for the difference between the saturation values and the second term accounts for the symmetry between them (a negative value will give a positive contribution). The factor A is a constant set by the user according to the relative importance of each objective. The DC transfer analysis has been used with the input signal varying from -10mV to 10mV [1], following the assumption of low range differential input. We have made two sets of experiments: one in which only the first term of the fitness evaluation function was employed and another in which both terms were employed, with the value of A set to 20.

We have used linear rank, one-point crossover and elitism throughout all the experiments. The population size was kept at the value of 40 individuals and the evolutionary algorithm ran for 200 generations. Each execution lasts around 2 hours in a SPARC Ultra 2 Sun workstation.

Figure 3 shows the fitness of the best genotype along the evolutionary process for a successful execution and Figures 4, 5 and 6 show some of the best evolved circuits, free of penalties, for the first set of experiments. As can be seen from these figures, the evolutionary system usually did not use all the 15 possible components in the evolved solutions, i.e., some genes may represent components that are not electrically connected to the output.

The fitness index shown in Figure 3 is proportional to the first term of the fitness equation. These circuits have also been simulated using SPICE, producing similar results.

Let us analyse the circuit of Figure 4. It has achieved values of v_s^- and v_s^+ equal to 12V and -0.6V respectively, and a DC gain of 44 dB. This circuit shows how evolution organised the components and voltage sources in order to obtain a conventional single transistor amplifier configuration. Transistor Q1 is the amplifier component, receiving the input signal in its base; the output of the

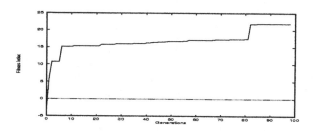

Fig. 3. Fitness of the best amplifier along the evolutionary process for a successful execution

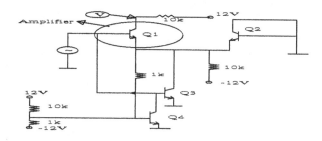

Fig. 4. Schematic of an Evolved Analogue Amplifier (First set of experiments)

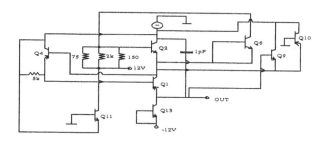

Fig. 5. Schematic of an Evolved Analogue Amplifier (First set of experiments)

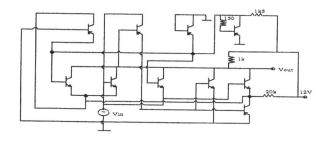

Fig. 6. Schematic of an Evolved Analogue Amplifier (First set of experiments)

amplifier is taken from the collector of transistor Q1. Transistors Q2, Q3 and Q4 operate as loads to the emitter of Q1: transistor Q2 has been configured as a diode working in its forward region and transistors Q3 and Q4 work in the reverse saturation region [9], which is not commonly used in analog design. This was the way found by evolution to reproduce a basic bipolar amplifier. It should be noted that the gain of 44 dB is the same of a single bipolar transistor with β equal to 100 (Q1 in the particular case).

The circuit of Figure 5 achieved values of v_s^- and v_s^+ equal to -3V and 0.5V respectively. This circuit presented a very high DC gain, around 130 dB, meaning that more than one stage is being used. The frequency response can be viewed in Figure 7. We placed a feedback capacitor to insert a pole in the amplifier, just for the sake of frequency response visualisation.

Finally, the circuit of Figure 6 presented values of v_s^- and v_s^+ equal to 0V and 12V respectively. It is interesting to note that evolution chose to use only the positive supply voltage in this circuit. This circuit exhibits a desirable behaviour for operational amplifiers, since it places the saturation voltages equal to the supply voltages (0 and 12 V in this case). The DC gain of this circuit is also 44 dB (meaning that there is just one amplification stage) and its power consumption is 0.19W.

Some of the evolved circuits have been implemented in reality, using the *BC 107 npn* transistor and *BC 177 pnp transistor* [12]. Figure 8 compares the DC transfer curve measured in reality to the one produced by the simulator for the circuit of Figure 6, showing that they are very close to each other. From this figure it can be seen that v_s^-(-10mV) and v_s^+(10mV) are around 0 and 12 Volts respectively. For large negative input values, it can be seen that the output follows the input. The implementability of the circuits derives from the precautions used during simulation. Figure 9 shows the implemented circuit in a prototyping board.

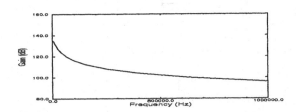

Fig. 7. Frequency response of the amplifier shown in Figure 5

When using the symmetry term in the fitness evaluation function, in a second set of experiments, we reduced the power supply values to -1.5V and +1.5V. When using high supply values, we have not obtained circuits that could place the saturation values near the positive and negative supply values, possibly be-

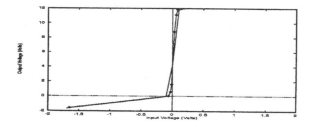

Fig. 8. DC response of the Evolved amplifier (measured values in points) shown in Figure 6

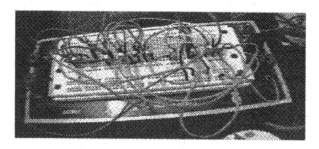

Fig. 9. Analogue Amplifier Implementation

cause the number of components has not been enough for the evolutionary process. Figure 10.a shows an evolved amplifier and Figure 10.b shows its DC response. The main advantages of this amplifier are its symmetry in the DC response and low power consumption (around 3 mW). Power supply values of 1.5V are typical of low power OpAmps.

It is important to clarify that, even though maximising the value of $|v_s^+ - v_s^-|$ is a way to maximise the DC gain, these two statistics are not always directly related, i.e., if a circuit A has a higher value of $|v_s^+ - v_s^-|$ than a circuit B, we can not affirm that it will also have a higher gain, since the measure of DC gain is related to the gradient of the linear portion of the curve shown in Figure 2. Nominal values of DC gain in commercial operational amplifiers usually range from 50dB to 200dB. As bipolar transistors are high gain devices, evolution can easily synthesise high gain structures, though these are not usually parsimonious ones. However, it has been verified that the task of arriving at the DC transfer curves of OpAmps, with a sharp slope at the origin of the DC transfer graph, is a harder one.

5 Intrinsic Evolution

This section describes preliminary results on experiments using the field programmable analog array MPAA020 of Motorola [10]. This board is based on

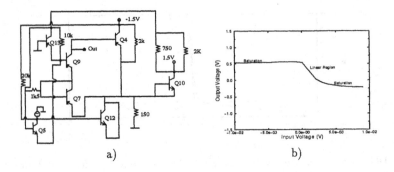

a)　　　　　　　　　　　　　　　　　b)

Fig. 10. Schematic and DC transfer curve of the low-power amplifier evolved in the second set of experiments

Switched-Capacitors technology, which is briefly introduced in this section. Despite the initial character of the experiments, the authors remark that this is pioneering study.

Switched-Capacitors circuits [6] are realised with the use of some basic building blocks, such as OpAmps, capacitors, switches and non-overlapping clocks. The operation of these circuits is based on the principle of the resistor equivalence of a switched capacitor [6]. This principle is illustrated in Figure 11, where ϕ_1 and ϕ_2 are the non-overlapping clocks. In circuit 11.a the average current is given by:

$$I_{avg} = \frac{C1(V1 - V2)}{T} \qquad (2)$$

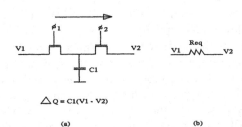

(a)　　　　　　　　　　　　　　(b)

Fig. 11. Switched capacitor / resistor equivalence principle [6]

where T is the clock period. This is equivalent to the resistor Req = T/C1 shown in Figure 11.b. Using this technology, integrators, filters, and oscillators may be designed. The analysis of these circuits usually involves the use of the Z-transform[6].

The VLSI chip used in our experiments is divided into 20 cells, each one containing an OpAmp and a set of programmable capacitors and switches [10].

This chip is configured by a total of 6,864 bits, which control circuit connectivity, capacitor values and other programmable features [10]. The aim of our experiments is to demonstrate that evolution can synthesise standard switched-capacitors circuits, such as oscillators, amplifiers, rectifiers and filters.

We report our initial experiment in the evolution of an oscillator. A human designed circuit implementing an oscillator is shown in Figure 12. This circuit uses two cells of the FPAA. A binary representation has been used in this test. We did not use the whole chip in this experiment; instead, we selected two cells to be programmed by our genotype, which is around 300 bits long. The genotype is downloaded into the board, configuring the circuit. Contrasting to the representation used in the previous experiments, this is not an explicit encoding of the analog circuit.

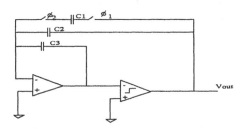

Fig. 12. Human Designed oscillator based on switched-capacitors technology [3]

The fitness evaluation function just tries to maximise the voltage difference between samples of the output generated by the genotypes, which corresponds to maximising the oscillator amplitude. This is shown in the following equation:

$$Fitness = \sum_{i=10}^{n} (s[i] - s[i - 10])^2 \qquad (3)$$

where $s[i]$ is the output sample at instant i. The number of samples per individual, n, and the sampling frequency have been set to 5,000 samples and 100 kHz respectively. The genetic algorithm had the following features: 40 individuals, 40 generations, one-point crossover, a high mutation rate of 7% per genotype bit and exponential rank selection. The experiment lasted a little less than 3 hours using a 166 MHz PC. The graph of Figure 13.a shows the best genotype fitness value along the generations and Figure 13.b displays the output of the best circuit, corresponding to a square wave of 3 Volts amplitude and frequency of 200 kHz (this wave could be captured by our fitness function, despite the aliasing problem). This initial result indicates that evolutionary systems are able to accomplish the task of evolving simple circuits based on this kind of technology.

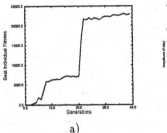 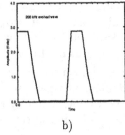

a)   b)

Fig. 13. Best individual fitness and evolved square wave in the intrinsic experiment

6 Conclusions

This article described experiments focusing on analog circuits evolution using extrinsic and intrinsic evaluation modes. Different techniques for analog circuits representation have been discussed and precautions in the use of simulators within evolutionary experiments have been presented. These precautions result in imposing constraints to the evolutionary algorithm due to physical limitations of semiconductor devices.

Our extrinsic experiments focused on the evolution of the DC transfer curve of operational amplifiers. We have obtained circuits whose DC responses were akin to the one of OpAmps and we have been able to analyse the way evolution manipulated components to arrive at some basic amplifier configurations. Some of the evolved circuits have been implemented in prototyping boards.

The continuation of the extrinsic experiments will be carried out in two directions: making the saturation voltages as symmetrical and close to the power supplies as possible, which is a requirement of some applications, by increasing the size of the genotypes; and making a more complete design plan of OpAmps, which may include bandwidth, power consumption and phase margin as additional objectives.

An experiment in intrinsic analog evolution has also been presented. Our intention is to study the differences between human-designed and evolved switched capacitor circuits. As demonstrated in the case of intrinsic evolution using FPGAs [15], we intend to arrive at circuits configurations smaller and less power consuming than the ones designed by humans.

7 Acknowledgements

The authors wish to thank Dr. Adrian Thompson for the collaboration and CAPES and Motorola for the support.

References

1. Bennett III, F. H., Koza, J. R., Andre, D., Keane, M. A., " Evolution of a 60 Decibel Op Amp Using Genetic Programming", Proceedings of the First International Con-

ference on Evolvable Systems (ICES96), LNCS 1259, pp. 455-469, Tsukuba, Japan, October, 1996.

2. DeGaris, H.,"Evolvable Hardware: Genetic Programming of a Darwin Machine", in Artificial Neural Nets and Genetic Algorithms, R.F. Albretch, C.R. Reeves, N.C. Steele (eds), Springer Verlag, NY, 1993.

3. "Easy Analog Design Software User's Manual", Motorola Inc. , 1997

4. Goldberg, D., "Genetic Algorithms in Search, Optimization and Machine Learning", Addison-Wesley Publishing Company, Inc., Reading, Massachusetts, 1989.

5. Grimbleby, J. B. , "Automatic Analogue Network Synthesis Using Genetic Algorithms", Proceedings of the First IEE/IEEE International Conference on Genetic Algorithms in Engineering Systems (GALESIAS - 95), pp.53-58, UK, 1995.

6. Johns, A. D., Martin, K., "Analog Integrated Circuit Design", John Wiley and Sons Inc., 1997

7. Koza J. R., Bennett III F. H., Andre, D. , Keane, M. A., "Four Problems for which a Computer Program Evolved By Genetic Programming is Competitive with Human Performance", Proc. of 1996 IEEE International Conference on Evolutionary Computation, IEEE Press., Pages 1-10.

8. Koza, John R., Andre, David, Bennett III, Forrest H, and Keane, Martin A. 1997. Design of a high-gain operational amplifier and other circuits by means of genetic programming. In Angeline, Peter J., Reynolds, Robert G., McDonnell, John R., and Eberhart, Russ (editors). Evolutionary Programming VI. 6th International Conference, EP97, Indianapolis, Indiana, USA, April 1997 Proceedings. Lecture Notes in Computer Science, Volume 1213. Berlin: Springer-Verlag. 125p;136.

9. Laker, K. R., Sansen, W., "Design of Analog Integrated Circuits and Systems", Mc. Graw-Hill Inc., 1994

10. Motorola Semiconductor Technical Data, "Advance Information Field Programmable Analog Array 20 Cell Version, MPAA020", Motorola Inc. , 1997

11. Sansen, W., " Notes on Low-Noise Analog CMOS and BiCMOS Design", Short Term Course promoted by the Imperial College, February, 25-27 , London, 1998

12. SGS Thomson Microelectronics, "Low Noise General Purpose Audio Amplifiers", pp. 1-7, 7-7, 1994.

13. Sipper, M, "EvoNet - The Network in Excellence in Evolutionary Computation: Report of the Working Group in Evolutionary Electronics (EvoElec)", November 25, 1997.

14. "SMASH User and Reference Manual", Dolphin Integration, France, 1993.

15. Thompson, A., "An Evolved Circuit, Intrinsic in Silicon, entwined with physics", Proceedings of the First International Conference on Evolvable Systems (ICES96), Tsukuba, LNCS 1259, pp. 390-405, Japan, October, 1996.

16. Thompson, A., "Hardware Evolution: Automatic Design of Electronic Circuits in Reconfigurable Hardware by Artificial Evolution ", PhD Thesis, University of Sussex, School of Cognitive and Computing Sciences, September, 1996.

17. Zebulum, R. S., Pacheco, M. A., Vellasco, M., "Comparison of Different Evolutionary Methodologies Applied to Electronic Filter Design", published in the Proceedings of the IEEE International Conference on Evolutionary Computation, to be held in Anchorage,Alaska, May 4-9, 1998.

Evolvable Hardware for Space Applications

Adrian Stoica, Alex Fukunaga, Ken Hayworth, Carlos Salazar-Lazaro

Center for Integrated Space Microsystems
Jet Propulsion Laboratory
California Institute of Technology
4800 Oak Grove Drive
Pasadena CA 91109, USA

Abstract. This paper focuses on characteristics and applications of evolvable hardware (EHW) to space systems. The motivation for looking at EHW originates in the need for more autonomous adaptive space systems. The idea of evolvable hardware becomes attractive for long missions when the hardware looses optimality, and uploading new software only partly alleviates the problem if the computing hardware becomes obsolete or the sensing hardware faces needs outside original design specifications. The paper reports the first intrinsic evolution on an analog ASIC (a custom analog neural chip), suggests evolution of dynamical systems in state-space representations, and demonstrates evolution of compression algorithms with results better than the best-known compression algorithms.

1 Introduction

Spacecraft autonomy plays a key role in future space missions. During remote missions, spacecraft are separated from Earth by distances that delay communications by many minutes (e.g. ~ 10 minutes one-way in communications with the Mars rover), which precludes real-time human operator control of the spacecraft. An intelligent, autonomous spacecraft must be able to cope with unexpected situations, and should be able to adapt to new environments. Spacecraft adaptation is largely controlled by on-board electronic hardware, hence a special need exists for adaptive electronic hardware.

Evolvable hardware (EHW) is adaptive hardware that self-organizes/reconfigures under the control of an evolutionary algorithm [1]. *Extrinsic* EHW refers to evolution in a software simulation (using models of the hardware behavior), followed by a download of the configuration of the most fitted solution to a programmable hardware. In *intrinsic* EHW the configuration bits are downloaded from the beginning to hardware, and the degree of adaptation/fitness is evaluated by observing the behavior of the real hardware. Successful evolution has been reported in simulations of analog (e.g. [2], [3]) and digital (e.g. [4]) circuits and in real digital hardware (e.g. [5] [6]). No intrinsic EHW on analog chips has been reported, an important reason being that the lack of commercial programmable analog chips suitable for EHW. The analog circuits evolved in simulations (e.g. [2], [3]) can not be extended directly to practical HW implementations. On the other hand, some researchers believe that the analog domain would be more suitable for evolution, and there were interpretations that even evolution on (digital) FPGA may have benefited from effect of analog underlying circuitry [5].

This paper addresses some EHW issues relevant to space applications. The paper is organized as follows: Section 2 discusses characteristic aspects of space-oriented EHW. Section 3 describes experiments in intrinsic evolution on application specific analog chips: a test in evolving a function approximator, and evolution of a vision-based tracking behavior for a mobile robot. Section 4 introduces a novel approach to EHW representing the system to be evolved in a behavioral AHDL (Analog Hardware Descriptive Language), in a state-space description. Section 4 presents an application of EHW to adaptive compression.

2   Space-oriented evolvable hardware

There are several characteristic aspects that need to be considered when addressing space-oriented EHW. It is very important to have a systems approach, understanding clearly that EHW is part of a bigger system for which optimality is sought. One needs to understand who/what provides the means for calculating a fitness function for candidate solutions, whether there is a target functionality or reward mechanism stored in some memory on-board, or reinforcement comes from the environment. Also, of most importance is to know how safe is EHW for the space system and also if evolution can provide a response in useful time.

The safety of space systems (such as satellites, spacecraft, planetary probes or rovers) being so critical, our current focus is on evolving adapted sensors and sensory information processing systems, rather than, for example, spacecraft control. The operations from the moment signals reach the sensors until a decision is made, or a coded signal is sent to ground, are fully inter-related and ultimately could be co-evolved in their ensemble to a global optimal signal processing efficiency. In practice, it may be simpler to consider them separately, and evolve independently. The operations could be, for example, signal acquisition (e.g. sensor adaptation in terms of sensor sensitivity domain/profile, focus of attention, etc.), signal pre-processing (e.g. filtering, amplification), extraction of information for on-board decisions (such as sensor-pointing), and preparation of a signal for transmission to Earth (e.g. compression).

The EHW directions we have explored aim to address some aspects from each of the above operations. We performed experiments in intrinsic evolution on analog ASICs, trying to understand more about intrinsic EHW and integration of such chips into higher level systems such as control of sensor arrays, antennas and solar panels, instrument pointing (in this sense we evolved circuits with desired I/O characteristic). We addressed the evolution of electronic circuits, which can be used for filtering or other signal transformations, exploring the design of evolvable CMOS chips based on transistor and elementary circuit blocks (current mirrors, differential pairs, etc) (which will be described in another paper). We addressed evolution of complex dynamic systems, which can be used to learn decision mechanisms or system behaviors. Thus, we developed a novel approach to EHW, which relies on a state-space representation of systems. We developed a simple test to explore the capability of evolving autonomous vision-based navigation. Finally, we approached evolution of algorithms for on-board signal processing, more precisely compression algorithms. In the context of space applications, compression is a very important problem because of the limited communications bandwidth between a spacecraft and the ground. EHW has already shown to be capable of deriving efficient adaptive compression [14].

3 Intrinsic evolution on programmable analog ASICs

This section reports results of intrinsic evolution on dedicated (special purpose) analog chips (ASIC), more precisely analog neural chips. The domain of evolutionary neural networks [5], as well as various analog neural chips have existed for several years, but no results on evolving on the chip have been reported. Previously, our group has explored other approaches for hardware-in-the-loop and on-chip learning, including gradient-descent approaches [7]. A main reason for performing intrinsic (hardware-in-the-loop) learning on an analog chip is that, unlike the digital case where very good models exist in advance, in analog there is always a slight discrepancy between a model and the physical implementation, and therefore a system evolved in software (extrinsic) may exhibit an offset behavior when downloaded to hardware. The tasks described in this section are small and should be regarded as demonstrating an idea rather than applications.

The chip used in the experiments, code-named NN-64, belongs to a family of programmable analog neural network chips developed at JPL [7] [8]. The chip consists of 64 neurons, each with 64 digitally programmable synapses and performs analog processing on analog input signals. The synapses have analog inputs received from chip inputs or from other neurons on the chip, which they multiply (using a multiplying DAC) with a digital weight, providing an analog signal to the somatic level. At the somatic level the analog contributions of the synapses are summed and passed through a sigmoid non-linearity, providing analog neural outputs. Signal processing from synaptic input to neural output takes ~250ns, while reprogramming the weights requires loading in rows of 64, 8 bits at a time, 64 rows for the full chip (or in random access). Loading at 33 MHz takes less than 2 microseconds per neuron, and about 120 microseconds for the full chip; the speed in the current setup where the download is controlled by software is about 3 orders of magnitude lower.

Test 1. The purpose of this test was to evolve on-chip a neural functional approximation. A feedforward, three layer 5-3-1 network was used to learn a simple function of one variable. The target was a bell shape Gaussian response at a linear increasing input. The genome was 23 bytes long, coding the values for the 23 8-bit synaptic weights. Each neuron was pre-biased to have a 2V output in the absence of the input signal. The fitness function was determined based on the sum of the squared errors between the calculated target function, and the circuit response, as measured at 15 input values. We used the Population-Based Incremental Learning (PBIL), an algorithm that "evolves" a probability vector that biases a randomized generation of bit strings representing candidate solutions, and has been found to be competitive with genetic algorithms for a wide variety of optimization problems [9]. A population of 200 networks was evolved for 160 generations. The result can be compared to the target in Fig. 1 (left). The response at a ramp signal is illustrated in the oscilloscope caption in Fig. 1 (right).

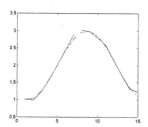 

Fig. 1. A function learned on the chip (intrinsic EHW): (left) closeness to target; (right) response on the oscilloscope.

Test 2. The purpose of this test was to evolve visuo-motor tracking behavior for a mobile robot. A single neuron mapped low-resolution visual images to steering controls. The video input was preprocessed to provide a low-resolution, 3x3 image. The neuron output was a value in the [-1, 1] interval. In terms of steering controls [-1, 1] mapped to [-90, 90] degrees turn in respect to robot's frontal direction (-90 signifying a 90 degrees anti-clockwise turn, +90 signifying a 90 degrees clockwise turn, etc.). A training set collected in a human-controlled driving session was simplified to obtain 12 training patterns like the ones shown in Fig. 2 (input: pixel image, output: steering value). Evolution took place on the chip (intrinsic), the fitness being measured against the stored desired behavior (stored training set). The problem can be seen as evolving the weights for a 9 to 1 neural function approximation. Again, we used PBIL with a population of 200 individuals for 160 generations. The resulting neural controller had an approximation error below 5% (on the training set), which proved sufficient for driving the robot around the track.

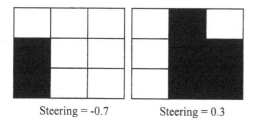 

Steering = -0.7   Steering = 0.3

Fig. 2. Examples from the training set used for learning, and Khepera robot following a marked trail.

4   Evolution of dynamical systems in state-space representations

The behavior of systems, including electronic systems, can be described in terms of an analog descriptive language. Different levels of design abstractions appear in an analog modeling hierarchy (see for example [10]): primitive (device), functional (macromodel), and behavioral (high-level language description) level. The representations commonly used for evolving hardware are primitive or functional. The approach briefly exposed here, and treated in more detail in [11], relates to a behavioral description: a state-space representation expressed by differential equations. Moreover, an intrinsic evolution is proposed, using specially designed hardware that implements this representation: an analog computing machine, which we built and tested in a simple prototype form. In brief the representation we refer to is the state-space representation:

$$\dot{\vec{q}}(t) = \vec{f}(\vec{q}(t); \vec{x}(t))$$
$$\vec{y}(t) = \vec{g}(\vec{q}(t); \vec{x}(t))$$

where x(t) is a vector of continuous signal values coming into the system, y(t) is a vector of continuous output signal values, and q(t) is a vector of continuous internal state values, the "memory" of the system. The functions f() and g() are vector valued and in general non-linear. Figure 3 illustrates an example of the equivalence between a circuit in its schematic description and the state-space representation, graphically displayed by drawing the vector field f().

The prototype programmable analog computer implements with enough flexibility the description in terms of differential equations. Fast context switching allows the state-space of a dynamic system to be decomposed into a lookup-table of smaller vector representations. A search is employed in terms of modifications of the vector field towards a target that ensures certain optimality. This technique called the "modeling clay" approach to bio-inspired hardware is described in detail in [11].

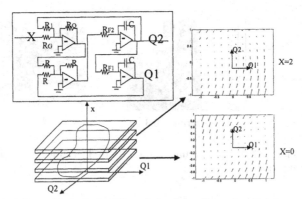

Fig. 3 An active filter circuit from [12] and the three-dimensional state-space of the circuit dynamics with two Q1xQ2 vector field planes plotted at different points along the x-axis

5 Evolution of algorithms for on-board signal processing: results in lossless compression

In space applications compression is necessary in order to enable the downlink of massive amounts of science data (images). (The application of EHW for image compression was pioneered by Salami et al [13]). Because image compression is extremely computationally intensive, a low-power, fast, hardware implementation of a compression algorithm is desirable. An EHW system could be used to automatically generate a hardware-based image compression algorithm specially adapted for the class of images captured by the spacecraft. Image compression for space communications can be approached both in intrinsic and extrinsic EHW mode. For example, suppose a deep space probe needs to send thousands of similar images (e.g., atmospheric images) from the mission target (say, Pluto) back to Earth. The spacecraft could send several exemplar images back to the ground, where an FPGA configuration adapted for the class of images is evolved and uploaded to the spacecraft (extrinsic EHW). Alternatively, the spacecraft could evolve image-specific compression strategies directly using on-board hardware (intrinsic EHW). The work presented in the following relates to the first alternative. A nonlinear model used by the compression algorithm is evolved, which can be then compiled to an FPGA configuration, and finally downloaded (up-link to the spacecraft) to the real FPGA.

A genetic programming (GP) system was developed to perform adaptive image compression based on predictive coding. Predictive coding uses a compact model of an image to predict pixel values of an image based on the values of neighboring pixels. A model of an image is a function model(x,y), which computes (predicts) the pixel value at coordinate (x,y) of an image, given the (known) values of some neighbors of pixel (x,y). Typically, when processing an image in raster scan order (left to right, top to bottom), neighbors are selected from the pixels above and to the left of the current pixel. To complete the compression, the error image (the differences between the predicted pixel value and the actual pixel value) is compressed using an entropy coding algorithm such as Huffman coding or arithmetic coding. If we transmit this compressed error signal as well as the model and all other peripheral information, then a receiver can reconstruct the original image by applying an analogous decoding procedure.

The GP system evolves s-expressions that represent nonlinear predictive models for lossless image compression. The error image is compressed using a Huffman encoder. Because the computational cost of evolving nonlinear predictive models using standard GP systems would be prohibitively expensive, we have implemented a highly efficient, genome-compiler GP system which compiles s-expressions into native (Sparc) machine code to enable the application of GP to this problem. The terminals used for genetic programming were the values of the four neighboring pixels (Image[x-1,y-1],Image[x,y-1], Image[x+1,y-1], Image[x-1,y]), and selected constant values: 1, 5, 10, 100. The functions used were the standard arithmetic functions (+,-,*, %), and MAX/MIN (which return the max/min of two arguments). A detailed presentation of this system and of the results obtained is reported in [14].

The system was evaluated comparing the size of the compressed files with a number of standard lossless compression algorithms on a set of gray scale images. The images used were science images of planetary surfaces taken from the NASA Galileo Mission image archives. The compression results (file size) of the following algorithms are shown in Table 1:

- evolved - the evolved predictive coding compression algorithm.
- CALIC - a state-of-the art lossless image compression.
- LOCO-I - recently selected as the new ISO JPEG-LS (lossless JPEG) baseline standard.
- gzip, compress, pack - standard Unix string compression utilities (*gzip* implements the Lempel-Ziv (LZ77) algorithm, *compress* implements the adaptive Lempel-Ziv-Welch (LZW) algorithm, and *pack* uses Huffman coding).
- szip - a software simulation of the Rice Chip, the current standard lossless compression hardware used by NASA.

It is important to note that in our experiments, a different model was evolved for each image. In contrast, the other approaches apply a single model to every image. Thus, the time to compress an image using the genetic programming approach is several orders of magnitude greater than the time it takes to compress an image using other methods. (This may be reduced if one can evolve models that perform well for a class of images, as opposed to models specialized for individual images). However, the time to decompress an image is competitive with other methods.

Table 1. Compression ratios of various compression techniques applied to set of test images.

Image Name	Original size	evolved	CALIC	LOCO-I	Com-press	gzip	pack	szip
Earth	72643	30380	31798	32932	42502	40908	55068	40585
Earth4	11246	5513	5631	5857	7441	6865	8072	7727
Earth6	20400	9288	10144	10488	11339	10925	13264	12793
Earth7	21039	10218	11183	11476	13117	12520	15551	13269
Earth8	19055	9594	10460	10716	11699	11350	13298	12465

The results obtained show that for science data images, an evolvable-hardware based image compression system is capable of achieving compression ratios superior to that of the best known lossless compression algorithms.

5 Summary

In this paper we have discussed characteristic aspects of space-oriented evolvable hardware, and identified a set of specific applications. The paper contains the first reported intrinsic analog EHW results. A novel approach to EHW based on a representation of systems in terms of state-space was introduced. An EHW based image compression system was described, which achieves compression ratios superior to that of the best known lossless compression algorithms.

Acknowledgements

The research described in this paper was performed at the Center for Integrated Space Microsystems, Jet Propulsion Laboratory, California Institute of Technology and was sponsored by the National Aeronautics and Space Administration. The authors wish to thank Drs. A. Thakoor, T. Daud, B. Toomarian, S. Thakoor, C. Assad for the ideas shared during discussions on evolvable hardware, and to the anonymous reviewers for their comments and suggestions.

References

1. De Garis, H. "Evolvable Hardware: Genetic Programming of a Darwin Machine". Int. Conf. on Artificial Neural Networks and Genetic Algorithms, Innsbruck, Austria, Springer Verlag, 1993
2. Grimbley, J. B. Automatic Analogue Network Synthesis using Genetic Algorithms, 1st IEE/IEEE Conf: Genetic Algorithms in Engineering Systems, UK, 1995
3. Koza, J., Bennett III, F. H., Lohn J., Dunlap, F., Keane M. A., and Andre, D. "Automated Synthesis of Computational Circuits Using Genetic Programming". In Proc. of Second Annual Genetic Programming Conference, Stanford July 13-16, 1997
4. Hemmi, H., Hikage, T. and Shimohara, K. AdAM: A Hardware Evolutionary System , In Proc. of ICEC, (193-196), 1997
5. Thompson, A. Silicon Evolution. In: Proceedings of Genetic Programming 1996 (GP96), J.R. Koza et al. (Eds), pages 444-452, MIT Press 1996
6. Higuchi, T., Murakawa, M., Iwata, M., Kajitani, I., Liu, W. and Salami, M. , "Evolvable Hardware at Function Level." In Proc. of ICEC, (187-192), 1997
7. Duong, T. A. et al., "Learning in neural networks: VLSI implementation strategies," In: Fuzzy Logic and Neural Network Handbook, Ed: C.H. Chen, McGraw-Hill, 1995
8. Eberhardt, S. et al, "Analog VLSI Neural Networks: Implementation Issues and Examples in Optimization and Supervised Learning," IEEE Trans. Indust. Electron. v39 (6):p. 552-564, Dec. 1992.
9. Baluja. I. Genetic Algorithms and Explicit Search Statistics. In Advances in Neural Information Processing Systems 9. Proceedings of the 1996 Conference. 1997. p.319-25
10. Stoica, A. On hardware evolvability and levels of granularity. Proc. of the International Conference "Intelligent Systems and Semiotics 97: A Learning Perspective, NIST, Gaithersburg, MD, Sept. 22-25, 1997
11. Hayworth, K., The "Modeling Clay" approach to bio-inspired electronic hardware, To appear in Proc. ICES98, 1998.
12. Horowitz, P., Winfield, H.: The Art of Electronics 2nd ed Cambridge Univ. Press 1989
13. Salami, M., Murakawa, M., Higuchi, T., Data compression based on evolvable hardware, Proc. Evolvable Systems Workshop, International Joint Conference on Artificial Intelligence, 1997
14. Fukunaga A, Stechert A. Evolving nonlinear predictive models for lossless image compression with genetic programming. To appear in Proceedings of 3rd Annual Genetic Programming Conference (GP-98) , Madison, Wisconsin USA , July 22 – 25, 1998

Embryonics: A Macroscopic View of the Cellular Architecture

Daniel Mange, André Stauffer, and Gianluca Tempesti

Logic Systems Laboratory, Swiss Federal Institute of Technology, IN-Ecublens, CH-1015 Lausanne, Switzerland. E-mail: {name.surname}@epfl.ch

Abstract. The ontogenetic development of living beings suggests the design of a new kind of multicellular automaton endowed with novel quasi-biological properties: self-repair and self-replication. In the framework of the Embryonics (embryonic electronics) project, we have developed such an automaton. Its macroscopic architecture is defined by three features: multicellular organization, cellular differentiation, and cellular division. Through a simple example, a stopwatch, we show that the artificial organism possesses the macroscopic properties of self-replication (cloning) and self-repair. In order to cope with the complexity of real problems, the cell will be decomposed into an array of smaller elements, the molecules, themselves defined by three features: multimolecular organization, self-test and self-repair, and finally cellular self-replication, which is the basis of the macroscopic process of cellular division. These microscopic properties are the subject of a companion paper [9].

1 Introduction

1.1 The POE model of bio-inspired systems

Recently, engineers have been allured by certain natural processes, giving birth to such domains as artificial neural networks, evolutionary computation, and embryonic electronics. In analogy to nature, the space of bio-inspired hardware systems can be partitioned along three axes: phylogeny, ontogeny, and epigenesis; we refer to this as the POE model [10](pp. 1-12). The phylogenetic axis involves evolution, the ontogenetic axis involves the development of a single individual from its own genetic material, essentially without environmental interactions, and the epigenetic axis involves learning through environmental interactions that take place after formation of the individual.

1.2 The ontogenetic axis

This paper is devoted to hardware implementations inspired by the ontogenetic processes of living beings. The main process involved in the ontogenetic axis can be summed up as growth, or construction. Ontogenetic hardware exhibits such features as replication and regeneration, which find their use in many applications. Replication can in fact be considered as a special case of growth - this process involves the creation of an identical organism by duplicating the genetic material of a mother entity onto a daughter one, thereby creating an exact clone.

Research on ontogenetic hardware systems began with von Neumann's work in the late 1940s on self-replicating machines. This line of research can be divided in two main stages:

- von Neumann [22] and others, Langton [6] and others, Reggia et al. [18], Tempesti [21], and Perrier et al. [16] developed self-replicating automata which are *unicellular* organisms: there is a single genome describing (and contained within) the entire machine.
- Inspired by Arbib [2], [3], Mange et al. [7], [10], Marchal et al. [11], Nussbaum et al. [12], Aarden et al. [1] and Ortega et al. [13], [14], [15], proposed a new architecture called *embryonics*, or embryonic electronics. Drawing inspiration from three features usually associated with the ontogenetic process of living organisms, namely, multicellular organization, cellular differentiation, and cellular division, they introduced a new cellular automaton complex enough for universal computation, yet simple enough for physical implementation through the use of commercially available digital circuits. The embryonics self-replicating machines are *multicellular* artificial organisms, in the sense that each of the several cells comprising the organism contains one copy of the complete genome.

1.3 Objectives and contents

Our final objective is the development of very large scale integrated circuits capable of self-replication and self-repair. These two properties seem particularly desirable for very complex artificial systems meant for hostile (nuclear plants) or inaccessible (space) environments. Self-replication allows the complete reconstruction of the original device in case of a major fault, while self-repair allows a partial reconstruction in case of a minor fault.

This paper is devoted to a macroscopic description of the Embryonics project. Section 2 describes the three architectural features of our artificial organisms: multicellular organization (the organism consists of an array of identical physical elements, the cells), cellular differentiation (each cell contains the complete blueprint of the organism, that is, its genome, and specializes depending on its position within the array), and cellular division (each mother cell generates one or two daughter cells). This last mechanism is the object of a formal description by an L-system. Section 3 shows that the multicellular organism thus defined is capable of self-replication (it can produce a copy of itself) and of self-repair (it can replace one or more faulty cells).

The microscopic study of the cell, which relies on three fundamental features: multimolecular organization (the cell is itself decomposed into an array of physically identical elements, the molecules), fault detection within each molecule and self-repair of the cell (through the replacement of the faulty molecules), and cellular self-replication (each group of molecules forming a mother cell is capable of replicating itself to produce a daughter cell and thus bring about the cellular division described at the macroscopic level) is described in a companion paper [9]. The outline of this paper constitutes the core of Section 4.

2 Embryonics' macroscopic features

In the framework of electronics, the environment in which our quasi-biological development occurs consists of a finite (but as large as desired) two-dimensional space of silicon. This space is divided into rows and columns whose intersections define the cells. Since such cells (small processors and their memory) have an identical physical structure, i.e., an identical set of logic operators and of connections, the cellular array is homogeneous. Only the state of a cell, i.e., the contents of its registers, can differentiate it from its neighbors.

2.1 Multicellular organization

The *multicellular organization* divides the artificial organism (ORG) into a finite number of cells (Figure 1), where each cell ($CELL$) realizes a unique function, described by a sub-program called the *gene* of the cell. The same organism can contain multiple cells of the same kind (in the same way as a living being can contain a large number of cells with the same function: nervous cells, skin cells, liver cells, etc.).

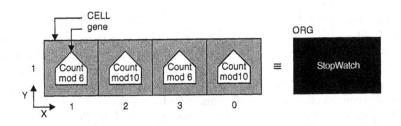

Fig. 1. Multicellular organization of StopWatch.

In this presentation, for clarity's sake, we will confine ourselves to a simple example of a one-dimensional artificial organism: a *StopWatch* implemented with four cells and featuring two distinct genes ("Countmod 10" for counting the units of seconds or minutes, "Countmod 6" for counting the tens of seconds or minutes); the design of these genes is described in detail elsewhere [10](pp. 204-216).

2.2 Cellular differentiation

Let us call *operative genome* (OG) the set of all the genes of an artificial organism, where each gene is a sub-program characterized by a set of instructions and by its position (its coordinates X, Y). Figure 1 then shows the operative genome of StopWatch, with the corresponding horizontal (X) and vertical (Y) coordinates. Let then each cell contain the entire operative genome (Figure 2a): depending on its position in the array, i.e., its place in the organism, each cell can interpret the operative genome and extract and execute the gene which configures it.

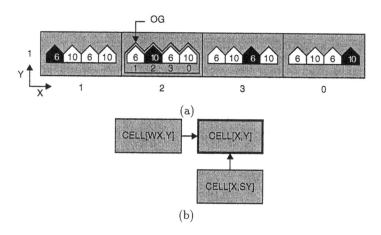

(a)

(b)

Fig. 2. Cellular differentiation of StopWatch. (a) Global organization; OG: operative genome (genes and coordinates). (b) Central cell $CELL[X, Y]$ with its west neighbor $CELL[WX, Y]$ and its south neighbor $CELL[X, SY]$.

In summary, storing the whole operative genome in each cell makes the cell universal: it can realize any gene of the operative genome, given the proper coordinates, and thus implement *cellular differentiation*.

In every artificial organism, any cell $CELL[X, Y]$ computes its coordinate X by incrementing the coordinate WX of its neighbor immediately to the west (Figure 2b). Likewise, it computes its coordinate Y by incrementing the coordinate SY of its neighbor immediately to the south. To verify the property of self-replication of the organism (see Subsection 3.1), the first, "mother cell" is distinguished by the coordinates $X, Y = 1, 1$, and the last cell is distinguished by the coordinates $X, Y = 0, 1$. In the StopWatch example, the computation of the coordinate X occurs modulo-4 (the organism has four cells on the X axis), while the computation of the coordinate Y, which plays no role outside of self-replication (see Subsection 3.1), occurs modulo-1 (the organism is one-dimensional). Any cell $CELL[OG, X, Y]$ can thus be formally defined by a program (its operative genome OG) and by its two coordinates X, Y. In the case of StopWatch, we have the program of Figure 3.

The artificial organism ORG, StopWatch, can be described as the concatenation of four cells ($CELL[OG, X, Y]$ with $X = 1, 2, 3, 0$ and $Y = 1$), and a set

```
X = (WX+1) mod 4
Y = (SY+1) mod 1
case of X:
    X = 1: Countmod 6  (10 minutes)
    X = 2: Countmod 10 (minutes)
    X = 3: Countmod 6  (10 seconds)
    X = 0: Countmod 10 (seconds)
```

Fig. 3. The operative genome OG of StopWatch.

of border conditions ($WX = 0$ to the west of the first cell $CELL[OG, 1, 1]$ and $SY = 1$ to the south of each of the four cells):

$$ORG = CELL[OG, 1, 1], CELL[OG, 2, 1], CELL[OG, 3, 1], CELL[OG, 0, 1] \qquad (1)$$

which, in our particular example, becomes:

$$StopWatch = Countmod\,6, Countmod\,10, Countmod\,6, Countmod\,10 \qquad (2)$$

2.3 Cellular division

At startup, the mother cell (Figure 4), arbitrarily defined as having the coordinate $X, Y = 1, 1$, holds the one and only copy of the operative genome. After time $t1$, the genome of the mother cell is copied into the neighboring (daughter) cells to the east (the second cell of the desired organism) and to the north (the first cell of the first copy of our original organism). The process then continues until the four cells of StopWatch are completely programmed: in our example, the furthest cell is programmed after time $t3$.

L-systems, originally conceived as a mathematical theory of plant development [17], [4], [5], [19], [20], are naturally suitable for modeling growth processes. The very simple case of the cellular division of StopWatch (Figure 4) can be described by the two-dimensional production of Figure 5a, where $\emptyset$ indicates an empty cell.

From the axiom of Figure 5b, we obtain, through the application of the production (Figure 5a), the successive derivations of Figure 5c, each denoting a step of the cellular division, and thus of the growth, of our cellular organism, StopWatch. We do indeed find, at time $t3$, a complete copy of the artificial organism described by expression (1).

2.4 Genotype, phenotype and ribotype

In biology, all ontogenetic development converts a linear genetic information, the DNA or *genotype*, into a protein (that is, a three-dimensional molecule which constitutes the *phenotype*). The genotype-phenotype transformation is performed by a third entity, the *ribosome*, in charge of decoding the DNA: it

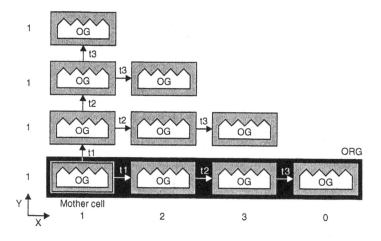

Fig. 4. Cellular division of StopWatch.

is the *ribotype*. The ribosome is, in fact, a special protein, and thus a three-dimensional structure belonging to the same family as the phenotype (Figure 6a)[8]. This relationship can be resumed by:

$$GENOTYPE + RIBOTYPE = PHENOTYPE \qquad (3)$$

or, to emphasize the kinship between ribotype and phenotype:

$$RIBOTYPE[GENOTYPE] = PHENOTYPE \qquad (4)$$

where *RIBOTYPE* can be considered as a function of the argument *GENOTYPE*.

Similarly, our operative genome *OG* represents the DNA, or genotype, of the artificial organism StopWatch. It is interpreted by multiple processors, the artificial cells *CELL*, which represent the counterpart of the ribotype. The phenotype, that is, the operation of our organism *ORG*, is the result of the computation executed in parallel by the cells *CELL* on the program *OG*. Relation (4) thus becomes, in our case:

$$\sum_{X=1}^{0} CELL[OG, X, 1] = ORG \qquad (5)$$

which, for StopWatch and according to expression (1) (Figure 6b), can be written:

$$CELL[OG, 1, 1], CELL[OG, 2, 1], CELL[OG, 3, 1], CELL[OG, 0, 1] = ORG \qquad (6)$$

```
                              ∅
                              CELL[OG,WX,(SY+1)mod1],∅
        ∅
        CELL[OG,WX,SY],∅ -> CELL[OG,WX,SY],CELL[OG,(WX+1)mod4,SY],∅
```

(a)

```
        ∅                       ∅
        CELL[OG,WX,SY],∅ = CELL[OG,1,1],∅
```

(b)

```
    ∅
    CELL[OG,1,1],∅
t1: CELL[OG,1,1],CELL[OG,2,1],∅

    ∅
    CELL[OG,1,1],∅
    CELL[OG,1,1],CELL[OG,2,1],∅
t2: CELL[OG,1,1],CELL[OG,2,1],CELL[OG,3,1],∅

    ∅
    CELL[OG,1,1],∅
    CELL[OG,1,1],CELL[OG,2,1],∅
    CELL[OG,1,1],CELL[OG,2,1],CELL[OG,3,1],∅
t3: CELL[OG,1,1],CELL[OG,2,1],CELL[OG,3,1],CELL[OG,0,1],∅
```

= ORG
(c)

Fig. 5. L-system model of StopWatch. (a) The production. (b) The axiom. (c) The cellular division derivation.

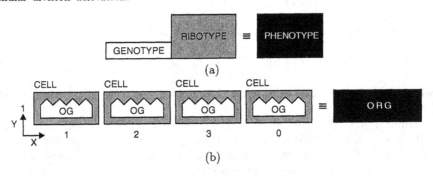

Fig. 6. Genotype-phenotype relationship. (a) The transformation. (b) StopWatch application.

3   Self-replication and self-repair as macroscopic properties

3.1   Organism's self-replication (cloning)

The *self-replication* of an artificial organism, i.e., the production of an exact copy of the original or "cloning", rests on two hypotheses:

– there exists a sufficient number of spare cells (unused cells at the right of the original organism, or at the upper side of the array), at least four in our example (to produce one copy);

– the calculation of the coordinates produces a cycle ($X = 1 \to 2 \to 3 \to 0 \to 1$ and $Y = 1 \to 1$ in Figure 7).

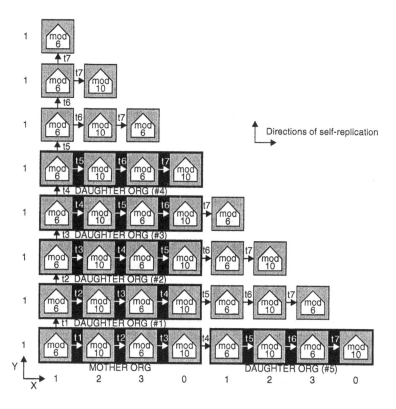

Fig. 7. Self-replication of a 4-cell StopWatch organism in an unlimited homogeneous array of cells.

As the same pattern of coordinates produces the same pattern of genes, self-replication can be easily accomplished if the microprogram of the operative genome OG, associated to the homogeneous array of cells, produces several occurences of the basic pattern of coordinates. In our example (Figure 7), both the repetition of the vertical coordinate pattern ($Y = 1 \to 1 \to 1 \to 1 \to 1 \to 1 \to 1 \to 1$) and of the horizontal coordinate pattern ($X = 1 \to 2 \to 3 \to 0 \to 1 \to 2 \to 3 \to 0$), associated to an unlimited array of cells, produce five copies, the *daughter organisms*, of the original or *mother organism*. Given a sufficiently large space, the self-replication process can be repeated for any number of specimens in the X and/or Y axes.

Formally, the computation of the different steps of the cellular division described by the L-system of Figure 5c will produce the following sequence of daughter organisms (Figure 7):

- after time $t4$: daughter organism $\sharp 1$, in the 2nd row;
- after time $t5$: daughter organism $\sharp 2$, in the 3rd row;
- after time $t6$: daughter organism $\sharp 3$, in the 4th row;
- after time $t7$: daughter organism $\sharp 4$, in the 5th row, and $\sharp 5$, in the 1st row.

3.2 Organism's self-repair

In order to demonstrate *self-repair*, we have decided to add spare cells to the right of the original unidimensional organism (Figure 8). These cells may be used not only for self-repair, but also for self-replication.

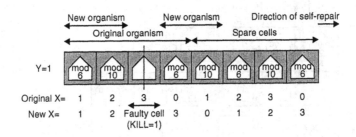

Fig. 8. Self-repair of a 4-cell StopWatch organism with four spare cells and one faulty cell.

The existence of a fault is detected by a $KILL$ signal which is calculated in each cell by a built-in self-test mechanism realized at the molecular level (see the companion paper [9]). The state $KILL = 1$ identifies the faulty cell, and the entire column (if any) to which the faulty cell belongs is considered faulty, and is deactivated (column $X = 3$ in Figure 8). All the functions (X coordinate and gene) of the cells at the right of the column $X = 2$ are shifted by one column to the right. Obviously, this process requires as many spare cells or columns, to the right of the array, as there are faulty cells or columns to repair (four spare cells tolerating four successive faulty cells in the unidimensional example of Figure 8). It also implies that the cell has the capability of bypassing the faulty column and shifting to the right all or part of the original cellular array.

With a sufficient number of cells, it is obviously possible to combine self-repair (or growth if any) in the X direction, and self-replication in both the X and Y directions.

4 Cell's microscopic features

In all living beings, the string of characters which makes up the DNA, i.e., the genome, is executed sequentially by a chemical processor, the *ribosome*. Drawing inspiration from this biological mechanism, we will realize each cell of our artificial organism by means of a small electronic processor, a *binary decision machine*, executing sequentially the instructions of our artificial genome, the operative genome OG. In analogy with the ribosome, which is itself decomposed into smaller parts, the molecules, we will embed our artificial cell into an array of programmable logic devices, an FPGA whose basic elements will be considered as our artificial molecules. The detailed design of this molecular architecture is the subject of the companion paper [9].

Acknowledgments

This work was supported by grants 20-42'270.94 and 20-49'375.96 from the Swiss National Science Foundation to which we express our gratitude.

References

1. A. C. Aarden, E. Blok, H. Bouma, and R. Schiphorst. Leven op silicium. Technical Report BSC-44N97, Faculteit den elektrotechniek, Universiteit Twente, 1997.
2. M. A. Arbib. Simple self-reproducing universal automata. *Information and Control*, 9:177–189, 1966.
3. M. A. Arbib. *Theories of Abstract Automata*. Prentice-Hall, Englewood Cliffs, N.J., 1969.
4. H. Kitano. Designing neural networks using genetic algorithms with graph generation system. *Complex Systems*, 4:461–476, 1990.
5. H. Kitano. Morphogenesis for evolvable systems. In E. Sanchez and M. Tomassini, editors, *Towards Evolvable Hardware*, volume 1062 of *Lecture Notes in Computer Science*, pages 99–117. Springer-Verlag, Heidelberg, 1996.
6. C. G. Langton. Self-reproduction in cellular automata. *Physica D*, 10:135–144, 1984.
7. D. Mange, D. Madon, A. Stauffer, and G. Tempesti. Von Neumann revisited: A Turing machine with self-repair and self-reproduction properties. *Robotics and Autonomous Systems*, 22(1):35–58, 1997.
8. D. Mange and M. Sipper. Von Neumann's quintessential message: Genotype + ribotype = phenotype. *Artificial Life*. (to appear).
9. D. Mange, A. Stauffer, and G. Tempesti. Embryonics: A microscopic view of the molecular architecture. In M. Sipper, D. Mange, and A. Perez, editors, *Proceedings of The Second International Conference on Evolvable Systems: From Biology to Hardware (ICES98)*, Lecture Notes in Computer Science. Springer-Verlag, Heidelberg, 1998.
10. D. Mange and M. Tomassini, editors. *Bio-Inspired Computing Machines*. Presses polytechniques et universitaires romandes, Lausanne, 1998.

11. P. Marchal, C. Piguet, D. Mange, A. Stauffer, and S. Durand. Embryological development on silicon. In R. A. Brooks and P. Maes, editors, *Artificial Life IV*, pages 365–370, Cambridge, Massachusetts, 1994. The MIT Press.

12. P. Nussbaum, P. Marchal, and C. Piguet. Functional organisms growing on silicon. In T. Higuchi, M. Iwata, and W. Liu, editors, *Proceedings of The First International Conference on Evolvable Systems: From Biology to Hardware (ICES96)*, volume 1259 of *Lecture Notes in Computer Science*, pages 139–151. Springer-Verlag, Heidelberg, 1997.

13. C. Ortega and A. Tyrrell. Design of a basic cell to construct embryonic arrays. In *IEE Proceedings on Computers and Digital Techniques*. (to appear).

14. C. Ortega and A. Tyrrell. Biologically inspired reconfigurable hardware for dependable applications. In *Proceedings of the Colloquium on Hardware Systems for Dependable Applications*. IEEE Professional Group A2, 1997.

15. C. Ortega and A. Tyrrell. Fault-tolerant systems: The way biology does it. In *Proceedings of the 23rd Euromicro Conference*. IEEE Computer Society Press, 1997.

16. J.-Y. Perrier, M. Sipper, and J. Zahnd. Toward a viable, self-reproducing universal computer. *Physica D*, 97:335–352, 1996.

17. P. Prusinkiewicz and A. Lindenmayer. *The Algorithmic Beauty of Plants*. Springer-Verlag, New York, 1990.

18. J. A. Reggia, S. L. Armentrout, H.-H. Chou, and Y. Peng. Simple systems that exhibit self-directed replication. *Science*, 259:1282–1287, February 1993.

19. A. Stauffer and M. Sipper. L-hardware: Modeling and implementing cellular development using L-systems. In D. Mange and M. Tomassini, editors, *Bio-Inspired Computing Machines*. Presses polytechniques et universitaires romandes, Lausanne, 1998.

20. A. Stauffer and M. Sipper. Modeling cellular development using L-systems. In M. Sipper, D. Mange, and A. Perez, editors, *Proceedings of The Second International Conference on Evolvable Systems: From Biology to Hardware (ICES98)*, Lecture Notes in Computer Science. Springer-Verlag, Heidelberg, 1998.

21. G. Tempesti. A new self-reproducing cellular automaton capable of construction and computation. In F. Morán, A. Moreno, J. J. Merelo, and P. Chacón, editors, *ECAL'95: Third European Conference on Artificial Life*, volume 929 of *Lecture Notes in Computer Science*, pages 555–563, Heidelberg, 1995. Springer-Verlag.

22. J. von Neumann. *Theory of Self-Reproducing Automata*. University of Illinois Press, Illinois, 1966. Edited and completed by A. W. Burks.

Embryonics: A Microscopic View of the Molecular Architecture

Daniel Mange, André Stauffer, and Gianluca Tempesti

Logic Systems Laboratory, Swiss Federal Institute of Technology, IN-Ecublens, CH-1015 Lausanne, Switzerland. E-mail: {name.surname}@epfl.ch

Abstract. The ontogenetic development of living beings suggests the design of a new kind of multicellular automaton endowed with novel quasi-biological properties: self-repair and self-replication. In the framework of the Embryonics (embryonic electronics) project, we have developed such an automaton. Its macroscopic architecture is defined by three features: multicellular organization, cellular differentiation, and cellular division which are described in a companion paper [5]. In order to cope with the complexity of real problems, the cell is itself decomposed into an array of smaller elements, the molecules, themselves defined by three features: multimolecular organization, self-test and self-repair, and finally cellular self-replication, which is the basis of the macroscopic process of cellular division. These microscopic properties are illustrated by the example of an up-down counter. Finally, we propose a design methodology based on three successive configurations of the basic molecular tissue, a novel FPGA. These configurations are analogous to the operation of three kinds of genetic information: the polymerase, ribosomic, and operative genomes.

1 Cell's microscopic features

1.1 Objectives and contents

In all living beings, the string of characters which makes up the DNA, i.e., the genome, is executed sequentially by a chemical processor, the *ribosome*. Drawing inspiration from this biological mechanism, we will realize each cell of an artificial organism by means of a small electronic processor, a *binary decision machine*, executing sequentially the instructions of our artificial genome, the operative genome OG. In analogy with the ribosome, which is itself decomposed into smaller parts, the molecules, we will embed our artificial cell into an array of programmable logic devices, an FPGA whose basic elements will be considered as our artificial molecules.

While the macroscopic description of the Embryonics project and the corresponding properties (self-repair and self-replication of the artificial organism) are described in a companion paper [5], Section 1 of this paper is dedicated to the microscopic study of the cell; this study relies on three fundamental features: multimolecular organization (the cell is itself decomposed into an array of physically identical elements, the molecules), fault detection within each molecule and self-repair of the cell (through the replacement of the faulty molecules), and cellular self-replication (each group of molecules forming a mother cell is capable of replicating itself to produce a daughter cell and thus bring about the cellular division described at the macroscopic level [5]). This last mechanism is the object of a formal description by an L-system.

Section 2 will finally propose a design methodology based on three successive configurations of the basic molecular tissue, a novel FPGA; these configurations are analogous to the operation of three kinds of genetic information: the *polymerase genome*, dividing the silicon space in order to realize the macroscopic cellular division, the *ribosomic genome*, building the binary decision machine which constitutes the core of the cell, and the *operative genome*, which defines the particular application.

1.2 Multimolecular organization

In order to implement any digital system, in particular the binary decision machine of our artificial cell, into a reconfigurable array, we require a methodology capable of generating, starting from a set of specifications, the configuration of a homogeneous network of elements, the *molecules*, each molecule defined by an identical architecture and a usually distinct state (the *molecule code* or *MOLCODE*).

To meet our requirements, we have selected a particular representation: the *ordered binary decision diagram* (OBDD). This representation, with its well-known intrinsic properties such as canonicity [1], was chosen for two main reasons:

- it is a graphical representation which exploits well the two-dimensional space and immediately suggests a physical realization on silicon;
- its structure leads to a natural decomposition into molecules realizing a logic test (a diamond), easily implemented by a multiplexer.

We will illustrate the handling of ordered binary decision diagrams through a simple example, an artificial cell realizing an up-down counter. Our choice will lead us to define a field-programmable gate array (FPGA) as a homogeneous multicellular array where each molecule contains a programmable multiplexer with one control variable, implementing precisely a logic test.

Let us consider the relization of the aboved-mentioned modulo-4 up-down counter, defined by the following sequences:

- for $M = 0 : Q1, Q0 = 00 \rightarrow 01 \rightarrow 10 \rightarrow 11 \rightarrow 00$ (counting up);
- for $M = 1 : Q1, Q0 = 00 \rightarrow 11 \rightarrow 10 \rightarrow 01 \rightarrow 00$ (counting down).

It can be verified that the two ordered binary decision diagrams $Q1$ and $Q0$ of Figure 1a (where each diamond represents a multiplexer, each square an input boolean value, and each diamond embedded in a square a 1-bit memory, i.e., a flip-flop) correspond to a possible realization of the counter [6](pp. 132-135, 239-240).

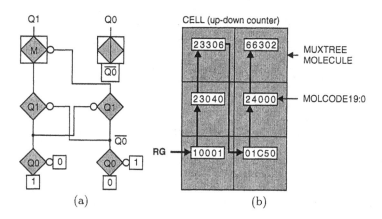

Fig. 1. Modulo-4 up-down counter. (a) Ordered binary decision diagram. (b) Multi-molecular implementation of the artificial cell with six MUXTREE molecules; RG: ribosomic genome (sum of the $MOLCODE$).

The reconfigurable molecule, henceforth referred as MUXTREE (for multi-plexer tree), consists essentially of a programmable multiplexer (with one control variable), a D-type flip-flop, and a switch block allowing all possible connections between two horizontal and two vertical long-distance busses. The behavior of a MUXTREE molecule, described in detail elsewhere [6](pp. 135-143), is com-pletely defined by a molecular code organized as a 20-bit data $MOLCODE19 : 0$, itself stored in a *configuration register* CREG.

The *multimolecular organization* divides finally the artificial cell, our up-down counter, into a finite number of molecules (six), where each molecule is defined by a unique configuration, its molecular code $MOLCODE19 : 0$. For clarity's sake, each $MOLCODE$ is represented in Figure 1b by five hexadecimal characters.

Let us call *ribosomic genome* (RG) the string of all the molecular codes of our artificial cell (Figure 1b), where each molecular code is a 20-bit or 5-hexadecimal characters word $MOLCODE19 : 0$:

$$RG = \sum MOLCODE19 : 0 = 10001, 23040, 23306, 01C50, 24000, 66302 \quad (1)$$

In conformance with the definitions of the companion paper [5](Subsection 2.4), the ribosomic genome RG represents the genotype of our cell. It is directly interpreted by the FPGA, that is, by the MUXTREE molecules that act as the equivalent of the ribotype. The phenotype describes the operation of the complete cell. Relation (4) of [5] becomes, in the general case of a cell:

$$\sum MOLECULE[RG] = CELL \quad (2)$$

and, in the particular case of our reversible counter:

$$\sum MUXTREE[RG] = \sum MUXTREE[10001, 23040, 23306, 01C50, 24000, 66302]$$
$$= up-down\ counter$$

1.3   Molecule's self-test and cell's self-repair

We have already described the capability of the complete organism to self-repair by sacrificing a complete column of each faulty cell [5](Subsection 3.2). This operation is analogous to the cicatrization process of living beings, the scar being represented by the column sacrificed by the artificial organism. If the cell is complex, this process is costly: it is thus indispensable to dispose of a second self-repair mechanism, situated at the molecular level. The biological inspiration is again immediate: the DNA's double helix offers a complete redundancy of the genomic information and allows the rectification of any base in an helix by comparison with the complementary base in the opposing helix. The specifications of the molecular self-repair system must include the following features:

- it must operate in real time;
- it must preserve the memorized values, that is, the state of the D-type flip-flop contained in each molecule;
- it must assure the automatic detection of a fault (self-test), its localization, and its repair (self-repair);
- it must involve an acceptable overhead;
- finally, in case of multiple faults (to many faulty molecules), it must generate a global signal $KILL = 1$ which activates the suppression of the cell and starts the self-repair process of the complete organism (see Subsection 3.2 of [5]).

All these constraints forced us to adopt a set of compromises with regard to the fault-detection capabilities of the system. A self-repairing MUXTREE molecule (or MUXTREE SR) can be divided into three parts (Figure 2) [12], [6](pp.249-258):

- the functional part of the molecule (the multiplexer and the internal flip-flop) is tested by space redundancy: the logic is duplicated (M1 and M2) and the outputs of the two elements compared to detect a fault; a third copy of the flip-flop was added to allow self-repair (i.e., to save the flip-flop state);
- the configuration register (CREG) is tested as the configuration is being entered (and thus not on-line); being implemented as a shift register, it can be tested using a special test sequence introduced in all elements in parallel before the actual configuration for the system;
- faults on the connections (and in the switch block SB) can be detected, but cannot be repaired, both because they cannot be localized to a particular connection, and because our self-repair system exploits the connections to

reconfigure the array; in the current system, therefore, we decided not to test the connections directly; this assumption is in accordance with the present state of the art [7].

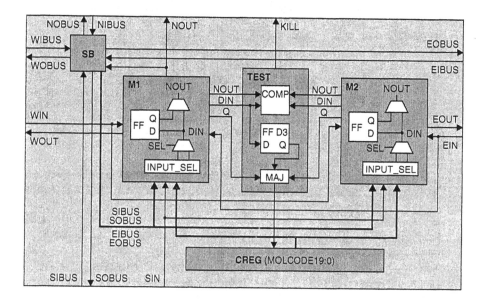

Fig. 2. A self-testing MUXTREE SR molecule using space redundancy.

The self-repair system had to meet the same constraints as those of the self-test system, and in particular the requirement that its additional logic be minimized. Exploiting the fact that the spare columns are distributed and that their frequency is programmable (Subsection 1.4), we limited the reconfiguration of the array to a single molecule per line between two spare columns (Figure 3). This allows us to minimize the amount of logic required for the reconfiguration of the array, while keeping a more than acceptable level of robustness. This mechanism is also in accordance with the present state of the art [9].

It should be added that if the self-repair capabilities of the MUXTREE SR molecular level is exceeded, a global $KILL$ signal is generated and the system will attempt to reconfigure at the higher (cellular) level, as mentioned in Subsection 3.2 of [5].

1.4 Cell's self-replication (cellular division): space divider

The macroscopic mechanism of cellular division (Subsection 2.3 and Figure 4 of [5]) implies the construction, from a unique mother cell, of a certain number of daughter cells identical to the mother cell. This construction represents a second

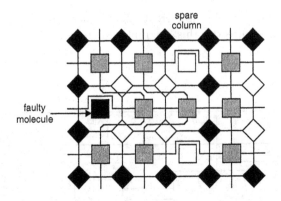

Fig. 3. The self-repair mechanism for an array of MUXTREE SR molecules.

example of self-replication (after that of the organism described in Subsection 3.1 of [5]). Performed by the molecular tissue MUXTREE SR, this self-replication will occur in two steps:

- first, a frontier will define, in the MUXTREE SR tissue, the cell's width, its height, and the number of spare columns;
- then, each molecule in each cell will receive its configuration, that is, its molecular code *MOLCODE* (the ribosomic genome *RG*).

The mechanism we have adopted to implement this process is to introduce a very simple *molecular automaton* or *space divider* (Figure 4), capable of creating a set of *boundaries* which partition the FPGA into blocks of molecules, each defining a cell. It then becomes possible to configure the entire array by entering the configuration of a single cell, the ribosomic genome *RG*, which will automatically be replicated within all the boundaries (cellular division). This molecular automaton is roughly inspired by the self-replicating Langton's loop [4]; its design is described elsewhere [6](pp. 240-249), [10].

The microscopic self-replication of a cell can be described by an L-system [8], [2], [3], [11], in the same manner as the macroscopic cellular division of the organism [5](Subsection 2.3). Our space divider is a two-dimensional molecular automaton and its structuring process is essentially a growth process, starting from the automaton molecule on the lower left-hand side, where the programming data are fed continuously. The representation of the symboles used in the developmental model is given in Figure 5a. The L-system description of the space divider starts with a single letter axiom which corresponds to a left branching apex. The productions applied to the axiom in order to obtain a cellular structure of 3 × 3 molecules are listed in Figure 5b; they fall into three categories:

- the branching signal propagation productions p1 to p4;
- the simple growth productions p5 to p14;

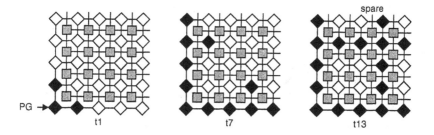

Fig. 4. Space divider: molecular automaton colonizing an array of MUXTREE SR molecules (each molecule of the space divider is a diamond; each MUXTREE SR molecule is a gray square); PG: polymerase genome.

– the branching growth productions p15 to p18.

Thirteen derivation steps of the developmental process are shown in Figure 5c, where the first character of the string to the left of the vertical separator is part of the program that is fed to the space divider: it is applied from the outside and corresponds to the left context of the first character after the vertical separator.

The molecular automaton finally implemented (Figure 5d) defines a division of the cellular space into squares of size 3×3 molecules (Figure 4 where each diamond corresponds to a square in Figure 5d).

A very interesting "bonus" of this system is that it becomes possible to use this automaton to define which columns of the array will be spare columns, used for self-repair (Figure 3). The frequency of these columns, and consequently the robustness of the system, is therefore entirely programmable, rather than hardwired, and can thus be set to meet the requirements of a single application.

Coming back to our original example of a modulo-4 up-down counter (Figure 1b), it should be obvious that the colonizing process of Figure 5d will generate a MUXTREE SR array able to implement our design (the six molecules of the counter) with a spare column to the right (three spare molecules).

The programming data of the space divider are equivalent to the *polymerase genome* (PG) of a living being, as they make possible the cellular division of the organism. Figures 5c and 5d detail the polymerase genome, which is described by a short cycle *iib*:

$$PG = iib, iib, iib, ... \qquad (3)$$

Let us call TISSUE the FPGA consisting of MUXTREE SR molecules, where each molecule includes a copy of the space divider. TISSUE represents then the lowest level hardware primitive in our hierarchy (it cannot be further decomposed). Coming back to the definitions of Subsection 2.4 of [5], the polymerase genome PG represents the genotype of the TISSUE FPGA, itself equivalent to the ribotype. The phenotype is the molecular tissue $\sum MOLECULE$, ready to

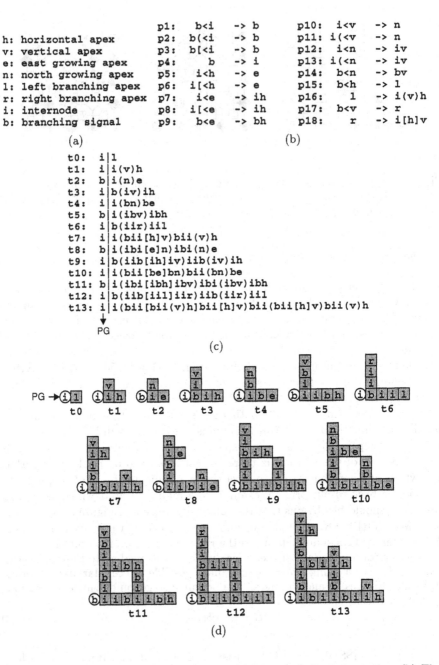

h: horizontal apex	p1:	b<i	-> b	p10:	i<v	-> n
v: vertical apex	p2:	b(<i	-> b	p11:	i(<v	-> n
e: east growing apex	p3:	b[<i	-> b	p12:	i<n	-> iv
n: north growing apex	p4:	b	-> i	p13:	i(<n	-> iv
l: left branching apex	p5:	i<h	-> e	p14:	b<n	-> bv
r: right branching apex	p6:	i[<h	-> e	p15:	b<h	-> l
i: internode	p7:	i<e	-> ih	p16:	l	-> i(v)h
b: branching signal	p8:	i[<e	-> ih	p17:	b<v	-> r
	p9:	b<e	-> bh	p18:	r	-> i[h]v

(a)   (b)

```
t0:  i|l
t1:  i|i(v)h
t2:  b|i(n)e
t3:  i|b(iv)ih
t4:  i|i(bn)be
t5:  b|i(ibv)ibh
t6:  i|b(iir)iil
t7:  i|i(bii[h]v)bii(v)h
t8:  b|i(ibi[e]n)ibi(n)e
t9:  i|b(iib[ih]iv)iib(iv)ih
t10: i|i(bii[be]bn)bii(bn)be
t11: b|i(ibi[ibh]ibv)ibi(ibv)ibh
t12: i|b(iib[iil]iir)iib(iir)iil
t13: i|i(bii[bii(v)h]bii[h]v)bii(bii[h]v)bii(v)h
```
            ↓
           PG

(c)

Fig. 5. L-system model of the space divider. (a) The symbol representation. (b) The production set. (c) A sample derivation; *PG*: polymerase genome. (d) The final molecular automaton realizing a space division.

receive its ribosomic genome. Relation (4) of [5] thus becomes, in the general case:

$$TISSUE[PG] = \sum MOLECULE \tag{4}$$

and, in the particular case of the space divider in our MUXTREE SR molecules:

$$TISSUE[iib, iib, iib, ...] = \sum MUXTREE \tag{5}$$

Finally, we will note that the process of molecular development of Figure 5d (the self-replication of the cell) represents a microscopic (that is, molecular level) description of the macroscopic process of cellular division of the organism [5](Figure 4). The construction of a daughter cell requires a total time t (Figure 4 of [5]), decomposed into 12 steps at the molecular level (Figure 5d).

2 Conclusion

2.1 The missing link

In Sections 2 and 3 of the companion paper [5], the macroscopic characteristics and properties of Embryonics have been illustrated by means of an artificial multicellular organism, a StopWatch, made up of four cells (without spares), each a small processor (a binary decision machine) with the associated memory necessary to store the operative genome. In Section 1 of this paper, the microscopic architecture and features of Embryonics' cells have been described through another, far simpler example: an up-down counter comprising six MUXTREE SR molecules (without spares).

The design and the implementation of a specimen of the StopWatch's cell into a regular array of MUXTREE SR molecules is a complex task which is beyond the scope of this paper; as an example, the realization of the basic cell of an even simpler multicellular organism, a modulo-60 counter with only two cells, requires 600 MUXTREE SR molecules (without spares) for both the binary decision machine and the associated memory (a 360-bit shift register) [6](pp. 258-265).

2.2 Design methodology and genome hierarchy

In our Embryonics project, the design of a multicellular automaton requires the following steps:

- the original specifications are mapped into a homogeneous array of cells (binary decision machines with their associated memory); the software (a microprogram) and the hardware (the architecture of the cell) are tailored according to the specific example (Turing machine, electronic watch, random number generator, etc.); in biological terms, this microprogram represents the *operative genome* (*OG*);

- the cell's hardware is implemented into a homogeneous array of molecules, the MUXTREE SR molecules; spare columns are introduced in order to improve the global reliability; our artificial cell being analogous to the ribosome of a natural cell, the string of the molecule codes can be seen as the *ribosomic genome (RG)*;
- the dimensions of the final molecular array, as well as the frequency of the spare columns, define the string of data required by the molecular automaton (the space divider that creates the boundaries between cells); as this information will allow the generation of all the daughter cells starting from the first mother cell, it can be considered as equivalent to the *polymerase genome (PG)*.

Given the basic TISSUE FPGA (i.e., the array of MUXTREE SR molecules, with a space divider automaton in each molecule), the corresponding programming has to take place in reverse order:

- the polymerase genome (PG) is injected in order to obtain the boundaries between cells:

$$TISSUE[PG] = \sum MOLECULE \tag{6}$$

- The ribosomic genome (RG) is injected in order to configure the array of MUXTREE SR molecules and obtain the final architecture of each cell:

$$\sum MOLECULE[RG] = CELL \tag{7}$$

- The operative genome (OG) is stored into the memory of each cell in order to make the cell ready to execute the specifications of the whole organism ORG:

$$\sum CELL[OG] = ORG \tag{8}$$

By replacing $CELL$ in expression (8) with the values derived from (2) and (4), we can finally show the sequence of three configurations that transforms the primitive FPGA TISSUE into an operative multicellular organism:

$$\sum ((TISSUE[PG])[RG])[OG] = ORG \tag{9}$$

Echoing biology, we have faced complexity by decomposing the organism into cells and then the cells into molecules. This decomposition implies multiple configuration steps: the polymerase genome organizes the space by defining the cells' boundaries, the ribosomic genome defines the architecture of each cell as an array of molecules, and finally the operative genome makes up the program which will be executed by the cellular processors to accomplish the required task. The Latin motto "divide and conquer" maintains its relevance even today.

Acknowledgments

This work was supported by grants 20-42'270.94 and 20-49'375.96 from the Swiss National Science Foundation to which we express our gratitude.

References

1. R. E. Bryant. Symbolic boolean manipulation with ordered binary-decision diagrams. *ACM Computing Surveys*, 24(3):293–318, 1992.
2. H. Kitano. Designing neural networks using genetic algorithms with graph generation system. *Complex Systems*, 4:461–476, 1990.
3. H. Kitano. Morphogenesis for evolvable systems. In E. Sanchez and M. Tomassini, editors, *Towards Evolvable Hardware*, volume 1062 of *Lecture Notes in Computer Science*, pages 99–117. Springer-Verlag, Heidelberg, 1996.
4. C. G. Langton. Self-reproduction in cellular automata. *Physica D*, 10:135–144, 1984.
5. D. Mange, A. Stauffer, and G. Tempesti. Embryonics: A macroscopic view of the cellular architecture. In M. Sipper, D. Mange, and A. Perez, editors, *Proceedings of The Second International Conference on Evolvable Systems: From Biology to Hardware (ICES98)*, Lecture Notes in Computer Science. Springer-Verlag, Heidelberg, 1998.
6. D. Mange and M. Tomassini, editors. *Bio-Inspired Computing Machines*. Presses polytechniques et universitaires romandes, Lausanne, 1998.
7. R. Negrini, M. G. Sami, and R. Stefanelli. *Fault Tolerance Through Reconfiguration in VLSI and WSI Arrays*. The MIT Press, Cambridge, 1989.
8. P. Prusinkiewicz and A. Lindenmayer. *The Algorithmic Beauty of Plants*. Springer-Verlag, New York, 1990.
9. A. Shibayama, H. Igura, M. Mizuno, and M. Yamashina. An autonomous reconfigurable cell array for fault-tolerant LSIs. In *Proceedings of the IEEE International Solid-State Circuits Conference*, pages 230–231 and 462, February 1997.
10. A. Stauffer and M. Sipper. L-hardware: Modeling and implementing cellular development using L-systems. In D. Mange and M. Tomassini, editors, *Bio-Inspired Computing Machines*. Presses polytechniques et universitaires romandes, Lausanne, 1998.
11. A. Stauffer and M. Sipper. Modeling cellular development using L-systems. In M. Sipper, D. Mange, and A. Perez, editors, *Proceedings of The Second International Conference on Evolvable Systems: From Biology to Hardware (ICES98)*, Lecture Notes in Computer Science. Springer-Verlag, Heidelberg, 1998.
12. G. Tempesti, D. Mange, and A. Stauffer. A robust multiplexer-based FPGA inspired by biological systems. *Journal of Systems Architecture*, 43(10):719–733, 1997.

Modeling Cellular Development Using L-Systems

André Stauffer and Moshe Sipper

Logic Systems Laboratory, Swiss Federal Institute of Technology, CH-1015 Lausanne, Switzerland. E-mail: {name.surname}@di.epfl.ch, Web: http://lslwww.epfl.ch.

Abstract. A fundamental process in nature is that of ontogeny, whereby a single mother cell—the zygote—gives rise, through successive divisions, to a complete multicellular organism. Over the years such developmental processes have been studied using different models, two of which shall be considered in this paper: L-systems and cellular automata. Each of these presents distinct advantages: L-systems are naturally suited to model growth processes, whereas if one wishes to consider physical aspects of the system, e.g., as pertaining to actual implementation in hardware, then an inherently spatial model is required—hence the cellular automaton. Our goals herein are: (1) to show how L-systems can be used to specify growing structures, and (2) to explore the relationship between L-systems and cellular automata. Specifically, we shall consider the case of membrane formation, whereby a grid of artificial molecules is divided into cells.

1 Introduction

A fundamental process in nature is that of ontogeny, whereby a single mother cell—the zygote—gives rise, through successive divisions, to a complete multicellular organism, possibly containing trillions of cells (e.g., in humans). Studying this process of cellular development is interesting both from a biological standpoint, wherein we wish to enhance our understanding of ontogeny in nature, as well as from an engineering standpoint, wherein we wish to build better machines, inspired by such natural processes [6].

Over the years such developmental processes have been studied using different models, two of which shall be considered in this paper: L-systems and cellular automata. Introduced almost three decades ago as a mathematical theory of plant development, L-systems capture the essence of growth processes [2]. Basically, an L-system is a string-rewriting grammar that is coupled with a graphical interpretation—the system can be used to churn out a plethora of finite strings that give rise (through the graphical interpretation) to one-, two-, or three-dimensional images.

Cellular automata (CA) are dynamical systems in which space and time are discrete. A cellular automaton consists of an array of cells, each of which can be in one of a finite number of possible states, updated synchronously in discrete time steps, according to a local, identical interaction rule. The state of a cell at the next time step is determined by the current states of a surrounding neighborhood of cells. This transition is usually specified in the form of a rule table, delineating the cell's next state for each possible neighborhood configuration. The cellular array (grid) is n-dimensional, where $n = 1, 2, 3$ is used in practice [10, 13]. A one-dimensional CA is illustrated in Figure 1 (based on Mitchell [7]).

Rule table: **Grid:**

neighborhood: 111 110 101 100 011 010 001 000
output bit: 1   1   1   0   1   0   0   0 $t = 0$ |0|1|1|0|1|0|1|1|0|1|1|0|0|1|1|

  $t = 1$ |1|1|1|1|0|1|1|1|1|1|1|0|0|1|1|

Fig. 1. Illustration of a one-dimensional, 2-state CA. The connectivity radius is $r = 1$, meaning that each cell has two neighbors, one to its immediate left and one to its immediate right. Grid size is $N = 15$. The rule table for updating the grid is shown to the left. The grid configuration over one time step is shown to the right. Spatially periodic boundary conditions are applied, meaning that the grid is viewed as a circle, with the leftmost and rightmost cells each acting as the other's neighbor.

Each of the above models presents distinct advantages. L-systems are naturally suited to model growth processes such as cellular development. On the other hand, if one wishes to consider physical aspects of the system, e.g., as pertaining to actual implementation in hardware, then an inherently spatial model is required—hence the CA.

Our goals herein are: (1) to show how L-systems can be used to specify growing structures, and (2) to explore the relationship between L-systems and CAs. Specifically, we shall consider the case of membrane formation, whereby a grid of artificial molecules is structured into cells. We begin in Section 2 with an introduction to L-systems. Section 3 demonstrates how a number of elemental developmental mechanisms, whose physical embodiment is that of a CA, can be described by L-system rewriting rules. Section 4 delineates the modeling of membrane formation using an L-system, followed by its implementation as a two-dimensional CA. Finally, we end with concluding remarks in Section 5.

2 L-systems

Lindenmayer systems—or L-systems for short—were originally conceived as a mathematical theory of plant development [2, 9]. The central concept of L-systems is that of rewriting, which is essentially a technique for defining complex objects by successively replacing parts of a simple initial object using a set of *rewriting rules* or *productions*. The most ubiquitous rewriting systems operate on character strings. Though such systems first appeared at the beginning of this century [9], they have been attracting wide interest as of the 1950s with Chomsky's work on formal grammars, who applied the concept of rewriting to describe the syntactic features of natural languages [1]. L-systems, introduced by Lindenmayer [2], are string-rewriting systems, whose essential difference from Chomsky grammars lies in the method of applying productions. In Chomsky grammars productions are applied sequentially, whereas in L-systems they are applied in parallel and simultaneously replace all letters in a given word. This difference reflects the biological motivation of L-systems, with productions intended to capture cell divisions in multicellular organisms, where many divisions may occur at the same time.

As a simple example, consider strings (words) built of two letters, A and B. Each letter is associated with a rewriting rule. The rule $A \to AB$ means that the letter A is to be replaced by the string AB, and the rule $B \to A$ means that the letter B is to be replaced by A [9]. The rewriting process starts from a distinguished string called the axiom. For example, let the axiom be the single letter B. In the first derivation step (the first step of rewriting), axiom B is replaced by A using production $B \to A$. In the second step, production $A \to AB$ is applied to replace A with AB. In the next derivation step both letters of the word AB are replaced simultaneously: A is replaced by AB and B is replaced by A. This process is shown in Figure 2 for four derivation steps.

Fig. 2. Example of a derivation in a context-free L-system. The set of productions, or rewriting rules is: $\{A \to AB, B \to A\}$. The process is shown for four derivation steps.

In the above example the productions are context-free, i.e., applicable regardless of the context in which the predecessor appears. However, production application may also depend on the predecessor's context, in which case the system is referred to as context-sensitive. This allows for interactions between different parts of the growing string (modeling, e.g., interactions between plant parts). Several types of context-sensitive L-systems exist, one of which we shall concentrate on herein. In addition to context-free productions (e.g., $A \to AB$), context-sensitive ones of the form $U<A>X \to DA$ are introduced, where the letter A (called the strict predecessor) can produce word DA if and only if A is preceded by letter U and followed by X. Thus, letters U and X form the context of A in this production. When the strict predecessor has a one-sided context, to the left or to the right, then only the $<$ or $>$ symbol is used, respectively (e.g., $U<A \to DA$ is a left-context rule and $A>X \to DA$ is a right-context one). Figure 3 demonstrates a context-sensitive L-system. We note that, defining a growth function as one describing the number of symbols in a word in terms of its derivation length, then this L-system exhibits square-root growth: after n derivation steps the length of the string (X symbols excluded) is $\lfloor \sqrt{n} \rfloor + 2$. Other growth functions can also be attained, including polynomial, sigmoidal, and exponential [9].

```
p1: U<A>A -> U
p2: U<A>X -> DA        0: XUAX       5: XADAAX
p3: A<A>D -> D         1: XADAX      6: XUAAAX
p4: X<A>D -> U         2: XUAAX      7: XAUAAX
p5:  U    -> A         3: XAUAX      8: XAAUAX
p6:  D    -> A         4: XAADAX     9: XAAADAX
```

(a)   (b)

Fig. 3. A context-sensitive L-system. (a) The production set. (b) A sample derivation. Note that if no rule applies to a given letter then that letter remains unchanged.

As noted above, L-systems were originally designed to model plant development. Thus, in addition to a grammar that produces finite strings over a given alphabet (as defined above), such a system is usually coupled with a graphical interpretation. Several such interpretations exist, one example of which is the so-called turtle interpretation, based on a LOGO-style turtle [9]. Here, the string produced by the L-system is considered to be a sequence of commands to a cursor (or "turtle") moving within a two- or three-dimensional space. Each symbol represents a simple command (e.g., **move forward, turn left, turn right**) such that interpretation of the string gives rise to an image.

In summary, there are two important aspects concerning L-systems, which shall serve us herein: (1) such a system gives rise to a growing, one-dimensional string of characters, (2) which can then be interpreted as a one-, two-, or three-dimensional image.

3   Using L-systems to describe cellular development

In this section we demonstrate how a number of basic components—or operations—related to cellular development can be modeled by L-systems. This developmental model involves four main processes or mechanisms: (1) simple growth, (2) branching growth, (3) signal propagation, and (4) signal divergence.

Simple growth arises from the application of productions p1 to p3 of Figure 4. In these productions the symbol a represents the apex and i the internode. The terminology introduced here is borrowed from the description of tree-like shapes [9]. A tree has edges that are labeled and directed. In the biological context, these edges are referred to as branch segments. A segment followed by at least one more segment is called an *internode*. A terminal segment (with no succeeding edges) is called an *apex*. In the productions, the ( ) and [ ] symbol pairs represent a left and right branch, respectively. These are used in so-called bracketed L-systems with the parentheses being a form of recursive application [9]: a string is interpreted from left to right to form the corresponding image. When a left bracket is encountered then the current position within the image is pushed onto a pushdown stack, with a right bracket signifying that a position is to be popped from the stack. Thus, one can model plants with branches, sub-branches, etc., or, in our case, create such constructs as multi-dimensional growing structures. The corresponding CA interpretation of a single derivation step of the simple growth productions is also given in Figure 4.

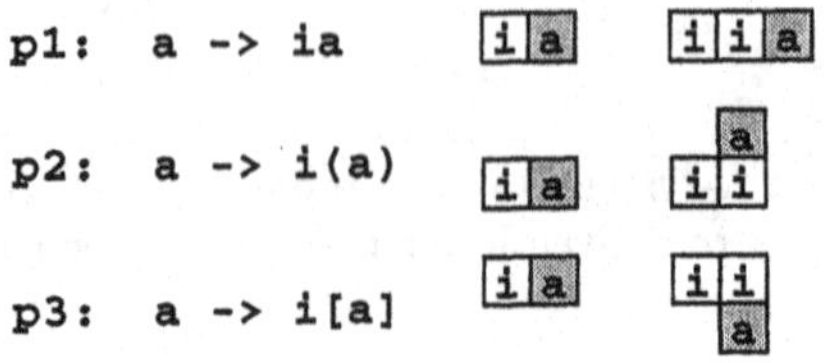

p1: a -> ia

p2: a -> i(a)

p3: a -> i[a]

Fig. 4. Some simple growing structures along with their CA interpretation. The ( ) and [ ] symbol pairs represent a left and right branch, respectively. As the cellular space considered is a two-dimensional grid, the branching angle is 90°.

Productions p1 to p4 of Figure 5 give rise to *branching growth*. The figure also depicts the CA interpretation of a single derivation step. Note that in both Figures 4 and 5, all productions are context-free.

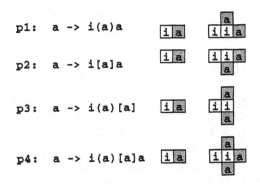

p1: a -> i(a)a

p2: a -> i[a]a

p3: a -> i(a)[a]

p4: a -> i(a)[a]a

Fig. 5. Some branching structures along with their CA interpretation.

Signal propagation of a given signal s is modeled by productions p1 and p2 of Figure 6. The context-sensitive production p1 means that internode i with a signal s to its left becomes an s, while application of the context-free production p2 transforms signal s to an internode i. Consequently, the CA interpretation of a single derivation step of these productions causes signal s to move one cell to the right (Figure 6).

Productions p1 to p10 of Figure 7 model *signal divergence*, whereby a given signal s divides into three signals t, u, and v. Some of these productions are

```
p1:   s<i -> s
p2:    s -> i
```

Fig. 6. Productions used to obtain signal propagation, along with their CA interpretation. The CA state s is propagated to the right.

```
p1:   s(<i  -> t
p2:   s<i   -> v
p3:    s    -> i

p4:   s[<i  -> u
p5:   s<i   -> v
p6:    s    -> i

p7:   s(<i  -> t
p8:   s[<i  -> u
p9:   s<i   -> v
p10:   s    -> i
```

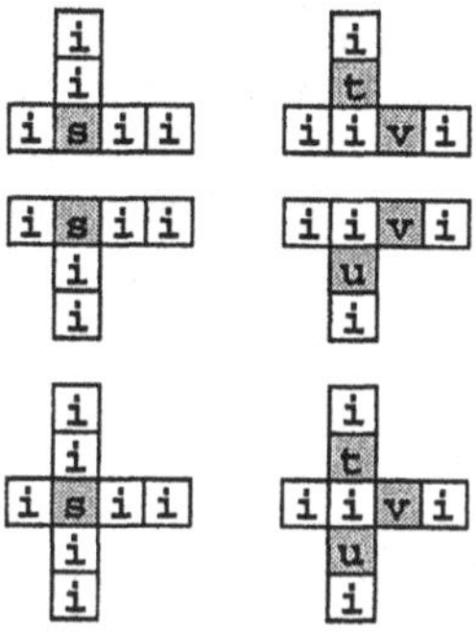

Fig. 7. Productions used to obtain signal divergence, along with their CA interpretation. These implement the cases of a given signal, i.e., CA state (denoted s) that breaks up to yield two or three new states (denoted t, u, and v).

context-sensitive. The corresponding CA interpretation of a single derivation step is also shown.

Using the components described above, as well as a number of others, we have previously shown how L-systems can be used to specify self-replicating structures, thereafter to be implemented as cellular systems [12]. The fabrication of artificial self-replicating machines has diverse applications, ranging from nanotechnology to space exploration. It is also an important aspect of the Embryonics project, described ahead. (A short survey of self-replication is provided in [8]; for detailed information see the online self-replication page at http://lslwww.epfl.ch/~moshes/selfrep/.) Below, we shall focus on another application of our approach, namely, membrane formation.

4 Membrane formation: The space divider

The Embryonics (Embryonic Electronics) project, under development at the Logic Systems Laboratory for the past four years, has as its ultimate objective the construction of large-scale integrated circuits, exhibiting properties such as self-repair (healing) and self-replication. It is based on three features usually associated with the ontogenetic process in living organisms, namely, multicellular organization, cellular differentiation, and cellular division [3–6]. An Embryonics "organism" consists of a multitude of artificial cells, which are in turn composed of yet finer elements, referred to as molecules. The first operational phase of such a system consists of structuring a *tabula rasa* "sea" of identical molecules into cells, which are defined by their borders, or membranes. The space divider, described in this section, is a two-dimensional CA whose purpose is to structure such a sea of molecules. The structure of this molecular sea ultimately consists of squares of functional molecules—the cells. The structuring process is essentially a growth process, starting from the bottom-left CA cell, where the programming data is continuously fed. (Note: the term cell is used here in two different contexts, as it originates from two different sources. In a CA, the cell is the elemental unit, equivalent to an Embryonics *molecule*, whose cell is one level up, i.e., composed of an ensemble of molecules. Thus, one must take care to distinguish between a CA cell and an Embryonics cell—this can be inferred from the text. Within the framework of the Embryonics project it would probably be better to refer to cellular automata as molecular automata, however, in order to conform to existing literature we shall retain the extant terminology.)

As noted, L-systems are naturally suited for modeling growth processes. Therefore, we first describe the space divider as an L-system developmental model. The graphical interpretation will be that of a CA. The representation of the symbols (letters) used in the developmental model of the space divider is given in Figure 8.

```
h: horizontal apex
v: vertical apex
e: east growing apex
n: north growing apex
l: left branching apex
r: right branching apex
i: internode
b: branching signal
```

Fig. 8. Symbol representation of the space divider L-system model.

The space divider L-system starts out with a single-letter axiom *l* which corresponds to a left-branching apex. The productions applied to the axiom in order to obtain the square structure are listed in Figure 9a. They fall into three categories: (1) the branching signal propagation productions p1 to p4, (2) the simple growth productions p5 to p14, and (3) the branching growth productions p15 to p18.

The first character of the string—to the left of the vertical separator—is part of the program that is fed to the space divider. This comprises the input to the system—an artificial "genome" that determines the structure that the space will take on (this genome also contains additional information related to the ultimate task that the "organism" will carry out—see references [3, 6] for further details). This input mechanism is a novel element that we have added

```
p1:   b<i    -> b        p10:   i<v    -> n
p2:   b(<i   -> b        p11:   i(<v   -> n
p3:   b[<i   -> b        p12:   i<n    -> iv
p4:     b    -> i        p13:   i(<n   -> iv
p5:   i<h    -> e        p14:   b<n    -> bv
p6:   i[<h   -> e        p15:   b<h    -> 1
p7:   i<e    -> ih       p16:     1    -> i(v)h
p8:   i[<e   -> ih       p17:   b<v    -> r
p9:   b<e    -> bh       p18:     r    -> i[h]v
```

(a)

```
 0:  i|1
 1:  i|i(v)h
 2:  b|i(n)e
 3:  i|b(iv)ih
 4:  i|i(bn)be
 5:  b|i(ibv)ibh
 6:  i|b(iir)iil
 7:  i|i(bii[h]v)bii(v)h
 8:  b|i(ibi[e]n)ibi(n)e
 9:  i|b(iib[ih]iv)iib(iv)ih
10:  i|i(bii[be]bn)bii(bn)be
11:  b|i(ibi[ibh]ibv)ibi(ibv)ibh
12:  i|b(iib[iil]iir)iib(iir)iil
13:  i|i(bii[bii(v)h]bii[h]v)bii(bii[h]v)bii(v)h
```

(b)

Fig. 9. Space divider L-system model. (a) The production set. (b) A sample derivation.
Note: when applying the above productions to derive a string like $x_1|x_2(x_3x_4[x_5]x_6)x_7$
then the context of a letter x_i depends upon its position. Thus, the context letters x_{i_l}
and x_{i_r} of $x_{i_l}<x_i>x_{i_r}$ are defined as follows: $x_1<x_2$, $x_2(<x_3$, $x_3<x_4$, $x_4[<x_5$, $x_4<x_6$,
and $x_2<x_7$. For example, when deriving the string $a|b(cd[e]f)g$, the context of f is not
the e to its immediate left but rather d. (Note that so as to simplify notation vertical
separators were omitted from the production set of (a).) Formally, L-systems where
the context of a letter can be further down the string are known as IL-systems or
(k,l)-systems, meaning that the left context is a word of length k and the right context
is a word of length l [9].

to the classical L-systems model, in which development from a given "seed"
(axiom) takes place with no external intervention. This feature gives rise to
a programmable cellular space, which can be programmed and structured via
a signal from an external source. The genome character (leftmost letter of the
string), input at each derivation step from the external source, comprises the left
context of the letter to the immediate right of the vertical separator. Thirteen
derivation steps of the developmental process are shown in Figure 9b. The CA
interpretation of this process defines a division of the cellular space into squares
of size 2 x 2 cells (Figure 10).

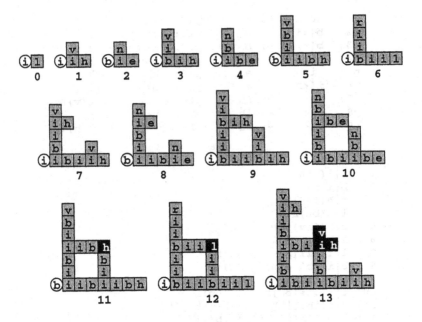

Fig. 10. CA interpretation of the space divider L-system model.

In Figure 10 the successive characters of the programming data (the genome) appear in the bottom-left circle. The membrane size, i.e., the number of cells per square side, is programmable and equals the number of internodes i between two successive branching signals b. The black squares of Figure 10 (at time steps 11, 12, and 13) correspond to the underlined string segments of Figure 9b (at derivation steps 11, 12, and 13, respectively). These represent two confluent branches, resulting from the closing of the square, which thus amount to the same CA state (or states).

The complete specification of the CA derived from the above L-system, as well as a hardware implementation using Field-Programmable Gate Arrays (FPGAs) is described by Stauffer and Sipper [11]. As noted, this comprises part of the Embryonics project, which is ultimately realized in hardware.

5   Concluding remarks

We have shown how L-systems can be used to specify growing structures, specifically concentrating on the case of membrane formation. The L-system is then transformed into a cellular automaton, an inherently spatial model which leads directly to a hardware implementation. The study of systems that exhibit growth is interesting both from a theoretical standpoint as well as from a practical one. This paper has shed light on the possible use of L-systems as an exploratory, and perhaps design tool within the realm of growth and development.

References

1. N. Chomsky. Three models for the description of language. *IRE Transactions on Information Theory*, 2(3):113–124, 1956.
2. A. Lindenmayer. Mathematical models for cellular interaction in development, Parts I and II. *Journal of Theoretical Biology*, 18:280–315, 1968.
3. D. Mange, D. Madon, A. Stauffer, and G. Tempesti. Von Neumann revisited: A Turing machine with self-repair and self-reproduction properties. *Robotics and Autonomous Systems*, 22(1):35–58, 1997.
4. D. Mange, A. Stauffer, and G. Tempesti. Embryonics: A macroscopic view of the cellular architecture. In M. Sipper, D. Mange, and A. Pérez-Uribe, editors, *Proceedings of The Second International Conference on Evolvable Systems: From Biology to Hardware (ICES98)*, Lecture Notes in Computer Science. Springer-Verlag, Heidelberg, 1998.
5. D. Mange, A. Stauffer, and G. Tempesti. Embryonics: A microscopic view of the molecular architecture. In M. Sipper, D. Mange, and A. Pérez-Uribe, editors, *Proceedings of The Second International Conference on Evolvable Systems: From Biology to Hardware (ICES98)*, Lecture Notes in Computer Science. Springer-Verlag, Heidelberg, 1998.
6. D. Mange and M. Tomassini, editors. *Bio-Inspired Computing Machines: Toward Novel Computational Architectures*. Presses Polytechniques et Universitaires Romandes, Lausanne, Switzerland, 1998.
7. M. Mitchell. *An Introduction to Genetic Algorithms*. MIT Press, Cambridge, MA, 1996.
8. J.-Y. Perrier, M. Sipper, and J. Zahnd. Toward a viable, self-reproducing universal computer. *Physica D*, 97:335–352, 1996.
9. P. Prusinkiewicz and A. Lindenmayer. *The Algorithmic Beauty of Plants*. Springer-Verlag, New York, 1990.
10. M. Sipper. *Evolution of Parallel Cellular Machines: The Cellular Programming Approach*. Springer-Verlag, Heidelberg, 1997.
11. A. Stauffer and M. Sipper. L-hardware: Modeling and implementing cellular development using L-systems. In D. Mange and M. Tomassini, editors, *Bio-Inspired Computing Machines: Toward Novel Computational Architectures*, pages 269–287. Presses Polytechniques et Universitaires Romandes, Lausanne, Switzerland, 1998.
12. A. Stauffer and M. Sipper. On the relationship between cellular automata and L-systems: The self-replication case. *Physica D*, 116(1-2):71–80, 1998.
13. T. Toffoli and N. Margolus. *Cellular Automata Machines*. The MIT Press, Cambridge, Massachusetts, 1987.

MUXTREE Revisited: Embryonics as a Reconfiguration Strategy in Fault-Tolerant Processor Arrays

César Ortega-Sánchez[1] and Andrew Tyrrell

Department of Electronics
University of York
York, YO10 5DD, UK
{cesar, amt}@ohm.york.ac.uk

Abstract: Embryonics' proposal is to construct arrays of processing elements with self-diagnosis and self-reconfiguration abilities able to tolerate the presence of failing cells in the same fashion as natural cellular systems do. Self-healing mechanisms found in nature and the implicit redundancy of cellular architectures constitute the foundations of embryonic systems' fault tolerance properties. It will be shown in this paper how by incorporating the biological concepts of chromosome and gene, the complexity of the MUXTREE embryonic architecture can be simplified, in comparison with the previous version. It is argued that by assuming a broader meaning for the concept of evolution it possible to classify embryonic arrays and other adaptable systems as evolvable.

1 Introduction

When a system (natural or man-made) reaches a certain level of complexity, it becomes very difficult to grasp all of its underlying dynamics, and therefore, it becomes less controllable and less reliable [1]. However, the needs of the modern individual are fulfilled using extremely complex systems. What would our society be without computers, satellites, medicines, mega-software and free market? Complex systems are the foundation of our life-style but they have become very difficult to design. Therefore, it is necessary to look for new methodologies and strategies to deal with complex systems. One approach is the refinement of traditional design techniques, but the techniques themselves are becoming too complex to be considered error-free. Evidently, we have to look somewhere else for the answers [2].

Nature offers to us some remarkable examples of how to deal with complexity and its associated unreliability. For example, the human body is one of the most complex systems ever known. Local failures are common, but the overall function of our organism is highly reliable because of the self-diagnosis and self-healing mechanisms that work ceaselessly throughout our bodies. These mechanisms are the result of

---

[1] This work has been partially supported by the Mexican Government under grants CONACYT-111183 and IIE-9611310226

millions of years of our gene's evolution. Evolving instead of designing seems to be an attractive alternative when dealing with complexity [3].

During the past few years the work done on evolvable systems has generated some remarkable results [4,5]. Genetic algorithms, neural networks, artificial brains, genetic engineering and evolvable hardware are just a few examples of this novel approach. What is in evolution that is so attractive for hundreds of engineers and scientists? The answer can be found in the characteristics that evolved systems possess, for example, adaptation, auto-regulation and learning.

Embryonics is the design of novel reconfigurable hardware inspired by mechanisms found in nature. We show in this paper how Embryonics can be used to incorporate fault tolerance characteristics, e.g. automatic diagnosis and reconfiguration, into processor arrays by using self-healing like mechanisms.

The paper is organised as follows. Section 2 offers a brief survey on the subjects of evolution and evolvable systems. The Phylogeny-Ontogeny-Epigenesis (POE) model is introduced as a way of classifying evolvable systems. In section 3 a short introduction to processor arrays and their associated fault tolerance techniques is given. The main characteristics of the Embryonics project are explained in section 4. In section 5 a particular embryonic architecture called MUXTREE is analysed and possible improvements are proposed. An example of how to implement a circuit using embryonic arrays is given in section 6. Conclusion are given in section 7.

2 Evolution and Evolvable Systems

To try to find a universal definition for evolution seems to be a fruitless search. The Oxford Dictionary of Current English defines evolution as "origination of species by development from earlier forms; gradual development of phenomenon, organism, etc." [6]. The first definition restricts evolution to natural (or artificial) selection; the second, uses the term as a synonym for development. Evolution, in a broad sense, is much more than that.

Anyone working on the field could give a personal definition for evolution. The term has different meanings for biologists, engineers, geneticists, software developers and priests. When a term is so difficult (or easy) to define, one way of explaining it is describing the properties of the objects possessing such attribute. Although it is not possible to achieve a general agreement on the definition of evolution, it is clear that evolvable systems are: adaptive, fault-tolerant, dynamic, complex in structure and behaviour, dependant on initial conditions (chaotic), capable of learning and made out of interacting simpler units. The list is not exhaustive but, incomplete as it is, it manifests a feature characteristic of evolution and evolvable systems: Change.

We perceive evolution as the continuous change of state in adaptive complex systems. Planetary systems, species and Economics although different in essence, all share the common attribute of perpetual change. Evolution is a sensory experience. It is us who, in an attempt to understand and predict the behaviour of such phenomena, try to find some coherent transition from one state to the next one. Change is the only objective attribute of dynamic complex systems, the rest is subjective interpretation.

It is possible now to state that evolution, in a broader sense, is the spontaneous and purposeful change of state in a dynamic complex system as a response to changes in the environment. Purpose is the subjective quality that we, as observers of the system, attribute to evolvable systems.

2.1 The POE Model

Sánchez et al. [7], proposed the Phylogeny-Ontogeny-Epigenesis model (POE model) as a framework to represent the three levels of organisation that can be distinguished in living beings.

Phylogeny embraces the evolution of species when genes are passed from one generation to the next. This is the kind of evolution usually associated with natural selection, where infrequent errors occurring during the copy of genes (mutations) originate new traits on the species. This traits, in some cases, allow the species to better adapt themselves to changes in the environment. Phylogeny presents evolution as species adaptation.

Ontogeny refers to the evolution (development, change of state) of multicellular organisms during their embryonic phase. When reproduction takes place, the new individuals are formed out of a single cell (the fertilised egg). During the following weeks after conception, the mother cell and all its offsprings copy themselves continuously following the "instructions" stored in their DNA (the genome). During this reproductive process cells differentiate to shape the different tissues, organs and limbs that characterise a complete healthy individual of a particular species. Ontogeny is evolution as embryonic development.

Epigenesis is the evolution of individuals' ability to respond to changes in the environment through learning. After birth every individual must adapt to the environment (society, geography, historical context) on which he/she has to live. To achieve adaptation every individual is born with a set of systems defined by the genome (nervous system, immune system and endocrine system), which suffer modifications when interacting with the outside world. Hence, epigenesis is evolution as learning. The POE model can be represented as three orthogonal axis, as shown in figure 1.

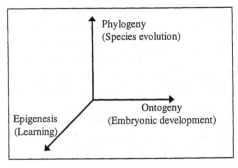

Fig. 1 The POE model for evolvable systems

Human-made evolvable systems can be classified using the POE model. The work on genetic algorithms and evolvable hardware can be placed on the phylogeny axis. Research on artificial intelligence and learning systems can be situated in the epigenesis axis. Works on self-reproducing cellular automata such as the Embryonics project would find a place in the ontogeny axis.

3 Processor Arrays

Processor arrays are a paradigm of computer architecture where multiple identical processing elements (cells) are interconnected in a regular pattern so that the processing power of the whole system (ideally) grows linearly with the number of processors. Communications are strictly local, i.e. each processor only communicates with its nearest neighbours. Examples of this kind of systems are:

Cellular Automata (CA).- In CAs the state of a cell is defined by the state of its nearest neighbours. There is a finite (usually small) number of possible states for a cell. The transition from one state to the next at discrete time steps in all the cells is what determines the array's behaviour and applications. CAs have been used for studying self-reproduction [8], and modelling of dynamic complex systems [9,10].

Systolic Arrays.- In systolic arrays the processing elements are designed to match a particular algorithm. Systolic means that synchronous pipelined computations take place along all dimensions of the array and result in very high computational throughput. The main applications for such computing machines are found in the fields of signal and image processing, pattern recognition, matrix arithmetic and graph algorithms [11].

Wavefront Arrays.- Wavefront arrays combine the systolic pipelining principle with the dataflow computing concept. They are data-driven and do not use a global clock, therefore, the requirement for correct timing in the systolic array is replaced by a requirement for correct sequencing in the wavefront array [12]. The applications of this architecture are practically the same as for systolic arrays.

Parallel structures, like those mentioned above, are good candidates for being implemented in silicon because of their regularity and relative simplicity in the processing unit. In recent years the idea of parallel computers on a chip has become feasible thanks to the advances in VLSI and WSI technologies. Nevertheless, production yields of VLSI circuits is far from being optimum; therefore, reconfiguration techniques have been explored for the past few years in order to provide VLSI processor arrays with fault tolerance.

3.1 Fault Tolerance Techniques in Processor Arrays

All fault tolerance techniques for hardware systems rely on the use of spare components to substitute failing elements. In the past, the cost associated with this redundancy has prevented the widespread use of fault-tolerant hardware. However, in the case of VLSI processor arrays, redundancy comes for free because not all available cells in the array are used on every application.

Fault tolerance in processor arrays implies the mapping of a logical array into a physical non-faulty array, i.e. every logical cell must have a correspondent physical cell [13]. When faults arise, a mechanism must be provided for reconfiguring the physical array such that the logical array can still be represented by the remaining non-faulty cells. All reconfiguring mechanisms are based on one of two types of redundancy: Time redundancy or hardware redundancy [14].

In time redundancy the tasks performed by faulty cells are distributed among its neighbours. In this scheme the application must allow a degradation in performance. When reconfiguration occurs, processors dedicate some time performing its own tasks and some performing faulty cells functions. Interconnection is a key issue. Besides, the algorithm being executed must be flexible enough so as to allow a simple and flexible division of tasks.

In hardware redundancy physical spare cells and links are used to replace the faulty ones. Therefore, reconfiguring algorithms must optimise the use of spares. In the ideal case a processor array with N spares must be able to tolerate N faulty cells; but, in practice, limitations on the interconnection capabilities of each cell prevents this goal from being achieved.

Most of hardware redundancy reconfiguration techniques rely on complex algorithms to re-assign physical resources to the elements of the logical array. In most cases these algorithms are executed by a central processor which also performs diagnosis functions and co-ordinates the reconfiguration of the physical array [15]. This approach has demonstrated to be effective, but its centralised nature makes it prone to collapse if the processor in charge of the fault tolerance functions fails.

An alternative approach is to distribute the diagnosis and reconfiguration mechanisms among all the cells in the array. In this way no central agent is necessary and the time response of the system improves. This mechanism resembles that found in natural cellular systems.

4   Embryology + Electronics = Embryonics

Embryonics was firstly proposed by Mange and Marchal [16,17] as a new family of field programmable gate arrays (FPGAs) inspired by the mechanism sustaining the development of multicellular organisms in nature.

4.1 The Central Dogma

The DNA is a ribbon of 2 billion characters which encode the ensemble of the genetic inheritance of the individual and, at the same time, the instructions for the construction and the operation of the complete organism. In this sense DNA can be both information and physical medium.

In any living being, every one of its constituent cells performs the same basic operation regardless of the particular function it is involved with, i.e. each cell interprets the DNA strand allocated in its nucleus to produce the proteins needed for the survival of the organism. Proteins are particular sequences of amino acids; such sequences are stored in the DNA as successions of nucleotide triplets (codons).

Protein synthesis imply two mechanisms: transcription and translation of the DNA. During transcription, the sequence stored in the DNA is copied by the enzyme RNA polymerase into messenger RNA (mRNA). During translation, mRNA is bond to ribosomes inside the cell where transfer RNA (tRNA) carrying amino acids are attached to the mRNA. The ribosome catalyses the bond between amino acids to build a molecule of the corresponding protein. When a cell reproduces, the offspring get a copy of its mother's DNA so that the complete process can be ceaselessly repeated. The flow of information from DNA to protein and from DNA of the parent to DNA of the offspring is known as the central dogma [18].

Although the DNA is identical in all the cells, only part of the strand is interpreted. Which part or parts of the DNA are interpreted will depend on the physical location of the cell with respect to its neighbours [19,20]. Figure 2 represents the way DNA's information is organised. Arrows indicate the direction in which complexity increases, e.g. a set of nucleotides forms a codon, etc.

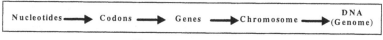

Fig. 2 Structure of DNA

The aim of Embryonics is to transport these basic structure to the 2-dimensional world of cellular arrays using specifically designed FPGAs as building blocks. An equivalent representation of the architecture of field-programmable processor arrays is shown in figure 3.

Fig. 3 Structure of a field-programmable processor array

4.2 Embryonics Architecture

Similarly to natural cellular systems, every one of the embryonic array's cells performs the same basic operation independently of the particular logic function it is involved with, i.e. each cell interprets one of the configuration registers allocated in its memory to perform the logic operations needed for the correct implementation of the system's specification. Which configuration register is selected will depend on the co-ordinates of the cell determined by those of its neighbours. Embryonic cellular arrays share the following properties with their biological counterparts [21]: Multicellular organisation (the logic blocks of an FPGA), cellular differentiation (every cell has a unique set of co-ordinates) and cellular division (every cell is configured by only one configuration register). Figure 4 shows the architecture of a generic embryonic system.

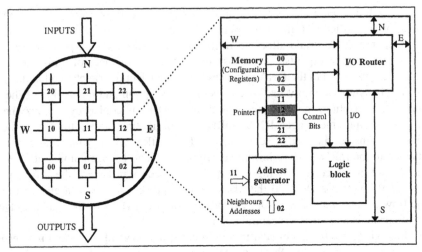

Fig. 4 Basic Components of an Embryonic System

The address generator assigns to each cell an individual set of co-ordinates, which depend exclusively on the co-ordinates of the nearest south and west neighbours. The logic block is controlled by a configuration register which is selected by the corresponding co-ordinates. When a fault is self-detected by a cell, it becomes transparent for the calculation of co-ordinates allowing another cell to take its co-ordinates and therefore its function. Digital data are transmitted from one cell to its neighbours through a North-East-West-South (NEWS) connection. The I/O router block allows the spread of information over all the array. This block is controlled by one section of the corresponding configuration register.

The architecture shown in figure 4 presents the following advantages:
- It is highly regular, which simplifies its implementation on silicon.
- The actual function of the logic block is independent from the function of the remaining blocks. This modularity can be exploited to produce a family of embryonic FPGAs, each member offering either a particular logic function, e.g. a binary selection function [21] or a microprogrammed architecture [22].
- Provided the architecture of the logic block is kept simple, it would be possible to implement built-in self test (BIST) logic to provide self-diagnosis without excessively incrementing the silicon area [21,23].

5 The MUXTREE Revisited

5.1 General Architecture

MUXTREE designates a particular implementation of embryonic FPGA where the processing element of the basic cell (logic block in figure 4), is a selector or multiplexer [21,24]. Multiplexers have the particular characteristic of being able to implement any node from an ordered binary decision diagram (OBDD), which in turn can represent any combinatorial or sequential logic function [25,26]. Therefore, the resulting architecture is ideal to be implemented as an FPGA. Nevertheless, the MUXTREE architecture, as shown in figure 4, has the following drawbacks:

- Each cell must be able to store the configuration registers of all the cells in the array. For example, in a 32x32 array, each cell must have 1,024 configuration registers, from which only one will be used. This level of redundancy prevents a practical and reliable implementation of the system.
- Depending on the reconfiguration technique chosen (row elimination, column elimination, neighbour substitution, etc.), only one of the co-ordinates is re-calculated when a fail is detected, therefore, it is possible to simplify the address generator block (see Fig. 6).

To overcome the first problem we propose a new architecture for the memory subsystem inspired by the concept of chromosome. The DNA strand is not a single double helix as the general concept might suggest, but it is divided into a number of sub-units called chromosomes.

In the same way, the genome or configuration program for an embryonic FPGA can be divided into smaller parts in order to concentrate the information closer to where it will be needed. For example, if the reconfiguration strategy followed is row replacement, then it would be convenient to divide the genome by columns because a failing cell can only be replaced by another on the same column. Figure 5 shows an example of the memory subsystem simplified by the chromosome analogy. Numbers in bold are the co-ordinates selecting the configuration register on each cell.

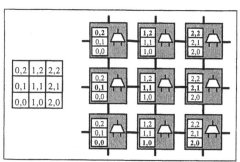

Fig. 5 Genome memory and its distribution in an embryonic array

This chromosome approach reduces the number of configuration registers per cell to the number needed in the original design divided by the number of columns. In the case of square arrays every cell contains only the square root of the number of registers needed in the original design, which represents substantial savings on silicon area.

Another characteristic of the previous version of the MUXTREE is that when a cell is self-detected faulty, it needed to broadcast a non-OK signal to all the cells in the same row and column so that they can recalculate their new co-ordinates and consequently, select a new configuration register. But close inspection of figure 5 reveals that when reconfiguration takes place the column co-ordinate remains constant.

Therefore, it is possible to simplify the reconfiguration mechanism by calculating row co-ordinates only and propagating non-OK signals exclusively to neighbours in the same row. Figure 6 illustrates these mechanisms.

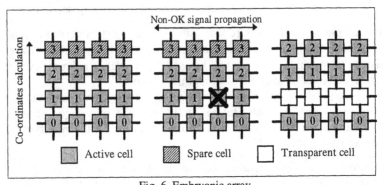

Fig. 6 Embryonic array
a) Co-ordinates calculation   b) Fault detection   c) Reconfiguration mechanism

One advantage of the proposed scheme is that it becomes possible to re-load independent chromosomes (configuration registers of a particular column), without affecting the others. This characteristic could be used to evolve embryonic arrays using genetic algorithms. The chromosomes in a genome would be the population exposed to crossover and mutation. Routing of circuits could be achieved following this approach.

5.2 Synthesis Method

The design cycle for implementing a system based on embryonic arrays is as follows:

1. Express the function to be implemented as a set of Boolean equations.
2. Obtain the OBDD for each one of the equations.
3. Replace each node of the OBDDs with a 2-1 multiplexer, using the node's variable as selection input and the outputs of the node as inputs to the multiplexer.
4. Map the network of multiplexers into an array of embryonic cells, assigning one multiplexer to each cell. This has been experienced to be the most difficult step due to the finite connectivity of the cells.
5. Obtain a configuration register for each cell in the array. The set of all configuration registers conforms the genome of the application.

After power-up all the cells in the array calculate their co-ordinates. At the same time external circuitry downloads the genome into the array column by column. The memory block was designed in such a way that chromosomes are downloaded to all the cells in the corresponding column at the same time, making this process faster.

6 Example

To illustrate embryonic arrays' properties the design of a programmable frequency divider is presented next. Figure 7 shows the circuit's block diagram. It is composed by a 3-bit selector which latches either the division factor **n**, or the next state of a 3-bit down-counter, according to the output of a zero detector. In this way, a 1-cycle wide pulse will be generated every **n** cycles of **F**. The output of the circuit is taken from the output of the zero detector. It will be high during one cycle of **F** when the down-counter reaches the 000 state.

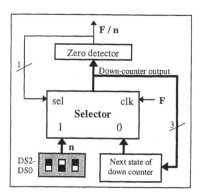

Fig. 7 Programmable frequency divider

Figure 8 shows the hardware implementation of the circuit shown in figure 7. Every multiplexer corresponds to a node in the corresponding OBDD. **A,B,C** are the outputs of the 3-bit down counter, **C** being the most significant bit. Multiplexers 1, 2 and 3 implement the selector block. These multiplexers update their outputs on the rising edge of **F**. **DS2**, **DS1** and **DS0** are used to set the value of **n**.

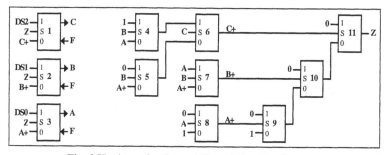

Fig. 8 Hardware implementation of circuit in figure 7.

Figure 9 shows how the circuits in figure 8 were mapped into an embryonic array. The numbers on each cell correspond with the numbers assigned to the multiplexers. Cells labelled **S** are spare cells, two rows for this example. Cells labelled **R** are routing cells. Routing cells are needed because every cell has direct connections only with its cardinal neighbours.

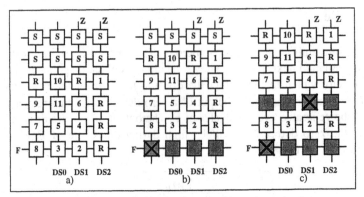

Fig. 9 Frequency divider implemented in embryonic array.
a) Without fails,   b) with one faulty cell,   c) with two faulty cells

Figure 10 shows the simulation results obtained for the frequency divider. Labels correspond with those of figure 8. **OK4** and **OK8** simulate fails in cells 4 and 8 respectively. Notice that when **OK** signals go to logic 0, there is a small time interval on which the output of the circuit remains constant, just before returning to normal behaviour. This is due to the reconfiguration process being carried out.

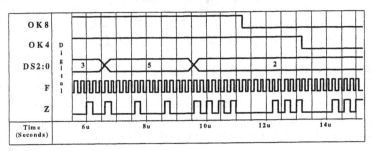

Fig. 10 Functional simulation of frequency divider

7 Conclusions

Evolution is perceived by the change of state in a dynamic complex system. In this broader sense, we can apply the attribute "evolvable" to numerous man-made systems whose configuration or response is influenced by the environment. In this way genetic algorithms, neural networks, artificial intelligence projects and Embryonics can all be classified as evolvable systems.

Embryonic arrays have demonstrated to be an attractive alternative for the design of biologically-inspired field programmable processor arrays with fault tolerance properties. We still are in the early stages of the Embryonics project, but the results obtained so far are encouraging to keep investigating on the application of biological concepts to the design of fault-tolerant engineering systems.

References

1. Paul G.: Beyond Humanity: CyberEvolution and future minds, Charles River Media, 1996
2. Avizienis A.: "Toward Systematic Design of Fault-Tolerant Systems", IEEE Computer, April, 1997, Computer Society Press, pp. 51-58
3. Kelly K.: Out of Control: The new Biology of machines, Fourth State-London, 1994
4. Sánchez E. et al.: (Eds.), Towards Evolvable Hardware: The evolutionary engineering approach, LNCS 1062, Springer-Verlag, 1996
5. Higuchi T. et al. (eds.): Evolvable Systems: From Biology to Hardware, LNCS, Springer-Verlag, 1997
6. The Oxford Dictionary of Current English, Oxford University Press, 1990
7. Sánchez E. et al.: "Phylogeny, Ontogeny and Epigenesis: three sources of biological inspiration for softening hardware", in Higuchi T. et al. (eds.), Evolvable Systems: From Biology to Hardware, Springer-Verlag, 1997
8. Langton C.: "Self-reproduction in Cellular Automata", Physica 10D, 1984, pp.135-144
9. Burks C.: "Towards Modelling DNA as Automata", Physica 10D, 1984, pp.157-167
10. Grassberger P.: "Chaos and Diffusion in Deterministic Cellular Automata", Physica 10D, 1984, pp. 145-156
11. Fortes J. et al.: "Systolic Arrays- From Concept to Implementation", IEEE Computer, July, 1987, pp.12-17
12. Kung S. et al.: "Wavefront Array Processors- Concept to Implementation", IEEE Computer, July, 1987, pp. 18-33
13. Grosspietsch K.: "Fault Tolerance in Highly Parallel Hardware Systems", IEEE Micro, Feb. 1994, pp.60-68
14. Chean M. et al.: "A Taxonomy of Reconfiguration Techniques for Fault-Tolerant Processor Arrays", Computer, January, 1990, pp. 55-69
15. Fortes J. et al.: "Gracefully Degradable Processor Arrays", Trans. on Computers, Vol.34-11, November, 1985, pp.1033-1043
16. Mange D. et al.: "Embryonics: A new family of coarse-grained FPGA with self-repair and self-reproduction properties", in Sanchez E. (Ed.), Towards Evolvable Hardware, LNCS 1062, Springer-Verlag, 1996, pp.197-220
17. Marchal P.: "Embryonics: The birth of synthetic life", in Sanchez E. (Ed.), Towards Evolvable Hardware, LNCS 1062, Springer-Verlag, 1996, pp.166-196
18. Murrell J.C. and Roberts L.M. (Eds.): Understanding Genetic Engineering, Ellis Horwood, Great Britain, 1989
19. Nüsslein-Volhard C.: "Gradients that Organize Embryo Development", Scientific American, August, 1996, pp.38-43
20. Gerhart J. and Kirschner M., Cells, Embryos and Evolution, Blackwell Science, 1997
21. Mange D. and Tomassini M. (Eds.), Bio-Inspired Computing Machines, Presses Polytechniques et Universitaires Romandes, Switzerland, 1998
22. Mange D., Madon D., Stauffer A. and Tempesti G.: "Von Neumann Revisited: A Turing Machine with Self-Repair and Self-Reproduction Properties", Robotics and Autonomous Systems, Vol.22-1, 1997, pp.35-38
23. Lala P.: Fault Tolerance and Fault Testable Hardware Design, Prentice-Hall, 1985
24. Ortega C. and Tyrrell A.: "Design of a Basic Cell to Construct Embryonic Arrays", IEE Procs. on Computers and Digital Techniques, May, 1998
25. Akers S.: "Binary Decision Diagrams", IEEE Trans. on Computers, Vol.27-6, June 1978
26. Liaw H. et al.: "On the OBDD-Representation of General Boolean Functions", IEEE Trans. on Computers, Vol.41-6, June, 1992, pp.661-664

Building Complex Systems
Using Developmental Process:
An Engineering Approach

Hiroaki Kitano

Sony Computer Science Laboratory
3-14-13 Higashi-Gotanda, Shinagawa
Tokyo 141 Japan
kitano@csl.sony.co.jp

Abstract. One of the central challenges of evolutionary computing and artificial life research is to establish a methodology for building very large complex system which has functional structures. Although it is increasingly recognized that the use of developmental process is the promising approach, none of existing method can create complex structures involving large scale repetitive subunits which characterize functional biological and artificial systems, such as brain, animal body, memory chips. In this paper, we present a powerful method of developing very complex structures based on a grammar-based approach. The introduction of novel meta-node and associated operations is the essential feature of the method. We demonstrate the strength of the method by actually developing the network topologies identical to human receptive fields of skin for touch stimuli and cerebeller cortex.

1   Introduction

This paper presents a method of morphogenesis-based system design for very complex systems. Since the first proposal to use developmental process for designing complex structure was made in 1990 [Kitano, 1990], numbers of methods have been proposed to incorporate developmental process in neural network and other structural design. As a result, the approach is now being recognized as an important methodology for designing highly complex structure. Nevertheless, we have not yet to achieve the level of complexity and functional structures using the developmental approach. None of the method proposed so far have shown to be scalable for truly large scale complex systems. One of the central reasons for the lack of ability in the existing method in creating complex structures is the absense of mechanism to generate very large scale repetitive structures.

In the real biological system, as well as in many engineering systems such as memory chips, very large scale repetitive structures play essential role. Human brain is composed of millions of columns, which are almost identical in their internal structure. Human retina, *Drosophila* eye, body segments, olfactory bulb, and many other organs entails vast number of repetitive structure. Even in artificial systems, memory array of DRAM, structure of conventional office buildings, grids of TFT flat dispay, and many other things are composed of largely repetitive structures in their essential part of the design.

In this paper, we present a framework of the method which enables developmental design of very large complex systems, which has large scale repetitive structures.

Obviously, there are two ways to attach this problem. The first approach is to closely simulate actual biological process of development. The second approach is to use well-defined formal approach. This would mean to use grammar-based method, instead of using metabolism-based method which are biologically more plausible. Both approaches has their merits, and should be investigated more in detail. In fact, we carry out research in both directions. In this paper, however, we concentrate the second approach to define engineering method, which is the grammar-based approach. Our research on more biologically feasible model shall be described elsewhere.

2   Current Methods

The developmental method of system design can be divided into two categories. The first group is based on grammar rewriting, originally proposed in [Kitano, 1990], and augmented by series of works including [Gruau and Whitley, 1993; Kitano, 1994b]. Karl Sims applied idea based on this type of model and successfully created impressive computer graphics image of artificial creatures which walk, swim, and chase in the virtual world [Sims, 1994]. Advantages of using grammar systems are well-definedness, transparency of the developmental process, and commonalities with other engineering methodologies. The second group is based on intraction of simulated gene products, as first proposed in [Kitano, 1994a; Kitano, 1995]. This line of research tries to closely model actual biological development, and are often computationally very expensive. As we have state in the previous section, we will concentrate our discussion on the grammar-based method in this paper.

The original idea of the grammar-based approach is to use the graph generation system, an extension of the Lyndenmayer system (L-system [Lindenmayer, 1968; Lindenmayer, 1971]) to develop structure of neural networks. It uses a set of rules which has 2 by 2 matrix on right-hand side of the rule. Each element in a matrix are simultaneously rewritten by corresponding rule. A final matrix defines connection topologies of the network. Gruau proposed a variation of the grammar-based model, named Cellular Encoding [Gruau and Whitley, 1993]. Cellular encoding applies a variation of genetic programming [Koza, 1993] to specifiy timing of cell division, cell types, and changes in links involving the cell. Unfortunately, these methods can not be easily scaled up to generate complex repetitive structures. This is because all operations are imposed on specific individual nodes, so that creation of large number of repetitive structures with similar internal topologies requires long chain of rules are coincidentally aligned to create such a structure. Such a scheme of generating repetitive strucutre us extremely fragile against mutation, and hardly be duplicated to expand complexity of the system.

Therefore, it is essential that any new method to achieve capabilities to generate very large complex systems. In the rest of paper, we describe a new method of grammar-based developmental system, which can generate very complex systems. It subsumes generative capability of all previsouly proposed grammar-based developmental models, including [Kitano, 1990; Gruau and Whitley, 1993]. We will demonstrate the power of the method by actually generating network topologies identical to subsystems of human neural systems.

3 The Framework

For the engineering method, we choose to use grammar-based rewriting rules to be the basis of the development system. The grammar-based model is much more computationally efficient than metabolism-based model as seen in [Kitano, 1994a], at the cost of biological plausibility. Also, it is more tractable and controllable, as basic developmental pathways are written in rewriting rules. The use of grammar-based model enable us to create a library of rules, which can develop substructures of larger systems. Designers may be able to use such a library in designing their system.

The major challenge in designing a system for development of complex and structured systems is how to create large scale repetitive structures. In our model, a new type of node called *Meta-node* is introduced, which represents cell assembly. Various new rules has been incorporated associated with meta-node.

In the biological systems, there are numbers of repetitive structures such as hyper-column in brain, instect eyes, etc. Recent studies in molecular biology provides us evidence which insists these structures are formed by diffusive components of gene products. In some case, the reaction diffusion system which creates the turing wave may play important role. While it is attractive approach to actually hybrid reaction-diffusion model into the grammar-based model, or to even use it as a main mechanism of pattern formation, we would avoid using these mechanism in this paper. First, the use of reaction-diffusion model is computationally expensive, since diffusion of substance has to be computed in three-dimension array for numbers of components. Second, the reaction-diffusion model can be included in the other model which will be described in the paper on biologically-feasible model of developmental design. The work described here takes an extreme end of formal and strcutured approach.

The other feature is that new model ensures rules to be expressive enough to over all possible change of nodes and links created as graphs.

3.1 Nodes

Three types of nodes and corresponding links are used in the formalism:

Precursor Node: Node which is not terminally differentiated.

Meta Node: Node which has array of precursor nodes.
Instance Node: Node which has already terminally differenciated.

Instead of using the reaction-diffusion model, we introduce meta-node. Meta-node is a node which has large number of nodes and their internal connections. The rule specifies number of nodes, node type, dimensions, and internal connections.

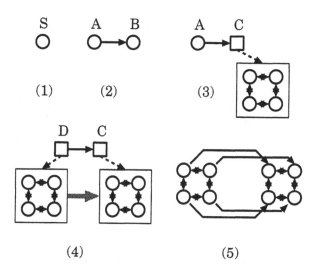

Fig. 1. Meta Node Expansion

Figure 1 shows a simple case of graph generation using meta-nodes. Initially, a graph starts from an initial symbol S [1]. Using the rewriting rule, it is expanded to two nodes connected by a link originated in A, destinating on B. Then, B is rewritten into C, which is a meta-node of 2 by 2 interconnected precursor-nodes inside. A is rewritten into D. At this time, a graph consolidation process takes place so that precursor nodes in each meta-node are directly connected.

Meta node has internal structure defined by its dimensions, composition cell type, and interconnection type.

3.2 Rewriting Grammar

Due to the introduction of meta-node, rewriting grammar needs to have rich feature set to express properties of meta-nodes and their interconnections. Table 1 shows data-structure for rewriting rule.

---

[1] Although current method uses bit-vectors to represent each node, we use alphabetical symbols in this section for ease of explanation.

	Node 1	Node 2	Link Change
	New Node State 1	New Node State 2	Internode Link
	X size	X size	Incoming Link
	Y size	Y size	Outgoing Link
	Intra Meta-Node Projection	Intra Meta-Node Projection	Inter Meta-Node Projection
	Projection Type	Projection Type	Projection Type
	-3N	-3N	-3N
Condition Part	-2N	-2N	-2N
Current Node State	-1N	-1N	-1N
Action Part	Self	Self	Self
Meta Node Flag	1N	1N	1N
	2N	2N	2N
	3N	3N	3N
	Alignment	Alignment	Alignment
	Pick 1	Pick 1	Pick 1
	BR	BR	BR
	DR	DR	DR
	CR	CR	CR
	RR	RR	RR

Table 1. Non-Compressed Data structure for a node division rule

Obviously, this rule representation is too big for genetic algorithms or other evolutionary approach to encode in an efficient manner. Also, each projection and meta-node features are directly written in each rule so that the projection acquired in one rule cannot be easily shared by other rules. Thus, actual implementation uses data structure as shown in Table 2 In the rule set shown in Table 2, rewriting rule only has pointers to data objects which defines featurs of projections and meta-nodes. When encoding this grammar system to the one dimensional string, chromosome should have three major regions; rewriting rule encoding regions, projections regions, and meta-node regions. Each region consists of fragments which encodes rule, projection, and node.

Links When a percursor node or an instance node divide, the question is what happen with links involved in the node. Figure 2 shows patterns of node division. In Figure 2-(A), inter-node connection patterns are shown. There are four possible connections after a node was divided. Figure 2-(B) and (C) shows connection with other nodes when there is an incoming link and an out-going link, respectively. Not counting the properties of each link, such as an activatory link or an inhibitory link, there are four basic connection types. Thus, when there are links from/to other node, there are 64 possible connection topologies involving two new nodes. How links involving the node changes after the division is specified in the rule.

	Projection
	Projection ID
Condition Part	-3N
Current Node State	-2N
Action Part	-1N
Meta Node Flag	Self
Node 1	1N
New Node State 1	2N
Meta-Node ID	3N
Projection ID	Alignment
Node 2	Pick 1
New Node State 2	BR
Meta-Node ID	DR
Projection ID	CR
Link Change	RR
Internode Link	
Incoming Link	Node
Outgoing Link	Node ID
Projection ID	X size
	Y size
	Projection ID

Table 2. Compressed Data structure for a node division rule

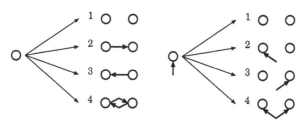

(A) Connection within two nodes   (B) Change of an incoming link

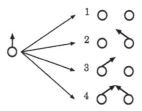

(C) Change of an outgoing link

Fig. 2. Divison Patterns

Projection The concept of projection is introduced to map nodes in two meta-nodes. Projection defines general characteristics of interconnection between nodes, that all nodes in specified meta-nodes shall follow. When the originating meta node and the target meta node is different, it defines inter-meta-node projections. When the originating meta-node and the target meta-node is same, it defines intra-meta-node projections. Numbers of projection types may be defined to capture complexity and diversity of interconnections in brain. Following two types are most convenient and, possibly, widely effective projection types.

Determinastic Projection Determinastic projection defines that for each node in the meta node, how interconnection shall be made to target meta-node in the determinastic way. Figure 3 shows some of these projections. for example, Projection `Self` link each node of the meta-node to a corresponding node in the target meta-node. 1N projection connects a node to the first neighbours (1N), nodes at one unit distance apart from the corresponding node, in the meta-node. 1.5N nodes are nodes at 45 degree unaligned from the 1N nodes from the source node, but are at 2 hop. Similarly, -1N projection means that each node in the meta-node recieves links from the 1N nodes in the target meta-node. Each projection can be independently defines so that complex projection such as `Self`, `1N`, `-2N` are possible.

When the target and the originating meta-node is same, and projection type is 1N, it creates mesh topology within the meta node. With node type specification on excitory and inhibitory links, it is possible to define `1N Activation`, `1.5N Inhibition`, which means commonly observed topologies that each node has out-going excitory links to the first neighbours, and recieves inhibitory links from slightly distant nodes.

Constraint Probabilistic Projection Constraint probabilistic projection (CPP) represents neural connections which exhibit certain level of regularities, but not fully determinastic. CPP an be defines by using four basic parameters; Base ratio (BR), divergence ratio (DR), Convergence Ratio (CR), and Recieving Ratio (RR). Distribution pattern (DP) defines whether projection to be made totally at random, or following some statistical distributions, such as Gaussian. Radius of Distribution (RD) and the Q of distribution (QD) defines the distance of the peak of the distribution from the originating node, and its sharpness.

Base ration defines a proportion of nodes in the meta node, that project links to target meta node. Divergence ratio is an average number of out-going links per a node. Convergence ratio is an average number of incoming links per a node in the target meta-node. Recieving ratio is a proportion of nodes in the target meta node that recieve projection from the originating meta node.

Suppose there is 100 nodes, a projection which is defined as $BR = 0.1$ and $DR = 100$, means that 10 out of 100 nodes can be a origin of divergent links, and each node will have 100 projecting links to nodes in the destination meta-node. Similarly, a projection defined as $RR = 1.0$ and $CR = 1000$ means that each node in destination meta-node will get 1000 projection from nodes in orginating meta-node.

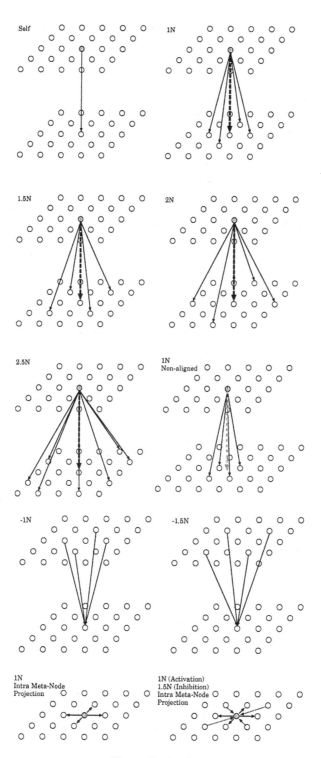

Fig. 3. Projection

226

4 Developmental Formation of Typical Neural Systems in Brain

The power of the proposed model shall be described by creating complex network topologies which actually exist in the human brain. In this paper, we demonstrate that the proposed model can generate topologies for (1) somatic sonsory systems, specifically a receptive field of skin for sensing touch stimulus, and (2) cerebeller cortex involving Purkinje cell, parallel fiber, mossy fiber, etc. These two neural subsystems are chosen to be representative of other neural subsystem, and their neural topologies are well-investigated [Kandel et al., 1991].

The receptive field of skin touch sensor neural system and cerebreller cortex is very large repetitive structures (Fig. 4). Skin's receptive field has strict layered architecture with array of idential and well aligned neural subsystems. On the other hand, cerebeller cortex has more randomness, although general structure can be viewed as collection of similar cell assembly. While highly determinastic model better suits for the skin receptive field, it cannot create cerebeller cortex.

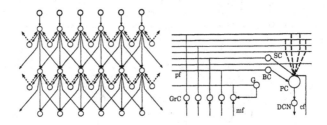

Fig. 4. Skin Neural System and Cerebellium

It should be noted, however, that the goal of the description below is to show that the proposed method can generate topologies exists in the human neural subsystems. It is not intended to claim nor imply that the order or mechanism of development simulate actual developmental process of human neural systems. It is only mean to show that our method can create, whatever the intermediate processes can be, topologies of neural networks with the level of complexity equivalent to human neural subsystems.

4.1 Somatic Sensor Receptive Field of Skin

Topology of receptive field of skin for touch stimuli can be created by using determinastic projections. Figure 5 illustrate process of meta-node division. First, the meta node A is created from the initial precursor node. This corresponds to detailed figure 6(A). There is no intra-meta-node connection in A. A split into B and C, and create Self link, as shown in figre 6(B). C divide into D and E. Self and 1N projection were created (Figure 6(C)). E also divide into F and G, create Self and 1N projection. Both D and

F divide into H, I and J, K, respectively, with inhibitory -1N non-aligned projections (Figure 6(E)). It is significant that the method can easily generate such a topology because this type of neural network topology, with slight variations, is widely observed in peripheral nervous systems, such as retina, auditory systems, and even in some of cortical regions.

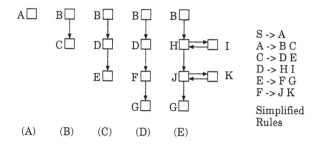

S -> A
A -> B C
C -> D E
D -> H I
E -> F G
F -> J K

Simplified
Rules

(A)   (B)   (C)   (D)   (E)

Fig. 5. Skin Neural System Formation

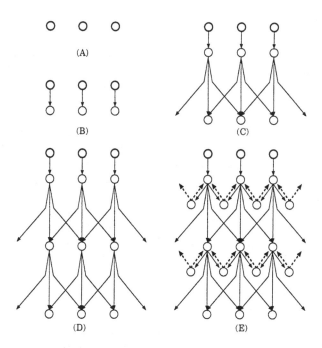

Fig. 6. Skin Neural System Formation

4.2 Cerebeller Cortex

Contrary to receptive fields, cerebeller cortex involves largely stochastic projections with high level of regularities. The formation of topologies of cerebeller cortex uses constraint probabilistic projections. First, a meta node with simple single incoming and out-going projection is created (figure 7(A)). For example, the dimension of the meta-node can be 100 by 100, having 10,000 nodes. This meta node split into two meta-nodes with constraint probabilistic projections with diverging nature (figure 7(B)). Property of the projection is BR = 1.0 (100%), DR = 500, CR = NA, and DR = 1.0 (100%). This means that for each node in Layer 1, there are 500 out-going links projecting randomly into nodes in the layer 2. Nodes in the layer 2 is about 5,000,000 nodes. A newly created meta-node again split to create additional meta-node with probabilistic projection with converging nature (figure 7(C)). There are about 2,500 nodes in the layer 3. This converging projection has property defines as CR = 80,000.0, RR = 1.0, BR = 1.0, DR = 2.0. Fianlly, each node in the meta-node at the top-most layer divide to create new nodes surrounding each node (figure 7(D)).

Although actual cerebellium circuit is more complexicated and modeling of dendrate and specific synaptic types would be needed, basic topologies can be created in this process.

5 Conclusion

In this paper, we proposed a new method of developmental design of complex system using grammar-based approach. The major contribution of the work described here is the introduction of meta node and associated projection scheme, which enable the grammar-based appraoch to generate very large scale repetitive and structured system. The power of the new method was demonstrated by actually generating neural network topologies in human neural subsystems; receptive field of skin somatic sensors and cerebeller cortex. Whether this method actually enables evolutoinary design of functional complex structures must await for actual experiments using physical and virtual agents interacting with real world, such as visual, audio, tactile sensing, actuators, and other spectrum of sensory and effector channels, what can be clearly stated at the end of this paper is that the method presented in this paper is, at this moment, the most powerful grammar-based model of developmental design of complex and structured systems.

References

[Gruau and Whitley, 1993] Gruau, F., and Whitley, D., "Adding Learning to the Cellular Development of NEural Networks: Evolution and the Baldwin Effect," *Evolutionary Computation*, 1(3): 213-233, 1993.

[Kandel et al., 1991] Kandel, E., Schwartz, J. and Jessell, T., *Principles of Neural Science: Third Edition*, Appleton & Lange, Norwalk, 1991.

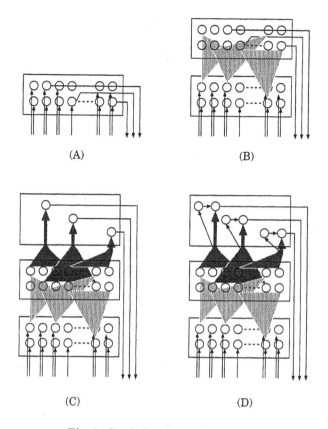

Fig. 7. Cerebeller Cortex Formation

[Kitano, 1995] Kitano, H., "A Simple Model of Neurogenesis and Cell Differentiation Based on Evolutionary Large-Scale Chaos", *Artificial Life,* 2: 79-99, 1995.

[Kitano, 1994a] Kitano, H., "Evolution of Metabolism for Morphogenesis," *Proc. of Alife-IV,* 1994.

[Kitano, 1994b] Kitano, H., "Neurogenetic Learning: An Intergrated Model of Designing and Training Neural Networks using Genetic Algorithms," *Physica D,* Aug. 1, 1994.

[Kitano, 1990] Kitano, H., "Designing Neural Network using Genetic Algorithms with Graph Generation System", *Complex System,* Vol. 4-4, 1990.

[Koza, 1993] Koza, J., *Genetic Programming,* The MIT Press, 1993

[Lindenmayer, 1968] Lindenmayer, A., "Mathematical Models for Cellular Interactions in Development," *J. theor. Biol.,* 18, 280-299, 1968.

[Lindenmayer, 1971] Lindenmayer, A., "Developmental Systems without Cellular interactions, their Languages and Grammars," *J. theor. Biol.,* 30, 455-484, 1971.

[Sims, 1994] Sims, K., "Evolving 3D Morphology and Behavior by Competition," *Proc. of Artificial Life IV,* Cambridge, MIT Press, 1994.

Evolving Batlike Pinnae for Target Localisation by an Echolocator

H. Peremans*, V. A. Walker, G. Papadopoulos, and J.C.T. Hallam

University of Edinburgh,
Department of Artificial Intelligence,
Forrest Hill 5, EH1 2QL Edinburgh, UK,
herbertp@dai.ed.ac.uk

Abstract. There is considerable evidence that pinna (external ear) shape plays a crucial role in the localisation, especially along the vertical dimension, of targets by echolocating animals. However, because of the complexity of the relation between pinna shape and localisation performance it is very difficult to design them so that the echolocator achieves specific localisation characteristics. Hence, we have developped a genetic algorithm (GA) which in conjunction with an acoustic echo simulator allows us to evolve desirable pinna shapes instead of having to design them. We use this method to evolve a rudimentary pinna that allows an echolocator, using a broadband call, to determine the vertical component of a target's location by comparing the measured intensities at different frequencies.

1 Introduction

By employing two laterally separated receivers, animals determine the (2D) bearing of a sound source based on inter-aural disparities of intensity (IID), arrival time (ITD) and/or phase. While the placement of two receivers on opposite sides of an acoustic perceptual system generates IIDs which are powerful lateralisation cues, adding pinnae to receivers creates directional cues which enable animals — using only two ears — to attain true 3D target localisation capability. In particular, bats which employ broadband calls are generally believed to localise targets in 3D by comparing IIDs across different frequencies [1–3]. That is, they exploit the fact that both position and size of the most audible portion of the frontal sound field changes with frequency, due to the passive acoustic properties of the head and pinnae.

In addition to its interest for biology, the investigation of pinna shape and its consequences for 3D target localisation is also showing great promise in providing very efficient solutions to the problems faced by robotic echolocation systems. In the past, there have been a number of proposals for ultrasonic sensor systems that can localise objects in 3D [4–6]. All of these systems make use of ITD's to calculate the position of a reflector in 3D. Consequently, they all tend to be

---

* H. Peremans was supported by a Marie Curie fellowship

rather big and use at least three receivers. In contrast to these approaches, we have shown elsewhere [7] how a binaural sonarhead, modelled on the bat's echolocation system, can perform 3D target localisation based on IID measurements only, facilitating the construction of sonarheads with very small inter-aural dimensions. However, the complexity of the relation between pinna shape and localisation performance makes it exceedingly difficult to design useful pinna shapes. Instead, we have developed a scheme based on a genetic algorithm that allows us to evolve desirable pinna shapes.

In the next section, we begin with an overview of the echolocation simulator employed in this work (Section 2). The genetic algorithm used to evolve the desired pinna shapes is described in Section 3. Section 4 provides a description of the experiments and, following the discussion of the results, Section 5 draws some conclusions and indicates possible further extensions of the work.

2 Echolocation simulator

The aim of the echolocation simulator is to simulate the acoustic signals picked up by an active echolocation system. To execute an experiment the user specifies the structure of the call to be emitted by setting the duration, the instantaneous amplitude and the instantaneous frequency. The user also specifies the positions and orientations of the transmitter and the receivers. It is possible to vary these parameters during the experiment to model the various active sensing strategies employed by bats. Finally, the positions of one or more point-like reflectors (from which complicated shapes can be assembled) are specified and 3D motion trajectories may be set for the targets as well as for the echolocator. Because of this transmitter, reflector and receiver motion the simulator has to take into account the Doppler shifts that occur during transmission, reflection and reception. Generating correct Doppler shifts is important because of their possible use as echolocation cues by bats.

The simulator then propagates the emitted cry through the environment until it reaches a receiver. The details of this process relevant for the current discussion, i.e. a stationary echolocator 'looking' at stationary targets, are given below, see [8] for a fuller account of the simulator's capabilities.

2.1 The propagation of acoustic signals within a simulated environment

As a first approximation to the true acoustic field created by the possibly moving transmitter in the presence of possibly moving reflectors, we assume that spherical wavefronts are generated by the transmitter. These wavefronts expand until they reach the target, at which point they are replaced by secondary spherically expanding wavefronst originating at the target which eventually reach the receivers [9]. Hence, all targets result in an echo at each receiver as long as they remain within the forward hemisphere defined with respect to the transmitter. Furthermore, we assume only a single echo is produced by each target, i.e. no secondary bounces of sound between targets are considered.

Propagation and absorption losses. While propagating, the amplitude of the call experiences spreading attenuation characterised by a $1/r$ dependency on the distance r travelled for both its outgoing and its incoming path. The amplitude of the call and echo is further attenuated by absorption losses in air. The absorption coefficient $\alpha(f)$ is frequency dependent and given by [10]
$\alpha(f) = 0.038f - 0.3$ dB/m.

Reflection losses. The targets are taken to be point reflectors that reflect diffusely, i.e. equal in all directions. To model scattering and absorption losses occurring during reflection, we allow the specification of a target dependent scattering coefficient σ, denoting the ratio of backscattered to incident pressure amplitudes.

In addition to the fixed reflection loss, targets with particular motions may introduce a time dependent reflection loss $\sigma(t)$ as well. This capability was added to model the occurrence of 'glints', i.e. high amplitude reflections corresponding with certain advantageous reflector orientations which characterise echoes reflected by the wings of the fluttering prey of many insectivorous bats.

2.2 Directivity of the transmitter and the receivers

Modelling the directivity, i.e. angular dependent sensitivity, of the transmitter and receivers is, for the purpose of this paper, the most important feature of the simulator. Because of the good correspondence with the behaviour of the transducer, i.e. the Polaroid transducer, used on our real sonarhead [7] the simulator models both the transmitter and the receivers as circular vibrating pistons.

Directivity due to finite size of transducer. When a source of finite size, rather than a point source, is used to generate sound, the amplitude of the generated pressure wave varies as a function of the off-axis angle. This variation in amplitude is due to interference between waves produced by different parts of the transducer surface. For particular angles this interference will be destructive, resulting in amplitude minima, for others it will be constructive, resulting in amplitude maxima, as shown in Fig. 1.

The directivity $H(f, \theta)$ specifies how the amplitude of the generated pressure wave depends upon the size of the transducer (radius $a = 1.4$ cm) and the off-axis angle θ when the piston is vibrating at frequency f [9]

$$H(f, \theta) = \frac{J_1(2\pi f a \sin(\theta)/c)}{2\pi f a \sin(\theta)/c},$$

with c the speed of sound and J_1 the first Bessel function. As can be seen from the polar response plotted in Fig. 1, the width of the main lobe decreases and the number of side lobes increases with increasing frequency.

The reciprocity theorem [11] states that the directivity of a transducer used as a transmitter is equal to that of the same transducer used as a receiver. Hence, the previous discussion applies equally well to the directivity due to the finite size of the receivers.

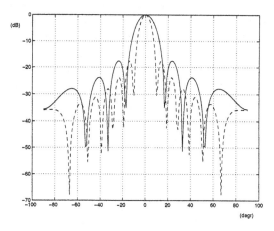

Fig. 1. The polar response of the transducer at 30kHz (solid line) and at 90kHz (dashed line).

Directivity due to pinnae. Another source of frequency dependent directional sensitivity is the presence of reflecting surfaces, i.e. the pinnae, around the receivers. As shown in Fig. 2, the presence of a reflecting surface generates a second, slightly longer, path through which the echo can reach the receiver. As

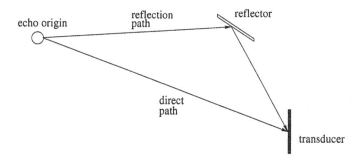

Fig. 2. The reflector provides an alternative path for the sound to reach the receiver.

before, interference between the two time shifted echoes will be constructive for particular angles and destructive for others depending on the ratio between the wavelength and the difference in the paths lengths of the direct and reflection paths.

In the simulator, pinnae are composed of a series of small reflectors placed around the receivers, each contributing one reflected echo to the composite echo. We assume that when multiple reflectors are present each reflector introduces a single additional echo path, i.e. no multiple reflections off the different reflectors are considered. For each target position, the simulator calculates the echo along the direct path and the echoes along the reflection paths. The sum of all these echoes is the final echo as picked up by the receiver.

The echoes along the reflection paths are calculated with respect to a virtual receiver, located at the mirror image of the real receiver in the reflecting surface [9], Fig. 3. Note that this simplification assumes perfect, i.e. specular, reflection

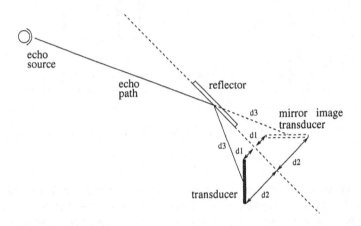

Fig. 3. The echo reflected onto the transducer by a reflector is equivalent to the echo that reaches an imaginary mirror-image of the receiving transducer.

and neglects the diffraction occurring at the edges of the finite size reflectors (L=same size as the receiver).

3   Evolving pinna shape for vertical localisation

As explained in the previous section, the shape of the pinna introduces frequency dependent spatial sensitivity. In this section we describe the GA used to explore the relationship between the shape of the pinna and the capacity of an echo-locator to localise targets. In particular we will concentrate on vertical target localisation, as pinna shape is believed to be mostly responsible for this capability [10]. As such, we had to decide on a representation scheme (genome), a generation and selection strategy and an optimisation criterion (fitness function).

3.1   The genome

in the work reported here, the structure of the pinna is approximated using at most three circular reflectors surrounding the receivers. Each

reflector's position and orientation is fully determined by 5 numbers. The first three represent the position of the center of the reflector's disc, while the remaining two are the azimuth and elevation angles of the vector normal to the reflector's surface.

Each gene's value varies within a predetermined space that samples the real space of possible positions and orientations. The position space is a cube, side 10 cm long, centred on the receiver. The orientation angles take values between -90° and +90° with a resolution of 2°.

3.2 The generation and selection strategy

The cross-over and mutation operators are the solution generating mechanisms of a GA. We used N-point cross-over: the two parents are broken into N+1 parts and a new child is formed by taking alternating sections from both parents. If cross-over doesn't occur, as determined by the cross-over probability, one of the parents is randomly chosen to be cloned. We have found it advantageous to introduce cross-over and mutation operators that treat certain parts of the genome separately, for instance, the part representing the position coordinates, the part representing the orientation angles, the part representing a particular reflector, etc.

In order to lead the search towards fitter solutions, genomes must be allocated reproductive opportunities depending on their fitness. We have used a tournament selection scheme wherein a set of genomes is randomly selected from the population and a fitness tournament is organised. The fittest genome participating in the tournament is selected with a given probability. If the fittest genome is not selected, then the second best is selected with the same probability, and so on. Finally, if no other genome has been selected, the last genome is.

3.3 The fitness function

For each pinna configuration, a fitness value is calculated, representing the genome's quality with respect to the optimisation criterion.

The fitness calculation proceeds in two steps. First, when a new configuration is created, an initial check is performed to ensure that the reflector configuration is a legal one: the reflectors do not occupy the same space and are not obscured by another reflector or receiver, see Fig. 4. The configurations for which some of the visibility trapezoids are non-empty are considered invalid and given a very bad fitness value. We prefer this approach to eliminating the invalid configurations from the population as that would mean losing the possibly interesting subsets of reflectors contained within them. Next, the sound field projected onto the receiver is calculated by the simulator described in the previous section and the fitness criterion is applied to this sound field.

As can be seen from Fig. 1, one can determine, apart from some ambiguity, the off-axis target angle θ by comparing the expected ratios of the echo intensities at different, at least two, frequencies with the actually measured ones. Elsewhere [7] we show that a similar scheme can be used to get complete 3D localisation information. Furthermore, the localisation accuracy and the elimination of the ambiguities depends critically upon how much the polar responses

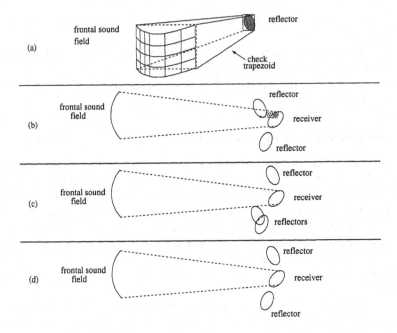

Fig. 4. Determining invalid pinnae configurations. (a) The frontal sound field is approximated by a square parallel to the xy plane. A reflector or receiver is approximated by the smallest rectangle that contains its projection onto the xy plane. The "check" trapezoid is derived by connecting the vertices of the two rectangles. (b) The "view" of the transducer is obscured by a reflector, (c) two reflectors touch, (d) a "valid" configuration.

at the different frequencies differ. Note that for the bare transducers the maximum sensitivity axes coincide, as can be seen from Fig. 1. Consequently, as the polar responses are circularly symmetric many positions in the frontal field would correspond with the same set of ratios between the intensities at different frequencies. Hence, in an attempt to evolve pinna shapes that would eliminate this high intrinsic ambiguity, we chose to judge the quality of a pinna shape, i.e. a reflector configuration, by the distance between the maximum sensitivity axes for the extremal frequencies in the frequency range considered. Since the calls we have been using for this work consist of a downward frequency sweep starting at 90kHz and ending at 30 kHz, the fitness criterion is given by the distance between the maximum sensitivity axes for these two frequencies.

To find the maximum sensitivity axis for a particular frequency the target's distance is kept constant and five sample target positions

are considered, corresponding to a square's vertices and centre, as shown in Fig. 5. For each point, the strength of the echo arriving in each ear is calculated and the point of maximum strength is determined. One of two

possible situations can occur:

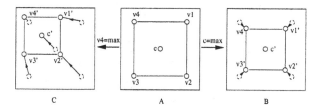

Fig. 5. Finding the maximum intensity point.

- the maximum strength echo corresponds with the central point, the size of the square side is decreased by 10% and the new square vertices are used as target positions;
- the maximum strength echo corresponds with one of the vertices of the square, the new square's centre is taken to be half-way between the old centre and the maximum strength vertex.

The initial square vertices are taken at $-20°$ and $20°$ azimuth and elevation. The procedure stops when the size of the square's sides corresponds to an angular deviation of less than $0.5°$. The centre of the final square is assumed to be the maximum strength point for that frequency.

This is a simple hill-climbing scheme and it works for the examined cases, since there were no significant local maxima.

4 Results and discussion

In the first experiment, the distance between the endpoints of the maximum sensitivity axes corresponding to the two extremal frequencies is used as a fitness value.

The results shown in Fig. 6 indicate that the GA is quite capable of evolving a reflector configuration or pinna shape that when added to the bare transducer drives the maximum sensitivity axes apart. As can be seen from the regions of the frontal sound field that the three reflectors focus onto the receiver, the GA accomplishes this by pointing the reflectors at the area out of the 90kHz main lobe but within the 30kHz one. Fig. 6 then shows that this strategy allows the GA to shift the maximal sensitivity axis of the 30kHz directivity pattern while keeping the axis of the 90kHz one fixed. Since the 30kHz main lobe is the largest one, it is easiest for the GA to shift the 30kHz maximum sensitivity axis by redistributing the energy that falls within this lobe but out of the 90 kHz main lobe. Note that this configuration evolved by the GA to maximise the distance between the maximal sensitivity axes, generates a large horizontal shift between the latter.

Therefore, in the second experiment, we start from the optimal configuration generated in the first experiment but this time only the vertical separation of the endpoints is used as a fitness criterion to ensure that the axes point to

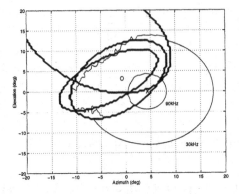

Fig. 6. The thick-line contours correspond to the regions of the sound field from which sound is reflected on the left receiver at both frequencies. The -6dB contours for 30 and 90kHz are also plotted (thin lines). O marks the 30kHz maximum intensity point and X marks the 90kHz one.

maximally different elevations. Alternatively, we could also look upon this second experiment as the second stage in the evolution process of the first experiment. As can be seen from the results shown in Figs. 7, the GA succeeds in further increasing the vertical separation between the two maximum sensitivity axes.

5 Conclusion

In this paper we have shown that it is possible to use a GA in combination with an acoustic echo simulation package to evolve pinnae that optimise particular components of an echolocation system. As an example of this technique's capabilities we have evolved a rudimentary pinna that allows an echolocator, using a broadband call, to determine the vertical component of a target's location by comparing the measured intensities at different frequencies.

Hence, this study proves the feasibility of using a GA in combination with an acoustic echo simulator to evolve pinnae shaped to optimise criteria relevant for an echolocating agent. As such, increasing the number of reflecting surfaces should allow us to explore the extent to which pinna shapes encountered in nature can be explained by various acoustic criteria. Furthermore, this would help us to actually build the pinnae evolved by the GA and mount them on our robotic sonar system to overcome some of the limitations of the commercially available transducers.

References

1. B. Lawrence and J. A. Simmons, "Echolocation in bats: The external ear and perception of the vertical position of targets.," *Science*, no. 218, pp. 481–483, 1982.

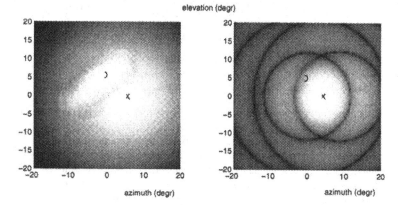

Fig. 7. Intensity map of the left acoustic map of the evolved solution when the call frequency is (a) 30kHz and (b) 90kHz. (O) marks the 30kHz maximum intensity point and (X) marks the 90kHz one, for a target placed at varying azimuth and elevation angles at a range of 0.5m.

2. Z. M. Fuzessery, D. J. Hartley, and J. J. Wenstrup, "Spatial processing within the mustache bat echolocating system: possible mechanisms for optimization.," *J. Comp. Physiol A*, no. 170, pp. 57–71, 1992.
3. M. Obrist, M. B. Fenton, J. Eger, and P. Schlegel, "What ears do for bats: A comparative study of pinna sound pressure transformation in chiroptera.," *J. Exp. Biol.*, no. 180, pp. 119–152, 1993.
4. R. Kuc, "Three dimensional tracking using qualitative sonar," *Robotics and Autonomous Systems*, vol. 11, pp. 213–219, 1993.
5. H. Akbarally and L. Kleeman, "A sonar sensor for accurate 3d target localisation and classification," in *Proceedings of the 1995 IEEE International Conference on Robotics and Automation*, (Nagoya), pp. 3003–3008, May 1995.
6. C. Delepaut, L. Vandendorpe, and C. Eugène, "Ultrasonic three-dimensional detection of obstacles for mobile robots," in *Proc. 8th Int. Conf. on Industrial Robot Technology*, (Brussels), pp. 483–490, 1986.
7. H. Peremans, A. Walker, and J. Hallam, "3d object localisation with a binaural sonarhead, inspirations from biology," in *1998 IEEE Int. Conf. on Robotics and Automation*, May 1998. in press.
8. V. A. Walker, *One tone, two ears, three dimensions: An investigation of qualitative echolocation strategies in synthetic bats and real robots*. PhD thesis, University of Edinburgh, 1997.
9. A. Pierce, *Acoustics*. New York: Acoustical Society of America, 1994.
10. R. Kuc, "Sensorimotor model of bat echolocation and prey capture," *Journal of the Acoustical Society of America*, vol. 96, pp. 1965–1978, 1994.
11. L. Kinsler, A. Frey, A. Coppens, and J. Sanders, *Fundamentals of acoustics*. New York: Wiley, 1982.

A Biologically Inspired Object Tracking System

Roger DuBois

Australian National University, Canberra ACT 2601, Australia

Abstract. The anatomy of the insect brain provides insights to neural architectures and visual processing algorithms which serve as blueprints for neuromimetic silicon chip designs. Selective attention reduces the amount of computation required by a biological system navigating through an information rich environment. In the insect visual system we see an example of task-specific sensor optimization for the detection of a topological invariant, the focus of expansion of the optic flow. A description of those regions of the optic lobe concerned with flow-field analysis is presented and this is followed by a description of a simple neural subsystem capable of detecting such a focus. This information is used in a feedback control system involving the peripheral sensors to gate object tracking and orienting systems. This robust and simple system is an ideal candidate for implementation in evolving silicon based vision systems.

1 Evolution: The Great Optimizer

Biological organisms have evolved a myriad of sensor optimizations to exploit information found in the natural environment. From the lowest phyla, some existing virtually unchanged for 70 million years to modern man, nature has achieved an elegance of design based upon the economy of energy. She does this by exploiting invariants from the multiple sources of information that may specify a given stimulus. Technological innovations frequently turn out to be mechanistic re-creations of existing biological functions. The concept of *bio-inspired systems* follows on naturally from this.

A typical notion about information in a perceptual system is that this information is correlated with affairs in the external world. We can imagine a huge covariance matrix describing all possible co-occurences between an internal 'image' and the external environment. If we restrict the entries in the covariance matrix to only nonzero, non-unity values, perceptual processing is inferential, where the premise for each inference is a mental construction rather than what is captured by the sensors and determined by the system architecture. This form of synthesis can be computationally expensive both for the biological organism and the designer of an autonomous robot. The challenge is to minimize the cost of this synthesis and perhaps, to eliminate it altogether. While this may seem a contemporary problem it should be remembered that all biological organisms evolve subject to similar demands for efficiency. If we consider evolution to be the optimization of information processing then, given the amount of time involved and great diversity of living systems, if an optimization for a particular task exists we can be reasonably sure that some organism will have evolved to take advantage of it. What optimization strategies might be useful?

1.1 Selective Attention

If we allow unity entries in the covariance matrix to represent invariant sources of information that specify an object in the external world, a biological system needs only to be constrained by evolutionary pressures to detect and attend to them. In the design of an autonomous robot we need to explicitly specify the invariants and then match the sensory subsytem to the detection task. For every invariant that may exist, perception becomes easier to understand and easier to implement on a machine. With sufficient numbers a perceptual system that does not rely *a priori* on synthetic model structures of the external world to work properly can be designed and built. Very sophisticated behaviour can be generated by systems which exhibit almost no measurable cognition. Take for instance the common fly: a high performance, visually guided animal, with limited neural architecture is capable of high speed navigation through optical flow-fields, collision avoidance and the detection and pursuit of small targets, with which it eventually mates. These are all desirable attributes for a robot! Reverse engineering of the more simple biological systems can be a profitable avenue of research.

Insects, flies for example, see each other and engage in various forms of high speed aerial pursuits prior to mating. The salient releasing mechanism for these orienting behaviours is the presence of a small target object in the animal's visual field. But by what mechanism does it focus its attention on the object of its desires in the presence of other distractors such as wind blown leaves or other insects.

1.2 Neuroanatomy — VLSI *in anima*

The primary visual system of the insect brain is highly structured, dominated by a retinotopic organization. It comprises three visual ganglia, namely the *lamina*, the *medulla* and the *lobula complex*. Each neuropil is strictly organized into columns and strata. Both the lamina and medulla are composed of parallel synaptic compartments, or columns, which exactly match in number the ommatidia in the retina [14][6][4][9].

Each column in the lamina (L in Fig. 1) known as an *optic cartridge*, receives inputs from a group of photoreceptors (R1-R6) that share the same visual axis as the overlying ommatidium. For a detailed description of the lamina synaptic pathways see [11]. The principal neurons of the lamina are concerned with encoding luminance contrast in environments having large intensity variations and exhibit high pass filter transfer characteristics. The lamina contains no motion-detection neurons and projects its filtered signals to the input (distal) layers of the medulla.

The medulla is a bistratified neuropil characterized by an extensive network of lateral interconnections and is the most peripheral structure in which movement detection takes place. The input layers of the medulla (m_i in Fig.1) are connected to the output layers (m_o) by arrays of short transmedullary interneurons [6]. These interneurons play a crucial role in gating signals through this structure.

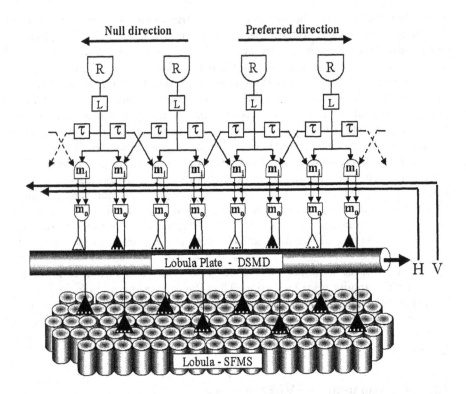

Fig. 1. Schematic representation of the insect motion detection system. The R units represent the photoreceptor array, the L units are lamina cells. The medulla is shown as a bi-stratified structure consisting of an input (mi) and an output (mo) side. DSMD is a wide-field, direction-selective motion-detecting neuron and SFMS is the small-field motion system and τ represents a delay line. Shown also is the feedback of horizontal (H) and vertical (V) flow-field information from the output region of the DSMDs to the peripheral medulla.

The initial stages of movement detection rely on sequence discrimination between spatially separate, retintopically organized signals having an asymmetric non-linear interaction between them. The computational elements are known as elementary movement detectors or EMD's [3] and it is this array of flickering elementary, motion signals that constitute the common currency for all visually guided behaviour in subsequent processing. These retinotopically organized signals are located in the proximal output layers of the medulla (m_o) but from here on the spatial organization of the signal becomes increasingly degraded as higher order properties of the visual world are computed by the third optic neuropil, the lobula complex.

The lobula complex consists of two complementary neuropils the *lobula* and the *lobula plate* each of which receive the parallel retinotopically organized flicker signals arriving from the output layers of the medulla.

In the lobula these spatially localized flickering signals are use to drive a position detection system whereby a position-sensitive torque signal is generated to orient the animal towards a particular target. These elemental small-field signals evoke very strong orienting responses, which increase in an dramatic manner as the spatial extent of the stimulating target decreases and on their own are perfectly adequate to direct visual flight. Observe for example the particularly impressive performance of a dragonfly in aerial pursuit of prey against an open sky.

However it must be stressed at this point that this sort of control, based purely on positional flicker information works *only in the absence of any other stimuli* or a moving textured background. The very high gain of the small-field system is inversely proportional to spatial extent of the stimulus and comes at the cost of a lateral inhibitory surround which rapidly shuts down the system in the presence of too many detractors. As I propose below, this operating limitation is overcome using the output from the complemetary wide-field lobula plate neurons.

The lobula plate contains large field *directionally-selective movement-detecting* (DSMD) neurons [4][5], which share a common network of presynaptic elements derived from the medulla. The DSMD neurons (Fig. 1) comprise several classes of fan-shaped, tangential cells, which integrate the EMD inputs and transmit the resulting signal, coding either global horizontal (H) or vertical (V) image motion via a centrally projecting axon to the optomotor systems controlling the flight musculature. Thus the average, global motion of the visual flow field is directly involved in flight control.

2   Engineering *in vivo* — Neurophysiology

Flies were immobilized with wax and presented with computer generated optic flow-fields. The optic lobe was explored with microelectrodes until neurons responsive to image movement were located. These neurons exhibited rapid (≈ 1 ms.) electrical depolarizations known as action potentials or 'spikes' whose occurence times were recorded as a sequence of pulses. The resulting point process appeared stochastic with Poisson statistics in the unstimulated animal. The statistical properties of this process changed with stimulation. Cells, with roughly circular receptive fields ($\approx 30°$), were recorded which exhibited classical bipolar optomotor responses, *i.e.* they exhibited continuous spontaneous spiking activity, modulated by movement of the visual field. Image movement in a preferred direction excited the response, while movement in the opposite, non-preferred, direction resulted in response inhibition.

If a cell's receptive field was centered on the focus of expansion of the flow-field such that the preferred and non-preferred motion vectors balanced, the excitation and inhibition cancelled and the cell's firing rate was no different from the normal unstimulated spontaneous rate *i.e.* a NULL signal. Thus, the combined influence of two cells of orthogonal directional preference and similar receptive fields would constitute a detector for the focus of expansion, *a topological invariant*.

2.1 Linear analysis

The recorded cells have an exceptional ability to resolve sudden changes in contrast frequency ($0.625°$ / 5.32 ms) and this behaviour shows rapid adaptation to the mean background image motion. The response to an arbitrary stimulus is reliably predicted by convolution of the stimulus history with the image velocity-impulse-response. These velocity kernels regarded as a linear transforming process, map the stimulus to the observed response. The transfer function G, estimated for each level of adaptation used was, in each case, 3rd-order:

$$G(s) = K \frac{(s - a)}{(s - b_1)(s - b_2)(s - b_3)}$$

As $s \to b_n$ the roots of the denominator polynomial, $G(s) \to \infty$ where each value b_n, is known as a pole of the transfer function and represents a characteristic natural frequency, or rate of change for the system. The poles completely describe a linear system. If the poles are plotted in the complex frequency plane (s-plane) we observe a remarkable feature: the root loci migrate along radial lines of constant damping ratio ($\xi \approx 0.5$) with increasing levels of adaptation. In control system theory a damping factor of 0.5 indicates a system is optimized for the best possible transient performance. Thus, the system not only measures the steady state velocity of the flow-field but also, by maintaining a constant damping ratio, encodes the velocity contrast as well.

3 Closing the Loop

The cells recorded from during these experiments had identical response characteristics to the classic lobula plate DSMD cells which convey flow-field information to the motor systems [8][10] but dual electrode studies and subsequent anatomical verification revealed them to be something entirely different.

Rather than being orthograde, centrally projecting afferent neurons they were discovered to be efferent *i.e.* they conveyed the horizontal or vertical flowfield information (H & V in Fig. 1) peripherally outwards from the midbrain to the input region of the medulla (m_i). These cells, known as the the *medullary tangentials* (MTH & MTV) receive their inputs from the output regions of the wide-field lobula plate cells coding global image motion, the DSMDs. In their turn these medullary tangentials relay these signals to the precise location where the transmedullary cells receive input from the lamina.

The transmedullary cells maintain the retinotopic organisation of the flicker signals arriving from the lamina as they relay information to the motion processing array in the medulla. The most likely candidates to receive this re-afferent signal from the lobula plate are the SUB, TM1 and TM5 transmedullary cells [14]. Interestingly these cells are the only transmedullary cells to have dendrites restricted to one group of terminals arising from a single optic cartridge and as a such are ideal candidates for inclusion in an small object detection system.

Thus we have a recurrent signal path, inward through the medulla on to the lobula complex and back to the medulla. This feedback from the lobula plate to the distal medulla gates signal transmission through the transmedullary cells which in turn feed into the position detecting system, the SFMS array. It is proposed that this arrangement constitutes a mechanism for static orientation and dynamic tracking of an object in both free-field and textured environments.

When the animal is stationary the (presumed inhibitory) NULL signal derived from the integrated output of the DSMD lobula plate neurons, is projected over the sensory array and every constituent high gain small-field element of the SFMS array spanning the entire visual field is gated open to receive input and provoke an orienting response.

However, when the animal moves through the environment its ego-motion creates visual flow-fields. When the receptive field of a particular MT cell is centered on a focus of expansion the feedback signal to that cell's projective field is locally NULL unlocking the high performance positional tracking system only along a receptive 'beam' aimed at those sensors directed at this focus. The two systems work in harmony since the lobula plate neurons are hardwired to the flight motor muscles specifically to balance the visual flow across each eye to enable steady flight.

This orienting and tracking behaviour is observed in flying insects on a hot day. When they are stationary they are receptive to any small object passing overhead and will chase it. If they are moving it is only when the object is caught in the glare of these perceptual headlights that it will be 'captured'.

3.1 Anima in silico

The insect visual system provides us with a neural architecture with which to construct a silicon based machine vision system capable of pursuit and obstacle avoidance. This 'smart-sensor' technology to be used in various autonomous devices or as adjuncts to piloted vehicles for line sensing, collision avoidance, missile tracking and so on.

Chip design is intimately linked with the choice of algorithm and several groups have made some progress with single-chip vision systems, some with more success than others. The innovative work of John Tanner and Carver Mead at Caltech on the *Correlating Optical Motion Detector* [15] led to several VLSI designs which were flawed by computationally expensive algorithms, based on the gradient scheme, [16][12] and the Marr and Hildreth zero-crossing detector [2]. Both of these schemes are excessively sensitive to noise since they rely on the calculation of spatial and temporal derivatives and at low ambient light intensity velocity estimates become problematic.

An approach taken by the Centre for GaAs VLSI technology and the Departments of Electrical & Electronic Engineering at the University of Adelaide, South Australia [1] based on the *Template Model* [7] is more promising. They have constructed a device whose input is an array of photodetectors interfaced with an array of graded refractive index (GRIN) lenses with parallel contrast differentiation. Changes in contrast are multiplexed for adjacent channels in space and in t_{i-1} sample time to form a 'Horridge template', an index into a lookup table providing information on object motion, direction, orientation, angular velocity and bearing. Several iterations of this design have already proved the feasability of the approach and the advantages of using gallium arsenide for low power, high speed operation, in low ambient light levels. It is proposed that the architecture/algorithm outlined above would be added easily to their design. Access to a spectrum of different image motion parameters also opens up the possibility of tuning the system to more subtle and salient features in the environment, *e.g.* objects moving with a particular angular velocity and so on.

References

1. Abbott, D., Yakoleff, A., Moini, A., Nguyen, X.T., Blanksby, A., Beare, R., Beaumont-SSmith. A. Kim, G., Bouzerdoum, A., Bogner, R.E., Eshraghi, K. (1995) Technology inspired obstacle avoidance - a technology independent program, Proc. SPIE, Int. Soc. Opt. Eng., 2591.

2. Bair, W. and Koch, C. (1989) Real-time motion detection using an analog VLSI zero-crossing chip. Proc. SPIE Visual Information Processing: From Neurons to Chips. Proc. SPIE, Vol 1473, pp. 59-65.

3. Buchner, E. (1976) Elementary movment detectors in an insect visual system, Biol. Cybern., 24: 85-101.

4. Hausen, K. (1982a) Motion sensitive interneurons in the optomotor system of the fly I. The horizontal cells: structure and signals, Biol. Cybern., 45: 67-79.

5. Hausen, K. (1982b) Motion sensitive interneurons in the optomotor system of the fly II. receptive field organization and response characteristics, Biol. Cybern., 46: 143-156.

6. Hausen, K. (1984) The lobula complex of the fly: structure, function and significance in visual behaviour. In M.A. Ali (Ed.), Photoreception and vision in invertebrates, Plenum, New York and London, pp. 523-560.

7. Horridge, G.A. (1991) Ratios of template responses as the basis of semivision. Phil. Trans. Roy. Soc. London B, Vol. 331, pp 189-197.

8. Ibbotson, M.R., Maddess, T. and DuBois, R.A. (1991) A system of insect neurons sensitive to horizontal and vertical image motion connects medualla to midbrain, J. Comp Physiol A,

9. Kirschfeld K (1972) The visual system of *Musca*: studies on structure and function. In: Wehner R. (ed) Information processing in the visual system of arthropods. Spinger, Berlin, pp 61-74.

10. Maddess, T., DuBois, R.A. and Ibbotson, M.R.(1991) Response properties and adaptation of neurons sensitive to image motion in the butterfly *Papilio aegus* , J. Comp Physiol A,

11. Meinertzhagen, L.A. and O'Neil S.D. (1991) Synaptic organization of columnar elements in the lamina of the wild type in *Drosophila melanogaster*, J. Comp. Neurol., pp. 232-236.

12. Moorea, A. and Koch, C. (1991) A multiplication based motion detection chip. In: Visual Information Processing: From Neurons to Chips. Proc. SPIE, Vol 1473, pp. 66-75.

13. Strausfeld, N.J. (1982) Functional neuroanatomy of the blowfly's visual system. In M. A. Ali (Ed.), Photoreception an dVision in Invertebrates, London New York: Plenum Press, pp. 483-522.

14. Strausfeld, N.J. (1989) Beneath the compound eye: neuroanatomical analysis and psychophysical correlates in the study of insect vision. In D.G. Stavenga and R.C Hardie (Eds.), Facets of Vision, Springer-Verlag, Berlin Heidelberg, pp. 360-390.

15. Tanner, J. and Mead, C. (1984) A correlating optical motion detector. Proc. Conferences on Advanced Research in VLSI, M.I.T., pp57-64.

16. Tanner, J. and Mead, C. (1986) An integrated analog optical motion sensor. In: VLSI Signal Processing II, pp.59-76.

The "Modeling Clay" Approach to Bio-inspired Electronic Hardware

Ken Hayworth

Center for Integrated Space Microsystems
Jet Propulsion Laboratory, Pasadena CA 91109, USA
Ken.J.Hayworth@jpl.nasa.gov

Abstract. The field of evolvable hardware or bio-inspired hardware holds the promise of automatically engineering complex electronic systems that remain adaptive during use. A growing number of experiments along these lines have been performed recently, mostly using off-the-shelf hardware or straightforward extensions of building blocks used by human engineers [1] [2]. In this paper we use the POE (Phylogeny Ontogeny Epigenesis) framework of bio-inspired hardware systems [3] and restrict evolutionary search and development considerations to pure hill-climbing search only, in order to develop some theory around evolution of electronic circuits. From this theory a new analog re-configurable hardware architecture is proposed for use in evolvable hardware. The hardware is a context switchable analog computer which can implement any general non-linear dynamic system on the level of the vector field representation. The optimization algorithm is a bio-inspired "molding" of the state-space description of the system. We call this novel hardware/ optimization algorithm platform the "modeling clay" approach to bio-inspired electronic hardware.

1   Introduction

This paper asks the question: "What general, re-configurable structure should a piece of evolvable analog hardware have that will allow it to evolve and adapt efficiently?" We seek the answer in biology.

The revolutionary point of Darwin's theory was quite simple: the slow accumulation of *small random improvements* could, in fact did, produce all the wonderful complexity of the biological world today. To a first, crude approximation, this evolution of complexity can be viewed as a type of hill-climbing search, greedily pursuing all changes that increase an individual's reproductive success. There is supporting evidence from biological simulations of evolution [6] that literal hill-climbing search can give rise to startling complexity, and this will be reviewed shortly. We do not wish to downplay the clear importance of crossover and the other complexities of population based evolutionary search techniques, we put these important questions aside so as to take advantage of the much simpler viewpoint which hill-climbing search to complexity allows. This focus on hill-climbing, when looked at within the POE framework of bio-inspired hardware, will direct us toward a new evolvable hardware architecture.

1.1 The POE framework

Sipper et al. [3] introduced the POE framework in which to compare evolvable hardware research directions. POE stands for Phylogeny Ontogeny Epigenesis representing natural selection, embryology, and learning respectively. For this paper we will need to somewhat expand on this framework, using the familiar language of evolutionary theory: "Genome space" is the space representing all possible genomes, and "phenome space", the space representing all possible creatures. If we, for the moment, **restrict** both evolutionary adaptive search and learning to use only greedy hill-climbing search techniques, the POE framework can be viewed as follows:

<u>Phylogeny</u> is the search over genome space using only information from the fitness measure of the corresponding point in phenome space. Evolution represents a hill-climbing search through this genome space. Genome space has a particular topology (a mathematical concept which simply allows us to talk about "connected" paths through these abstract spaces) which for an asexually reproducing creature, may be considered with respect to Hamming distance. Any ancestral tree will be seen to have followed a connected path with respect to the topology of this space.

<u>Ontogeny</u> is the explicit mapping of points in genome space to points in phenome space. Phenome space can be thought of as the space of all possible behaviors. A topology and dimensionality can be imposed on this space as well. Looked at this way, the genome-to-phenome mapping (ontogeny) is a mapping of the entire genome space into a subspace manifold of the phenome space. The key point of this realization is that now the hill-climbing search in genome space can also be viewed as a type of search through phenome space. This is an important point since it is the phenome's fitness that is directing the search in genome space.

We can think of the phenome space as being a road map and the search algorithm as driving around in town. If the genome-to-phenome mapping is "topology preserving" then driving to your destination is easy. The less topology preserving the genome to phenome mapping is, the more the road map will lead you astray. (This observation is usually talked about in terms of the "fitness landscape" of the genome space, an equivalent notion but less suited for the following discussion. A smooth fitness landscape is in most ways equivalent to a topology preserving genome-to-phenome mapping, however a fitness landscape based model is always tied to a particular problem. Since we are designing a general piece of analog hardware that will be well suited to a variety of tasks, we must take this more general view.)

<u>Epigenesis</u> can be thought of as a hill-climbing search over the parameters of the individual phenotype during its lifetime. For instance this might be neural connections which are determined by the genes but whose values are not. The analysis of this type of learning will require thinking in terms of yet another space, which we will call the "connection space", pushing the neural analogy, of that particular individual phenotype. A summary is in order:

- Ontogeny is a mapping of genome space into a sub-space manifold of phenome space
- The genome-to-phenome mapping should be topology preserving to allow efficient evolutionary hill-climbing search

- Epigenesis is mediated by a similar hill-climbing search, this time in an individual's "connection space"
- The connection-to-phenome mapping should be topology preserving to allow for efficient adaptation of the individual during its lifetime
- During epigenesis, the individual's point in phenome space, which starts at birth embedded in the sub-space manifold of the genome mapping, is allowed to wander from this manifold into the much larger phenome space accessible through the learning search

1.2 An example of an efficient evolutionary system

In biology, the genome to phenome mapping is the cell differentiation, cell migration, and cell population growth of biological embryology.[5] The genome encodes for proteins which provide the building blocks for cells, but more importantly many serve as regulatory proteins which, by their concentration, determine cell growth rates and the like. Proteins also regulate other proteins via gene expression, therefore, for each effect in cell behavior leading to embryological development, there is a good possibility that many dozens of genes affect this growth parameter.

Because of the use of regulatory cascades, mutations in specific genes may often produce only quantitative effects like bigger bones or more flattened noses. The reason is that knocking out or altering the shape of one of the proteins in a regulatory network may only affect the *concentration* of the end product, not its form.

One may object that qualitative change cannot occur with only quantitative step changes of concentrations of growth mediating chemicals and cell population sizes. A wonderful simulation, using only hill-climbing search, of the evolution of a vertebrate eye [6] brilliantly sets this objection to rest. In fact, all of Paleontology is based on the assumption that ancestral species can be identified by tracing quantitative shape and size changes in fossil data. The tree of life sketched out under these assumptions has "leaves" for all the creatures on earth today, for example humans and bats. Following their "branches" back in time through mostly quantitative size and shape changes in fossilized bone fragments, shows how our 5 fingered hand, with two bone forearm and single bone upper arm, and the bat's delicate webbed wing are connected through a series of quantitative morphings to our common ancestor's limb. Looking only at skeletal structures (which leave fossils) the entire tree of life is a history of gentle quantitative morphing of structural elements, with some division and fusing, which over eons give rise to the amazing qualitative changes apparent in today's world. It is ubiquitous examples like these which support the hill-climbing restriction which we are imposing on this discussion. Note how the structure of the experiment in [6] (simulated embryology of the eye through cell population growth rates) seems to work well with a hill-climbing technique. This intuition is what we will use in the "modeling clay" approach to electronic evolvable hardware.

"[I]f every form which has ever lived on this earth were suddenly to reappear… it would be quite impossible to give definitions by which each group could be distinguished from other groups, as all would blend together by steps as fine as those between the finest existing varieties…"
-Charles Darwin, as quoted in D. Dennett's Darwin's Dangerous Idea p85

2 Bio-inspired electronic circuitry

The term "evolvable hardware" brings to mind electronic components like transistors, resistors, and capacitors, automatically rewiring themselves in response to evolutionary pressure or a reinforcement signal. This vision has driven many researchers to attempt evolvable hardware platforms based on automatic switch-based routing, connecting components or higher level functional blocks [1][2]. It is worth noting that this methodology is an extension of how a human engineer designs circuits. A human engineer starts with discrete components or functional blocks, and has the end goal in mind. This is quite different from what biological evolution does. The switching in and out of components, in general, does **not** lead to small incremental changes in behavior.

"Bio-inspired" does not simply mean using a GA on an off-the-shelf piece of hardware. It means designing a piece of electronic hardware which has the special properties biological organisms have which makes evolution of complexity possible in the natural world. A few of these special properties were outlined in the previous section. It is time to try to apply these to the field of electronic circuits after we develop one more construct.

2.1 Pseudo-phenome space

The discussion of the extended POE framework to include hill-climbing search in the various "spaces" may have left the reader a little uneasy, like somehow the "hard part" of the problem was missed. If evolution and learning is simply going toward your goal in phenome space, then what is so hard? Here is the rub; phenome space is the space of all behaviors, an inherently ill-defined concept. Imposing a topology on this space appropriate to, for instance, the six-legged robot walking control problem would require that all methods of locomotion (tripod-gate, caterpillar gate, etc.) lie "close" to each other in this phenome space. (i.e. have connected paths between each other which do not pass through less efficient ways of walking) The prescription from the framework above would then be to "simply" find a genome representation and genome-to-phenome mapping which preserves this topology. A more general problem solving evolvable hardware platform would require a phenome space topology that not only is appropriate for six-legged walking behavior, but also a universe of different behaviors. This is not only difficult but is in principle impossible, and it reaches to the core of how we define "ill-posed" problems.

We believe, however, that the intuition provided by the framework is sound, and that in practice one can design bio-inspired, efficient, evolvable hardware platforms by attempting to approach as close as possible to this ideal. The key to doing this is to define a "pseudo-phenome" space, which is as close to the ideal of behavior topology as possible, and which allows definitions of genome space and connection space mappings preserving the character of this topology. Hill-climbing search is no longer "guaranteed" to reach the optimal solution, but then again, biology has many examples of organisms trapped on local maxima such as whales that have to surface periodically to breathe because of their land-dwelling ancestry. [4] In this paper the pseudo-phenome space representation suggested for bio-inspired electronic circuits is the inherently geometric state-space representation of a dynamic system.

2.2 State-space

The *behavior* of an electronic circuit can be described in many ways, but one is universal and precise, the language of differential equations. Any circuit topology can be represented in the following format called state-space form:

$$\dot{\vec{q}}(t) = \vec{f}(\vec{q}(t); \vec{x}(t)) \qquad (1)$$

$$\vec{y}(t) = \vec{g}(\vec{q}(t); \vec{x}(t))$$

Where x(t) is a vector of continuous signal values coming into the system, y(t) is a vector of continuous output signal values, and q(t) is a vector of continuous internal state values, the "memory" of the system. The functions f() and g() are vector valued and in general non-linear.

Take, for instance, the circuit in fig 1, which is an active filter design taken out of the book The Art of Electronics [8]. The system is linear and time invariant, has one input and one output. In state-space form the system equations are:

$$\begin{bmatrix} \dot{q}_1 \\ \dot{q}_2 \end{bmatrix} = \begin{bmatrix} 0 & -\dfrac{1}{R_{F1}C} \\ \dfrac{1}{R_{F2}C} & -\dfrac{1}{R_{F2}C}\dfrac{R_Q}{R_1} \end{bmatrix} \begin{bmatrix} q_1 \\ q_2 \end{bmatrix} + \begin{bmatrix} 0 \\ -\dfrac{1}{R_{F2}C}\dfrac{R_Q}{R_G} \end{bmatrix} x \qquad (2)$$

$$y = q_2$$

The dynamics of this circuit can be visualized geometrically by constructing a "state-space" which has axes representing the current voltage values of the capacitor states q1 and q2, and the value of the input variable x. As the physical system's currents and voltages change in response to the physics of the circuit connectivity, the point in state-space traces a deterministic curve through the space. At each point in state-space, a vector can be drawn representing dq/dt, this is done by evaluating the matrix equation above at each point in the state space. This produces a fully specified geometric description of the circuit behavior that is in all ways *equivalent* to the differential equation description or the component netlist description.

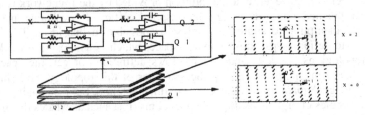

Fig. 1 Active filter circuit from The Art of Electronics p278 [8] along with the three-dimensional state-space of the circuit dynamics with two Q1xQ2 vector field planes plotted at different points along the x-axis

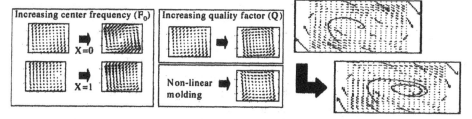

Fig. 2 (Left) Relation between circuit behavior and vector field symmetric molding. (Right) effects of mutation on vector field representation

The space of all state-spaces: Recall that we are interested in creating a pseudo-phenome space in which a natural topology, corresponding roughly to behavior of electronic circuits, exists. The space of all state-spaces (of a particular dimension) meets these requirements. This space is isomorphic to the space of all vector-valued functions, so there is a ready distance metric and topology in which to define local search. Following a connected path through the space of all state-spaces should lead you through mostly smooth changes in system behavior, providing more reliable information on which direction to go to reach a desired behavior.

2.3 The "modeling clay" approach defined

With the introduction of the space of all state-spaces as the pseudo-phenome space, we are now ready to apply the extended POE model to bio-inspired electronic circuit theory. "Mutations" will take the form of gentle morphings of the existing vector field description of the circuit (fig. 2). During life, adaptations will take the form of more local morphings of the vector field in response to reinforcement signals.

Let the genome description be a coarse grained parameterization of common state-space entities. In the bandpass circuit example, the tightness of winding of the vector field, and the stretch of the vector field along each of the axes are examples of linear transformations on the vector field. This is not a necessary restriction, but allows us to prove that the resulting genome-to-phenome mapping is a topology preserving mapping from the genome space (a parameterization of the space of linear operators on vector fields) to a sub-space manifold of our pseudo-phenome space: the space of all state-spaces. Since this mapping is a linear transformation from the genome space to the pseudo-phenome space, hill-climbing search for a particular behavior in this parameter genome space will succeed if the search would succeed in the sub-space manifold of the pseudo-phenome space directly. Recall that the pseudo-phenome space was designed so that this is likely the case.

This ontogeny gives us a phenotype embedded in the sub-space manifold of the pseudo-phenome space. In epigenesis, we can use efficient hill-climbing search to travel outward from this "birth" manifold if we define a "connection space" and another topology-preserving mapping. Break the state space into 1000 blocks, 10 on a side in the three-dimensional bounded state-space. Let the connection space be "control vectors" in each of these boxes, and the connection-to-pseudophenome mapping be a smooth spline interpolant of the resulting vector field defined by the control points. From spline theory, we see that this is also a linear transformation from connection space onto a set of basis vectors in the pseudo-phenome space. Now reinforcement learning can direct the hill-climbing on the connection space resulting in highly non-linear but smooth stretching and twisting of the dynamic system's vector field. Thus the final "adult" system can behave much different than the "child" system which was specified by only a few parameters.

We call this the **"modeling clay"** approach because the phylogeny step produces a sculptor's "rough" through coarse molding and twisting of the vector field, while the epigenesis step details that "rough" with finer, point-directed molding and stretching of the vector field. *The symmetry constraints inherent in the ontogeny mapping, and the spline smoothing enforced in the connection-mapping, make sure that the vector field "behaves like clay" in that local deformations are smoothed out locally, but do not propagate to ruin the hard won beauty which may exist in more distant parts of the sculpture.* Compare this to Nilsson and Pelger's eye evolution Simulation [6]. We have simply found a way of applying the analogy of cell population growth-based *physical* deformation to electronic circuit behavior.

3   Hardware implementation of the "modeling clay"

The "modeling clay" approach above requires a re-configurable hardware platform which can be programmed at the hardware level with a vector field description of the desired virtual circuit. We have recently constructed a simple prototype analog computer that embodies the functionality described in equation #1 above. A wide range of two state variable dynamic systems can be implemented on this computer. A test of the analog computer to evolve a target single variable transfer function by the "modeling clay" approach was successful. More ambitious tests will require refinement of the hardware described below.

The hardware consists of boards of discrete opamp chips, implementing the integration functions, summers, constant amplifiers (MDAC's), as well analog multipliers. These are common in all analog computers. What is special about this analog computer is that it is designed to be "context switchable", an idea taken from the field of CNN's [10]. Each of the coefficients of the cubic terms is embodied by an MDAC (Multiplying DtoA converter), each of which are in turn driven by a local digital memory store of 2 Kbytes. This allows the entire analog computer's dynamics to change (all function coefficients change) via a single broadcast addressing into the local memories. The computer is programmed with a vector field description of the desired circuit dynamics by breaking the state space of the system into disjoint regions each with its own vector field representation determined by unique coefficient values. The coefficient values of all the regions are stored in the local MDAC memories, with each region having a *common address* across all the local memory chips. As the analog state on the capacitors of the integrator opamps trace their paths through the state space of the virtual system, comparators can watch for when the state leaves a certain region of state space and crosses into another. When this occurs, the coefficients of that region are loaded into each particular MDAC from its private digital memory. Taken together, this is an analog computer architecture that allows for virtual dynamic systems of any size and complexity limited only by memory size. The actual coefficient storage can be implemented using the CMAC [9] compression and smoothing algorithm, or a caching virtual memory, both of which allow compression of empty regions in the dynamics in order to prevent the memory size from growing exponentially with the dimensionality of the state-space.

Also, an analog storage stack (not yet implemented) can be included to store and pass the values of the analog state variables. The analog stack allows multiple dynamic systems to interact on the same re-configurable analog hardware.

The key aspect of this context-switchable analog computer, which is desirable for the "modeling clay" approach to evolvable hardware, is that its *native language* is the vector field representation of dynamic systems, as programmed in the coefficient memories. Hardware-in-the-loop phylogeny and epigenesis can be accomplished by allowing the hill-climbing searches to manipulate this memory.

4 Conclusions

Keeping true to Darwin's vision of the accumulation of small improvements leading to complexity has inspired the design of a novel, bio-inspired electronic hardware. This hardware is a context switchable analog computer which compactly implements an enormous variety of virtual circuit dynamics by replacing the "component routing" paradigm, with a new "vector field molding" paradigm. It is hoped that this work will further intuition in other cases of bio-inspired engineering.

Acknowledgements
The research described in this paper was performed at the Center for Integrated Space Microsystems, Jet Propulsion Laboratory, California Institute of Technology. It was sponsored by the National Aeronautics and Space Administration.

References

1. Thompson, A. et al.: Unconstrained evolution and hard consequences, in <u>Towards Evolvable Hardware</u> Sanchez & Tomassini eds., Springer-Verlag Berlin 1996 p136-165
2. Higuchi, T. et al.: Evolvable hardware and its applications to pattern recognition and fault-tolerant systems, in <u>Towards Evolvable Hardware</u> Sanchez & Tomassini eds., Springer-Verlag Berlin Heidelberg 1996 p118-135
3. Sipper, M. et al.: The POE model of bio-inspired hardware systems, in Genetic Programming 1997 proceedings p510-511
4. Dawkins, R.: <u>Climbing Mount Improbable</u> W. W. Norton & Company NY,NY 1996
5. Wolpert, L. <u>The Triumph of the Embryo</u> Oxford University Press 1991
6. Nilsson, D., Pelger, S.: A pessimistic estimate of the time required for an eye to evolve, Proceedings of the Royal Society of London, B, 256, p53-8
7. Dennett, D.: <u>Darwin's Dangerous Idea</u> Simon & Schuster 1995
8. Horowitz, P., Winfield, H.: <u>The Art of Electronics</u> 2nd ed Cambridge Univ. Press 1989
9. Albus, J. S.: A new approach to manipulator control: the Cerebellar Model Articulation Controller (CMAC), Trans. of the ASME Sept.1975, p220-7
10. Roska, T., Chua, L.: The CNN Universal Machine, IEEE Trans. On Circuits and Systems-II: Analog and Digital Signal Processing, Vol 40, No 3, March 1993 p163-172

A "Spike Interval Information Coding" Representation for ATR's CAM-Brain Machine (CBM)

Michael Korkin[1], Norberto Eiji Nawa[2,3] and Hugo de Garis[2]

[1] Genobyte, Inc., 1503 Spruce Street, Suite 3, Boulder CO 80302, USA
korkin@genobyte.com, http://www.genobyte.com
[2] Evolutionary Systems Dept., ATR - Human Information Processing Research Laboratories, 2-2 Hikari-dai, Seika-cho, Soraku-gun, Kyoto 619-0288, Japan
{xnawa,degaris}@hip.atr.co.jp, http://www.hip.atr.co.jp/{~xnawa,~degaris}
[3] Furo-cho, Chikusa-ku, Bio-Electronics Lab., Graduate School of Engineering, Nagoya University, Nagoya 464-8603, Japan

Abstract. This paper reports on ongoing attempts to find an efficient and effective representation for the binary signaling of ATR's CAM-Brain Machine (CBM), using the so-called "CoDi-1Bit" model. The CBM is an Field Programmable Gate Array (FPGA) based hardware accelerator which updates 3D cellular automata (CA) cells at the rate of 100 billion a second, allowing a complete run of a genetic algorithm with tens of thousands of CA based neural net circuit growths and hardware compiled fitness evaluations, all in about 1 second. It is hoped that using such a device, it will become practical to evolve 10,000s of neural net modules and then to assemble them into humanly defined RAM based artificial brain architectures which can be run by the CBM in real time to control robots, e.g. a robot kitten. Before large numbers of modules can be assembled together, it is essential that the individual modules have a good functionality and evolvability. The "CoDi-1Bit" CA based neural network model uses 1 bit binary signaling, so a representation needs to be chosen based on this fact. This paper introduces and discusses the merits and demerits of a representation that we call "Spike Interval Information Coding" (SIIC). Simulation results using the SIIC representation method to evolve time dependent waveforms and simple functional modules are presented. The results indicate the suitability of the SIIC representation method to decode the bit streams generated by the CA based neural networks.

1 Introduction

The CAM-Brain Project ("CAM" stands for "Cellular Automata Machine" [10]) at the Advanced Telecommunications Research Laboratories (ATR) aims to construct a large-scale brain-like neural network system. If the project succeeds and our expectations are fulfilled, these "artificial brains" will have a large number of potential applications in several different fields, from "smart" domestic appliances to speech processing and robot control. Of course, up to now, this is pure speculation, and we admit there is still a long way to go before we can talk in more concrete terms. However, we believe that to realize a system that possesses a level of functionality and structural complexity similar to real biological brains, the most appropriate way, if not the only way, is to evolve them, as happened in nature.

The fundamental approach of the CAM-Brain Project is the growth/evolution of large-scale neural networks. Since the dawn of the Project [2], Cellular Automata (CA) have been chosen as the medium in which to grow the neural networks. CAs meet the requirements of generality and especially scalability, necessary for simulating large-scale systems. Moreover, the parallel nature of CAs allows their transposition into hardware, where higher speeds can be achieved. The CA based neural model initially used [3] suffered from an explosion of states and transition rules that blocked any attempt to implement it in hardware. Due to this problem, a new model called "CoDi" (from COllect and DIstribute) was proposed [4], greatly simplifying the system and for the first time allowing the implementation of the system in special hardware, namely XC6264 FPGAs (described below) which update the CA space at a rate of 100 billion cell a second, and should be able to perform a complete run of a genetic algorithm (with 10,000s of circuit growths and evaluations) in about a second.

Advances in hardware technology led to the development of devices called Field Programmable Gate-Arrays (FPGAs). FPGAs are hardware devices that can be reconfigured in run-time to perform different logic functions, wedding the flexibility of software with the speeds of hardware. This motivated the design/construction of a specific computer, called the CAM-Brain Machine (CBM) [6], for the evolution of neural networks under the CoDi model. The CBM will grow 16,000 neural network modules of roughly 10,000 3D CA cells each, updating 100 billion cells/second, a speedup of 500 times compared to the MIT machine "CAM8" [10] that the Project had been using previously to update CA cells quickly. The combination of biologically inspired algorithms with reprogrammable hardware devices comprises the field called "evolvable hardware", or "evolutionary electronics" (among other terms). In this field, several attempts of evolving behaviors in FPGA-based systems have been performed [5, 7, 9].

The experiments reported in this paper aim to tackle a fundamental issue of the project: how to interpret the signals that come from the neural network modules? The CoDi-1Bit model evolves neural networks whose signals that traverse the connections are digital, i.e. binary 0's and 1's. The question is, what kind of representation schemes should be used in order to extract useful information from the signals output by the neural networks modules. This paper proposes a "Spike Interval Informaction Coding" (SIIC) representation, based on results from the field of neuroscience reported in [8], which deals with the theory of information encoding in natural neural systems. Experiments evolving time dependent waveforms and simple functional modules were carried out. The results are encouraging so far, and indicate that the SIIC is a suitable representation scheme for the CoDi model.

2 The CoDi-1 Bit Cellular Automata Based Neural Net Model

This section gives an overview of the "CoDi" neural network model implemented in the CAM-Brain Machine hardware. The model is called "CoDi" due to the "COllect and DIstribute" nature of its neural signals. CoDi is a simplified CA-based neural network model developed at ATR in the summer of 1996 with two goals in mind. One was to make neural network functioning much simpler compared to the older CAM-Brain model developed in 1993 and 1994 [1, 2], so as to be able to implement the model directly in electronics and thus to evolve neural net modules at electronic speeds.

In order to evolve one neural network module, a population of modules is run through a genetic algorithm for several hundred generations. Each module evaluation consists of growing a new set of axonic and dendritic trees which interconnect the neurons in the 3D cellular automata space, then running the module to evaluate its performance (fitness).

The CoDi model [4] operates as a 3D cellular automata. Each cell is a cube which has six neighbor cells, one for each of its faces. By loading a different phenotype code into a cell, it can be reconfigured as a neuron, an axon, or a dendrite. Neurons are configurable on a coarser grid, namely one per block of 2*2*3 CA cells. In a neuron cell, five (of its six) connections are dendritic inputs, and one is an axonic output. An accumulator sums incoming signals and fires an output signal when a threshold is exceeded. Each of the inputs can perform an inhibitory or an excitatory function (depending on the neurons chromosome) and either adds to or subtracts from the accumulator. The neuron cell's output (axon) can be oriented in 6 different ways in the 3D space.

A dendrite cell also has maximum five inputs and one output, to collect signals from other cells. The incoming signals are passed to the output according to a given function. For instance, if the logic OR function is used, the output is active whenever at least one of the inputs is active. If an XOR function is used, the output is active when only a single input is active. Two or more active inputs block each other. The XOR dendrite is more plausible from the biological point of view. A similar phenomenon occurs in real dendrites in animals. An axon cell is the opposite of a dendrite. It has 1 input and maximum 5 outputs, and distributes signals to its neighbors. Before the growth begins, the module space consists of blank cells, which are used to grow new sets of dendritic and axonic trees during the growth phase. Blank cells perform no function in an evolved neural network. Figure 1 shows a schematic of the three types of cells.

As the growth starts, each neuron continuously sends growth signals to the surrounding blank cells, alternating between "grow dendrite" (sent to the neuron's dendritic connections) and "grow axon" (sent to the axonic connection). A blank cell which receives a growth signal becomes a dendrite cell, or an axon cell, and further propagates the growth signal, being continuously sent by a neuron, to other blank cells. The direction of the propagation is guided by the growth instructions attached to the cell. These local instructions indicate the directions that the growth signal should be propagated to and consists of a bit for each face of the cube cell. The growth signal is propagated to these directions whose corresponding the bit is set to 1 (except the direction where the signal comes from).

This mechanism allows the growth of a complex 3D system of branching dendritic and axonic trees, with each tree having one neuron cell associated with it. The trees can conduct signals between the neurons to perform complex spatio-temporal functions. The end-product of the growth phase is a phenotype bitstring which encodes the type and spatial orientation of each cell.

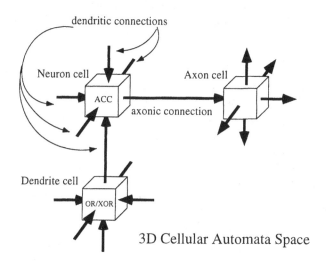

Fig. 1. Neuron, dendrite and axon cells in the CA space.

3 CAM-Brain Machine (CBM)

This section briefly describes the hardware implementation of the above CoDi-1Bit model, allowing CoDi neural net modules to be grown in hardware.

The CAM-Brain Machine (CBM) was especially designed to support the growth and signaling of neural networks built by the CoDi model. The CBM should fulfill the needs for high speeds, when simulating large-scale binary neural networks, a necessary condition when one is concerned with performing real-time control. The hardware core is implemented in XC6264 FPGA chips, in which the neural networks will actually grow. A host machine will provide the necessary interface to interact with the hardware core. It is planned that the CBM will be used to grow 16,000 neural networks modules, each with approximately 10,000 cells. The modules will be organized in architectures defined in advance, so several neural network modules will be interconnected to form a functional unity. For a complete description of the CBM, refer to [6].

4 The Spike Interval Information Coding (SIIC) Representation

The CoDi-1Bit model builds neural networks that output streams of 0's and 1's. However, it is arguable that if one wants to evolve modules of higher levels of complexity a more sophisticated scheme of information representation is necessary. Moreover, this representation scheme should be both efficient in conveying information and highly evolvable. In this context, the "Spike Interval Information Coding" (SIIC) representation was proposed.

In order to test the feasibility and the limitations of the new representation, several experiments were performed. The SIIC representation is inspired by the ideas presented in the book "Spikes: exploring the neural code", by Rieke et al. [8]. The book presents a novel hypothesis to explain how sensory signals are encoded in the action potentials or spikes that traverse neural systems. The classical theory of information encoded in neural signals was initiated by the work of E. D. Adrian, which strongly influenced the neuroscience community in the following years. Adrian's basic idea was that information about the intensity of the stimulus is encoded in the rate of spikes it generates. The rate of spikes can be calculated by counting the number of spikes in a fixed time window, following the beginning of the stimulus.

However, the work of Rieke et al. provides a new explanation. It claims that, if the classical theory is correct, the information rates would be inefficient. Moreover, it observes that "(...) single neurons can transmit large amounts of information, on the order of several bits per spike.". So the information should be contained not in the rates, which is a kind of averaging, but in the spikes trains themselves. The book [8] develops the theory and mathematical background to support its claims; since this theory is beyond the scope of this paper, it will not be developed any further. Here, we simply take part of it, namely, given a train of spikes, how should one decode it, in order to obtain useful information?

The procedure presented in the book has a different motivation from the one used in this paper. In the book, the aim is to find an appropriate method to construct an estimation of the analog stimulus signal from the spike train. Whereas, at the current stage of our research, the motivation is to find a method to extract information from the 0-1 bits that are output from the neural network modules evolved in the CBM. Considering that a 0 represents the absence of a spike, and a 1 represents the presence of a spike, the problem is quite similar. In earlier experiments, we found that the CoDi model evolved well for the cases of single fixed position outputs, however, we were unable to achieve satisfactory results nor a suitable representation method when using multiple non fixed position outputs (the so-called "unary" representation, where if N output surface neurons were firing at a given moment, the number N was being represented).

The spikes approach seems to be more suitable for the CoDi model in this sense, since it works with single fixed position outputs.

4.1 The SIIC Decoding Method

The procedure for decoding a spike train (the sequence of 0-1 bits), named SIIC, consists of convolving it with a special *convolution filter*, as shown in Fig. 2. The result obtained is called *estimated signal*, which is a time-dependent signal that is output from the neural network module to be evaluated by a fitness function or to be used in a subsequent process. The convolution process is discrete, since the convolution filter, the spikes trains and the results are discrete. The estimated signal is a digital representation of an analog signal, sampled at discrete time points, corresponding to the clock ticks.

The filter used in our experiments is shown in Figure 2. It is a digitized approximation of a curve presented in [8]. The SIIC decoding process is as follows:

1. Collect m bits from the output of the neural network module.
2. The estimated signal must have size $n \leq m$. Every point of the estimated signal is mapped to a point of the stream of bits collected in the previous step.
3. The filter and the bit stream are overlapped. The first point of the filter corresponds to the first bit of the stream and the same for the subsequent points.
4. To calculate the first point of the estimated signal, convolve the filter and the output stream, i.e. sum the values of the points of the filter where the correspondent point in the output stream is a 1.
5. The obtained value corresponds to the first value of the estimated signal. Then shift the filter to the next bit, so the first point of the filter corresponds to the second bit of the output stream, and repeat the procedure described in the previous step.
6. Repeat the procedure described in the two previous steps to calculate all the points of the estimated signal.

First, in order to test the potential of the SIIC, a simple experiment was undertaken. The objective was to check whether a simple GA would be able to design a stream of bits that, when decoded using the process described above, would generate a sinusoidal wave. Doing this, we could have a better idea of the potential of the SIIC itself without any interference from the CoDi model. The objective function to be approximated was a simple sine wave. The GA had to minimize the sum of the squares of the errors of each point of the estimated signal. The chromosomes had length of 336 bits, corresponding to the stream of bits to be decoded. The size of the stream of output bits does not necessarily have the same size as the estimated signal. It should only have at least the same size, or greater. The filter used is shown in Figure 2. The obtained sinusoidal wave after 2000 generations is shown in Figure 3 as the dashed line and the desired curve as the solid line.

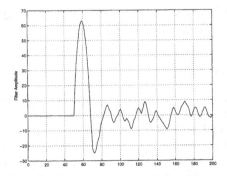

Fig. 2. Decoding filter for the spike trains.

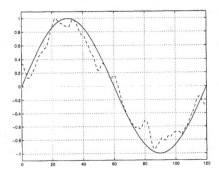

Fig. 3. Sinusoidal wave generated by a simple GA.

The result of this simple experiment shows that, at least for the case of evolving sinusoidal waves, the SIIC method performs well. The next step was to realize experiments integrated with the CoDi model of evolution of neural networks, which are described in the following sections.

4.2 CoDi and SIIC

The next step was to check the issue of the evolvability of the CoDi model, i.e. would the CoDi model be able to generate neural network modules whose binary outputs, decoded by the SIIC method, satisfy a given fitness function? Two experiments were performed. The first task was the generation of time dependent signals, sinusoidal waveforms, and the second, the construction of simple functional modules.

Sine Waves, Single and Multiple Periods First, we attempted to evolve a single period of a sine wave (Figure 4). The solid line is the target sine wave, and the dashed line is the obtained wave after 600 generations. The output stream collected from the neural network modules had length equal to 300 bits. The filter used in the convolution is the one shown in Figure 2, and the length of the estimated signal was 120 bits. The population had 30 chromosomes and the CA space size was 24*24*18. 48 input points (neurons) were chosen from one of the faces of the cubic CA space and were constantly firing in order to bring a high level of activity to the module. The output signal was collected from a point in the opposite face of the inputs. The estimated signal was normalized, so its discrete time points would have values of the same order of a unitary sine wave. The lower figure shows the actual stream of spikes that generated the sinusoidal waveform. The vertical black bars represent the stream of spikes through time. The absence of spikes in the start is due to the time the spikes take to traverse the CA space, since the input points and the output points were located in opposite faces. In these experiments, first the whole stream of bits was collected and then the SIIC method was applied to obtain the estimated signal. In a real application, in order to minimize the time delays, the decoding by the SIIC method should start as soon as the necessary number of bits is available. Also, for the sake of performance the filter should be shortened as much as possible. Assuming that the filter has a length of around 50 bits (this would be if the filter shown in figure 2 was truncated to the first positive and negative peaks, which are the relevant portions), the relative position in time of the stream of spikes and the estimated signals would be as shown in the following figures.

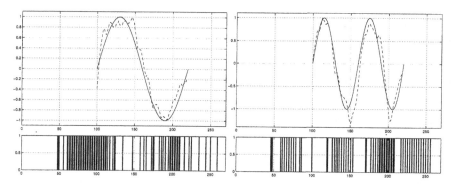

Fig. 4. Single period (left) and two periods of a sinusoidal wave (right) generated by the CoDi model and SIIC method. The lower figures show the actual spikes that generated the waveforms.

As an extension of the previous experiments, two, three and four periods of a sine wave were evolved. The length of the estimated signal was kept to 120 bits and the output stream to 300 bits. The obtained result are shown in Figure 4 and 5, where the dashed lines are the obtained waveforms and the solid lines the desired waves. It is interesting to notice the periodic nature of the spikes that generate periodic waveforms.

The number of periods were gradually increased in order to check the highest frequency of a sinusoidal waveform that could be evolved. Limited to a time

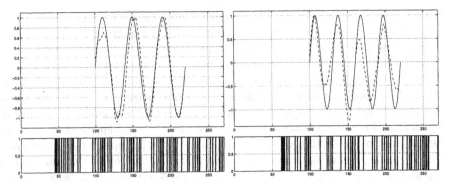

Fig. 5. Three periods (left) and four periods of a sinusoidal wave (right) generated by the CoDi model and SIIC method. The lower figures show the actual spikes that generated the waveforms.

window of 120 bits and using the filter shown in 2, the output considerably degraded for 8 periods of sine (Figure 6).

Sum of Sines and Cosines The last experiment of the first session was to evolve the following wave:

$$y(t) = \sin(2\pi t) + 0.3 \times \sin(4\pi t) + 0.6 \times \cos(6\pi t) + 0.2 \times \cos(14\pi t) \qquad (1)$$

The equation has no special meaning, serving merely as a more complex test case. The obtained result is shown in Figure 6.

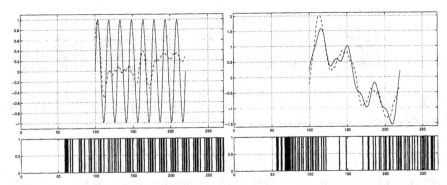

Fig. 6. Eight periods of a sinusoidal wave (left) and a sum of sines and cosines (right) generated by the CoDi model and SIIC method.

Overall, the results obtained in the experiments described above showed that, at least for the case of evolving sinusoidal waveforms, the SIIC method could achieve better results than the more simplistic schemes tried before (e.q. unitary representation, incremental up/down counter representation, Gray representation). The results indicate the ability of the CoDi-SIIC scheme to evolve neural network modules that can output time dependent signals at least this complex.

A Switchable Dual Function Module Once it became clear that a fixed position single point based representation was more evolvable than a stochastic multipoint representation (i.e. "unary"), our thoughts turned to the idea of trying to evolve a module whose behaviors could be placed under switchable control, i.e. a module with dual functionality, which could be switched from one behavior to the other depending on whether a "control" input was activated or not.

More specifically, the two fixed position input points "IN" and "SWITCH" of the input surface ($z = 0$), were chosen to be (8,8,0) and (16,16,0) for a cuboid of 24*24*18 3D CA cells, with a fixed output point of (11,12,9). If the output point was not an axon, fitness was defined to be zero.

Two experiments were run on the same module (same CoDi-1Bit circuit). In both experiments, the IN input was firing at every clock tick. In the first experiment, the SWITCH input was off for every clock tick. In the second experiment, the same chromosome was used to regrow the same module, and the SWITCH input fired for every clock tick. The module was evolved to give a very active output (lots of 1's) if SWITCH was off, but a low output (few 1's) if SWITCH was on. That is, SWITCH acts as an inhibitor.

The outputs in the two cases are shown below, firstly with SWITCH off, then on. Over 90 clock ticks, the first output had 42 more 1's than the second output.

SWITCH off	SWITCH on
000000000000000000000000000000	000000000000000000000000000000
000000011001111111111111111111	000000100001000001000001000001
111111111111111111111111111111	100000100001000001000001000001

The number of 1's in the two outputs were labeled as Sum_1 and Sum_2 respectively. The fitness definition finally settled upon, was:

$$IF(Sum_1 > Sum_2)$$
$$fitness = 10000 * (Sum_1 - Sum_2) + 0.001 * (Sum_1 + Sum_2)$$
$$IF(Sum_1 < Sum_2)$$
$$fitness = 100 * (Sum_2 - Sum_1) + 0.001 * (Sum_1 + Sum_2)$$

The term $0.001 * (Sum_1 + Sum_2)$ was used to encourage circuits to give nonzero output at the output point. The terms $100 * (Sum_2 - Sum_1)$ and $10000 * (Sum_1 - Sum_2)$ encouraged differences in the two outputs, with a strong preference for the first case to give more 1's in the output.

This result was very encouraging because it shows that controllable multifunction modules, at least like this switchable function, are evolvable with the

CoDi-1Bit model. Such modules will be very useful when the time comes to evolve modules to be placed in "artificial brain" architectures.

A Pattern Detector Module With a slight modification of the code used to evolve the above module, a pattern detector module was evolved which is capable of distinguishing between two square wave inputs, of 111000111000... and 11111000001111100000... In this case, no switch input was used. Two experiments were run. In the first, the input was the 6 clock tick cycle square wave input, applied at the fixed input point (8,8,0). In the second experiment, the circuit was regrown with the same chromosome and the 10 clock tick cycle square wave input was applied to the same fixed input point. The fitness definition was the same as above. Results are shown below. Over 90 clock ticks, the first output had 46 more 1's than the second output.

Input Wave 111000111000...   Input Wave 11111000001111100000...
Output: Output:

00000000000000000000000100110   00000000000000000000000000000010
11101111111110111111111111111111   00100010001000100010001000011000
11111111111111111111111111111111   10001000100010001000100010001000

Since the CoDi modules seem capable of evolving such detectors, it may be possible to evolve modules which are capable of detecting a specific phoneme analog input (e.g. the spike train (bit string) which when convolved with a particular convolution function gives the time dependent analog signal). In a manner similar to the above, one could input the signal in the first experiment, and a random signal in the second, and evolve the phoneme detector. Maybe one could evolve a set of detectors, one for each phoneme, for example.

5 Conclusions

This paper presented a "Spike Interval Information Coding" (SIIC) representation to decode the binary outputs of the neural network modules built by the CoDi model and to be run in the CAM-Brain Machine. The results obtained indicate that the SIIC method and the CoDi model could evolve time dependent waveforms, as simple as sinusoids, with fair levels of accuracy. The decoding process uses a filter, that is convolved with a spike train to generate the estimated signal. In the experiments described in this paper, the filter was defined in advance and remained unchanged through the evolutionary process. Since the shape of the filter greatly influences the evolvability of the system, we feel that it could be evolved too. Moreover, further clarifications about the decoding method itself are necessary.

The evolution of a switchable function, although a simple one, indicates the possibility of building more complex controllable functions. Until now, all the experiments were performed with constantly firing inputs, in order to bring a high level of activity to the modules. Controllable functions are necessary in any context. The development of a methodology to evolve controllable functions is necessary, in order to allow the evolution of more complex functional modules.

Ultimately, the aim of the CAM-Brain Project is to make "artificial brains", using the CBM to evolve large numbers (10,000) of CoDi modules very quickly, and then assemble them into humanly defined artificial brain architectures, such as a controller for a robot kitten. Future research will aim the evolution of multiple neural network systems, namely, which modules to evolve and in what kind of architectures to assemble these modules in order to obtain the desired behaviors.

References

1. Hugo de Garis. Artificial life: Growing an artificial brain with a million neural net modules inside a trillion cell cellular automata machine. In *4th. Int.Symposium on Micro Machine and Human Science*, October 1993.
2. Hugo de Garis. The CAM-Brain project: The genetic programming of a billion neuron which grows/evolves at electronics speeds in a cellular automata machine. In *Proceedings ALIFE IV, Fourth Int.Conf.on Artificial Life*, July 1994.
3. Hugo de Garis. CAM-Brain: ATR's billion neuron artificial brain project: A three year progress report. In *Proceedings of AROB'96, Int.Conf.on Artificial Life and Robotics*, February 1996.
4. Felix Gers and Hugo de Garis. CAM-Brain: A new model for ATR's cellular automata based artificial brain project. In *Proceedings of ICES'96, Int.Conf.on Evolvable Systems*, October 1996.
5. Didier Keymeulen, Marc Durantez, Kenji Konaka, Yasuo Kuniyoshi, and Tetsuya Higuchi. A evolutionary robot navigation system using a gate-level evolvable hardware. In *Proceedings of ICES'96, Int.Conf.on Evolvable Systems*, October 1996.
6. Michael Korkin, Hugo de Garis, Felix Gers, and Hitoshi Hemmi. CBM (CAM-Brain Machine): A hardware tool which evolves a neural net module in a fraction of a second and runs a million neuron artificial brain in real time. In John R. Koza, Kalyanmoy Deb, Marco Dorigo, David B. Fogel, Max Garzon, Hitoshi Iba, and Rick L. Riolo, editors, *Genetic Programming 1997: Proceedings of the Second Annual Conference*, July 1997.
7. John R. Koza, Forrest H. Bennett III, David Andre, and Martin A. Keane. Reuse, parameterized reuse, and hierarchical reuse of substructures in evolving electrical circuits using genetic programming. In *Proceedings of ICES'96, Int.Conf.on Evolvable Systems*, October 1996.
8. Fred Rieke, David Warland, Rob de Ruyter van Steveninck, and William Bialek. *Spikes: exploring the neural code*. MIT Press/Bradford Books, Cambridge, MA, 1997.
9. Adrian Thompson. An evolved circuit, intrinsic in silicon, entwined with physics. In *Proceedings of ICES'96, Int.Conf.on Evolvable Systems*, October 1996.
10. T. Toffoli and N. Margolus. *Cellular Automata Machines*. MIT Press, Cambridge, MA, 1987.

Learning in Genetic Algorithms

Erol Gelenbe
Department of Electrical
and Computer Engineering
Duke University
Durham, N.C. 27708-0291
erol@ee.duke.edu

Abstract Learning in artificial neural networks is often cast as the problem of "teaching" a set of stimulus-response (or input-output) pairs to an appropriate mathematical model which abstracts certain known properties of neural networks. A paradigm which has been developed independently of neural network models are genetic algorithms (GA). In this paper we introduce a mathematical framework concerning the manner in which genetic algorithms can learn, and show that gradient descent can be used in this frameork as well. In order to develop this theory, we use a class of stochastic genetic algorithms (GA) based on a population of chromosomes with mutation and crossover, as well as fitness, which we have described earlier in [18].

1. Introduction

Genetic Algorithms (GA) introdced by Holland [2] were inspired by the manner in which a set of chromosomes of different types undergoes transformations (see for instance Muehlenbein [6] and Goldberg [7]). The basic primitives associated with these algorithms are an initialization of the population of chromosomes, transformation operations, such as crossover -- which replaces a subset of the chromosomes by a new subset, mutation -- which simply replaces some chromosome by another, and fitness -- which determines how each type of chromosome will survive or be duplicated. Most studies and applications of genetic algorithms use variants of these operations. For instance crossover is typically limited to replacing two chromosomes by two other chromosomes. Fitness may determine simply the probability that a chromosome within a particular population is allowed to appear in the next generation. There is a vast literature on genetic algorithms and their applications including [1, 2, 3, 6,

7, 10, 11, 15, 16, 17]. One aspect of GAs which has not been significantly developed concerns mathematical tractability, and GA applications are mainly based on Montecarlo simulation. Yet mathematical tools, if further developed, could make it easier to design GAs with a specific functional behavior, and also to test GAs more rapidly that what would be possible via a lengthy Montecarlo simulation.

We model a finite set of _types_ of chromosomes simply denoted by the integers {1, ... , n}. At a given instant of time, our system will contain a _finite but unbounded number of chromosomes_ of any type, and these numbers will be denoted by some vector $k = (k_1,...,k_n)$. Thus each k_i is some non-negative integer representing the number of chromosomes of type i which are currently present. In more precise terms, the state of the system at time t is the vector $k(t) = (k_1(t), ... ,k_n(t))$ whose value is $k = (k_1,...,k_n)$ with $k_i \geq 0$, i=1,...,n. In this model the _total population size of chromosomes_ at any time t ,

$\Sigma^n_{i=1} k_i(t)$, is _unbounded_, while the _total number of types of chromosomes is finite_ (n). This system evolves under the following set of elementary rules :

(a) A chromosome of type i can be spontaneously created (_birth_),
(b) A chromosome of type i can survive or it may not survive ; this relates to the _fitness_ criterion of GA models,
(c) A chromosome of type i may be transformed into a chromosome of type j (_mutation_),
(d) A chromosome of type i may combine with a chromosome of type j (here j need not necessarily be different from i) to produce a chromosome of type k (_simple crossover_). In that case, the two originating chromosomes (of types i and j) disappear and are replaced by the type k chromosome. Note that we allow i and j to be the same, and indeed we may also have that k is the same as i or j (or both),
(e) A chromosome of type i may combine with a chromosome of type j, and both may be destroyed; we will call this the _destruction_ (or joint destruction) operation.

Note that this corresponds to a model in which _two_ individuals of any generation can only give rise to _a single_ individual in the next generation (Rule (d)). While Rule (e) appears to be novel in genetic algorith models but is plausible if one considers combinations of chromosomes which are not viable. Thus if we did not have Rule (a) the population of chromosomes would eventually become extinct, even if we did not have Rule (b), i.e. death. Therefore Rule (a) is needed to cover cases where we wish the population either to tend to some probabilistic steady-state or to grow indefinitiely.

Now that we have stated what is allowed to happen, we also have to indicate how these things happen and when. All events are considered over infinitesimal time periods Δt, and therefore occur within infinitesimally small intervals of time $[t, t + \Delta t)$. Due to Rule (a), the probability that one chromosome of type i is created in the interval $[t, t + \Delta t)$ is denoted by $\Lambda_i \Delta t$. The probability that such an event will not occur is $(1 - \Lambda_i \Delta t)$. Each type of chromosome has a specfic level of activity, describing how fast it gets involved in changes. We introduce this idea into our GA model, since we think it may be of use in applications and because it may also very well correspond to biological reality. This is quantified by an activity rate $a_i > 0$ which may be different for each type i of chromosome. Mutation, fitness and crossover are related to the activity rate as described below. Thus chromosomes with higher activity rates will tend to mutate or crossover more frequently. Clearly, this activity rate may be modulated by other effects such as the fitness, or the mutation probability. Thus the probability that the chromosome "is not fit", as well as the activity rate, will determine its capabilities with respect to survival, mutation and crossover.

The probability that a chromosome of type i is fit will be denoted by f_i. In the interval $[t, t + \Delta t)$ it will die and disappear from the system with probability $(1- f_i)a_i \Delta t$. On the other hand, a chromosome of type i will proceed to mutate or carry out a crossover only if it is fit and survives and this occurs with probability $f_i a_i \Delta t$. Notice that the activity rate also influences the destruction or death rate of chromosomes which are not fit.

Concerning Rule (c), the probability that a chromosome of type i mutates into one of type j is m_{ij}. Therefore in the interval $[t, t + \Delta t)$ the probability that it mutates into one of type j is is $f_i a_i m_{ij} \Delta t$, while the probability that this does not happen is simply $(1 - f_i a_i m_{ij} \Delta t)$. Clearly, for this to occur at all, there must be some chromosome of type i in the system. In the next section we will show how this definitions of fitness relates to the usual definition [3, 4, 7, 8] which is a metric relative to the population size as a whole, rather than to a single individual. Concerning Rule (d) for crossover, <u>the probability that a chromosome of type i combines with one of type j, to yield a chromosome of type v</u> is c_{ijv}. However for this to occur, the chromosome of type i must be activated and be fit as well. Thus, the probability that it will occur in the interval $[t , t + \Delta t)$ is $f_i a_i c_{ijv} \Delta t$, while the probability that this does not happen is $(1 - f_i a_i c_{ijv} \Delta t)$. Finally, for Rule (e) for mutual destruction we have that <u>the probability that a chromosome of type i combines with one of type j, to mutually destroy</u> is d_{ij} . For this to occur, the chromosome of type i must be activated and be fit as well. The probability that it will occur in the interval $[t , t + \Delta t)$ is $f_i a_i d_{ij} \Delta t$, while the probability that this does not happen is $(1 - f_i a_i d_{ij} \Delta t)$. Of course, a fit chromosome may also neither mutate nor take the initiative of combining with another chromosome for a crossover, and choose to remain for the time being in the same type with probability s_i

so that $\Sigma_{j=1}^{n} (m_{ij} + d_{ij}) + \Sigma_{j,v=1}^{n} c_{ijv} + s_i = 1$. The term $f_i a_i$ which will later appear in many expressions related to this model can be intuitively be interpreted as the mathematical counterpart of the type i chromosome being "fit and active". Note that because we introduce the notion of activation rate, and because we require a chromosome to be fit in order to mutate or do a crossover, our model distinguishes the case of a type i chromosome taking the initiative of combining with a type j chromosome, from the case where the initiative begins at a type i chromosome. These two cases can be made equivalent by setting $f_i a_i c_{ijv} = f_j a_j c_{jiv}$. The parameters Λ_i , f_i , a_i , m_{ij} , d_{ij} , c_{ijv} , s_i characterize the manner in which creation, fitness, activity, mutation and crossover rules are being applied by the GA. Note that if the crossover operation were not present in ths model, we would be merely dealing with a generalized birth and death process whose properties and solutions are well known [4,5]. The crossover operation, even in its limited present form, distinguishes the model discussed in this paper from these conventional multidimensional birth and death models. In the sequel we will present an algorithm based on gradient descent for selecting the fitness parameter for a GA which is being designed to solve an optimization problem. This is similar to learning algorithms for recurrent random neural networks [14]. Once a GA is set up with a given ensemble of parameters, it is then typically run as a Montecarlo (or sometimes deterministic) simulation to evaluate its behavior and effect. Though short term evolutions and behavior is of interest, statistical significance is generally only available if one considers the long run or steady-state behavior of the model. We will presently show that for the GA model we have defined above, the long run behavior need not be deduced from a simulation, since it is directly available from the Theorem we prove below. Let us denote $p(k,t)$ = Prob $[k(t) =k]$, i.e. the probability that the GA is in some state k at time t, assuming some arbitrary initial condition. Clearly, if we know $p(k,t)$ for all t, then we have full knowledge of system behavior. For large enough t, however, the stationary or long term behavior can provide enough information. This is precisely what we will be dealing with here, in that we will provide formulas to easily calculate the steady-state probabilities for $k = (k_1,...,k_n)$:

$$p(k) \equiv \lim_{t->\infty} P[k(t)=k] . \tag{1}$$

Theorem 1 The steady-state probability characterizing completely the GAs behavior is independent of the intial state configuration $k(0)$. It is given by the composite geometric distribution :

$$p(k_1,...,k_n) = \Pi^n_{i=1} (1 - \rho_i) \rho_i^{k_i} , \tag{2}$$

where the ρ_i , i=1, ... , n are the unique solution to the following system of equations :

$$\rho_i = \frac{ \{ \Lambda_i + \Sigma^n_{j=1} \rho_j f_j a_j m_{ji} + \Sigma^n_{u,v=1} \rho_u f_u a_u \rho_v c_{uvi} \} }{ \{ a_i(1- f_i s_i) + \Sigma^n_{j,v=1} \rho_j f_j a_j c_{jiv} + \Sigma^n_j \rho_j f_j a_j d_{ji} \} } \tag{3}$$

This expression has an intuitive explanation. The denominator contains all those terms which result in the <u>creation</u> of chromosomes of type i, while the denominator contains the terms which result in the <u>depletion</u> of type i chromosomes. Expression (2) is a remarkable and unusual result since it states that the joint probability of system state can be written as the product of the marginal probabilities related to each individual population of chromosomes, and that the distribution of the number of chromsomes of each type obeys a geometric distribution.

Remark 1 (Relative Fitness) We will relate the parameter f_i to another measures of fitness : it is the <u>relative fitness</u> of the population of Type i relative to the fitness of the population as a whole and we represent it by F_i . From (3) we can easily show that the average number of individuals of Type i is given by : $N_i = \rho_i /(1 - \rho_i)$, Similarly, the total average number of individuals can be calculated as: $N = \Sigma_{i=1}^{n} \rho_i /(1 - \rho_i)$. Then the average fitness of the population is $f = \Sigma_{j=1}^{n} f_j N_j/N$ and the relative fitness of the chromosomes of Type i is $F_i = f_i /f$.

Remark 2 (Fictitious chromosome types and Rule (e)) This second remark will simplify the subsequent analysis. Let us add a "fictitious" chromosome type, call it type I, so that I is not in the set $\{1, ... , n\}$. This will be a type of chromosome into which one can <u>enter</u> only by crossover so that :

$$c_{ij\,I} \text{ and } c_{ji\,I} \text{ may be non-zero for } i,j = 1, ... , n .$$

However we are not be able to <u>leave</u> this chromsome type in any way, i.e. : $m_{i\,I} = 0$, $m_{Ii} = 0$, $c_{Iji} = 0$, $c_{jIi} = 0$, for all $i,j = 1 , ... , n$ and $\Lambda_I = 0$, so that from (3) we necessarily have $\rho_I = 0$, since all terms in the numerator are null. However if we set $d_{ji} = c_{ji\,I}$ for each j,i we end up with :

$$\rho_i = \frac{\{ \Lambda_i + \Sigma_{j=1}^{n} \rho_j f_j a_j m_{ji} + \Sigma_{u,v=1}^{n} \rho_u f_u a_u \rho_v c_{uvi} \}}{\{ a_i(1- f_i s_i) + \Sigma_{j,v=1}^{n} \rho_j f_j a_j c_{jiv} + \Sigma_{j}^{n} \rho_j f_j a_j c_{ji\,I} \}}$$

or

$$\rho_i = \frac{\{ \Lambda_i + \Sigma^n_{j=1} \, \rho_j f_j a_j \, m_{ji} + \Sigma^n_{u,v=1} \, \rho_u f_u a_u \, \rho_v c_{uvi} \}}{\{ a_i(1 - f_i \, s_i) + \Sigma^{n+1}_{j,v=1} \, \rho_j \, f_j a_j \, c_{jiv} \}}$$

where we use the (n+1)-st chromosome type to denote the type I chromosome. This simple development indicates that the model with an additional fictitious chromosome type but no chromosome destruction (Rule (e)) is equivalent to the model with a Rule (e). Therefore the proof of Theorem 1 will be carried out as if Rule (e) did not exist, without loss of generality.

2. Designing a GA for a specific problem

In this section we turn to the issue of the choice of parameters for a specific optimization problem. Suppose that this problem's solution can be expressed by a set of binary variables X_i, i=1, ..., n which we encode by using the ρ_i so that $\rho_i = P[X_i = 1]$. Suppose also that we are seeking a solution $\{X_1, ..., X_n\}$ which minimizes some cost function $C(X_1, ..., X_n)$. The coresponding cost function in terms of the GA will be $C(\rho_1, ..., \rho_n)$. Let us limit first our attention to the choice of the fitness parameter f_k, k =1,...,n although the same principles can be used for any other of the parameters of the GA model we have introduced. Then we suggest that this choice be made by using the standard gradient algorithm because it will lead to a local minimum of C' with respect to the set of parameters under consideration:

$$f_k \Leftarrow f_k - \eta \, \partial C / \partial f_k \qquad (6)$$

for some positive constant learning rate η . We need to compute the derivative term, for which we use the chain rule: $\partial C / \partial f_k = \Sigma_i [\partial C / \partial \rho_i] \, \partial \rho_i / \partial f_k$. Clearly the issue then is to be able to compute $\partial \rho_i / \partial f_k$ for any i and k. We shall see below that this can be achieved simply by solving a linear system of equations (8) at each step of the iteration . To simplify matters, assume that $m_{ii} = c_{iiv} = d_{ii} = 0$ so that we eliminate all unnecessary transformations of the set of chromosomes. Let us rewrite (3) as $\rho_i = N_i / D_i$, where

$$N_i = \{ \Lambda_i + \Sigma^n_{j=1} \rho_j f_j a_j m_{ji} + \Sigma^n_{u,v=1} \rho_u f_u a_u \rho_v c_{uvi} \}$$

$$D_i = \{ a_i(1 - f_i s_i) + \Sigma^n_{j,v=1} \rho_j f_j a_j c_{jiv} + \Sigma^n_j \rho_j f_j a_j d_{ji} \}$$

We can then write:

$$\rho'_i = \partial \rho_i / \partial f_k = D_i^{-1} \{ N'_i - \rho_i D'_i \} \tag{7}$$

where N'_i and D'_i are the partial derivatives of the numerator and denominator of (3) with respect to f_k . We then have :

$$N'_i = \Sigma^n_{j=1} \rho'_j f_j a_j m_{ji} + \rho_k a_k m_{ki} + \Sigma^n_{v=1} \rho_k a_k \rho_v c_{kvi}$$

$$+ \Sigma^n_{u,v=1} (\rho'_u \rho_v + \rho_u \rho'_v) f_u a_u c_{uvi}$$

$$D'_i = \Sigma^n_{j,v=1} \rho'_j f_j a_j c_{jiv} + \Sigma^n_{v=1} \rho_k a_k c_{kiv} + \Sigma^n_{j=1} \rho'_j f_j a_j d_{ji}$$

$$+ \rho_k a_k d_{ki} - a_i s_i 1[i=k]$$

As a consequence, we rearrange terms in the above expressions to write:

$$N'_i - \rho_i D'_i = \Sigma^n_{j=1} \rho'_j f_j a_j [m_{ji} + \Sigma^n_{v=1} \rho_v c_{jvi} - \rho_i d_{ji}$$

$$- \rho_i \Sigma^n_{v=1} c_{jiv}] + \Sigma^n_{j=1} \Sigma^n_{u=1} \rho'_j \rho_u f_u a_u c_{uji}$$

$$- \rho_i \Sigma^n_{v=1} \rho_k a_k c_{kiv} - \rho_i \rho_k a_k d_{ki} + \rho_i a_i s_i 1[i=k]$$

Let us denote :

$$M_{ij} = D_j^{-1} f_i a_i \{ m_{ij} + \Sigma^n_{v=1} [\rho_v c_{ivj} - \rho_j c_{ijv}] - \rho_j d_{ij} \}$$

$$+ D_j^{-1} \Sigma^n_{u=1} \rho_u f_u a_u c_{uij}$$

We then express the derivative ρ' of the vector $\rho = (\rho_1 , \dots , \rho_n)$ as

$$\rho' = \rho' M + d(k) , \text{ or } \rho' = d(k)[I - M]^{-1} \tag{8}$$

where the matrix $M = [M_{ij}]$ is given above and $d(k) = (d_1(k), \ldots, d_n(k))$ is given by :

$$d_i(k) = -\rho_i D_i^{-1} \{\Sigma^n_{v=1} \rho_k a_k c_{kiv} + \rho_k a_k d_{ki}- a_i s_i 1[i=k]\} \quad (9)$$

It is interesting to see however that M does not depend on k, whereas $d(k)$ does of course. Thus $[I - M]^{-1}$ needs to be computed only once for all the f_k at each step of the gradient algorithm.

Remark 3 For each value of ρ, M and hence $[I - M]^{-1}$ can be calculated just once to obtain the derivatives of ρ_i the with respect to all of the f_k.

Remark 4 We now summarize the proposed gradient algorithm for the fitness parameters. For each iteration step (6) :

1. First calculate the ρ_i from the previous values of the f_i, for i=1,...,n .
2. Calculate the matrix M from ρ and the previous values of the f_i. Obtain the inverse $[I - M]^{-1}$. Note that the same M will be used for updating all the f_k.
3. Calculate d^k for each k =1, ... , n . Use (8) to calculate $\partial \rho_i / \partial f_k$ for each k.
4. Use (6), (7) and $\rho' = d(k)[I - M]^{-1}$ to calculate the new values of the f_k.
5. Stop if the difference between the previous and the current value of C' is sufficiently small.

Remark 5 Once the GA's parameters have been selected in this manner, the GA's solution is considered to provide a solution for the optimization problem at hand.
Let us now consider the problem of choosing the crossover parameters using a gradient algorithm. As in the previous section, we are suggesting that the crossover parameters be calculated on the basis of a gradient descent, where C is the cost function that the GA is supposed to minimize:

$$c_{uvk} \Leftarrow c_{uvk} - \gamma \, \partial C / \partial c_{uvk}$$

where γ is a small positice constant. We will have to compute the derivative of C with respect to each of the c_{uvk}, which will reduce in fact to calculating $\partial \rho_i / \partial c_{uvk}$, $i=1,\ldots,n$:

$$\partial C / \partial c_{uvk} = \Sigma^n_{i=1} \, \partial C / \partial \rho_i \cdot \partial \rho_i / \partial c_{uvk} \qquad (10)$$

Let $\rho^*_i = \partial \rho_i / \partial c_{uvk} = D_i^{-1} \{ N^*_i - \rho_i D^*_i \}$, where N^*_i, D^*_i are the partial derivatives of the numerator and denominator of (3) with respect to c_{uvk}. We have :

$$N^*_i = \Sigma^n_{j=1} \rho^*_j f_j a_j m_{ji} + \Sigma^n_{u,v=1} \rho^*_u f_u a_u \rho_v c_{uvi}$$
$$+ 1[i=k] \, \rho_u f_u a_u \rho_v$$

$$D^*_i = \Sigma^n_{j,v=1} \rho^*_j f_j a_j c_{jiv} + \Sigma^n_j \rho^*_j f_j a_j d_{ji} + 1[i=v] \, \rho_u f_u a_u$$

which leads to

$$\rho^*_i = D_i^{-1} \{ \Sigma^n_{j=1} \rho^*_j f_j a_j [m_{ji} - \rho_i d_{ji} + \Sigma^n_{v=1} (\rho_v c_{jvi} - \rho_i c_{jiv})]$$
$$+ 1[i=k] \, \rho_u f_u a_u \rho_v - \rho_i 1[i=v] \, \rho_u f_u a_u \}.$$

Again, we may simply express the derivatives ρ^* of the vector ρ with respect to c_{uvk} as te solution of a linear system :

$$\rho^* = \rho^* V + g(uvk) , \text{ or } \rho^* = g(uvk) [I - V]^{-1} , \qquad (11)$$

where the n by n matrix $V = [V_{ij}]$ is given by :

$$V_{ij} = D_j^{-1} f_i a_i [m_{ij} - \rho_j d_{ij} + \Sigma^n_{v=1} (\rho_v c_{ivj} - \rho_j c_{ijv})]$$

$$= M_{ij} - D_j^{-1} \Sigma^n_{u=1} \rho_u f_u a_u c_{uij}$$

and does not depend on the specific choice of the parameter c_{uvk} with respect to which we are conducting the gradient descent. This is an important remark because at each step of the gradient descent, one single matrix inversion $[I - V]^{-1}$ can be calculated for all n^3 different crossover parameters when these parameters are being updated with the gradient algorithm. Finally, the vector $g(uvk) = (g_1(uvk), \dots, g_n(uvk))$ is given by : $g_i(uvk) = D_i^{-1} \{1[i=k] \rho_u f_u a_u \rho_v - \rho_i 1[i=v] \rho_u f_u a_u\}$.

Remark 4 We now summarize the proposed gradient algorithm for the fitness parameters. For each iteration step of the gradient algorithm :

1. First calculate the vecot ρ from the previous values of the c_{uvk} , for $i=1,\dots,n$.
2. Calculate the matrix V from ρ and the previous values of the c_{uvk} . Obtain the inverse $[I - V]^{-1}$. Note that the same V will be used for updating all the c_{uvk} .
3. Calculate $g(uvk)$. Use (11) to calculate $\partial \rho_i / \partial c_{uvk}$ for each (u,v,k) .
4. Use (10), (11) and $\rho^* = g(uvk) [I - V]^{-1}$ obtained in Step 3 to calculate the new values of the c_{uvk} .
5. Stop if the difference between the previous and the current value of C is sufficiently small. Otherwise repeat these steps.

References

1. Bharucha-Reid, A.T. "Elements of the Theory of Markov Processes and their Applications," McGraw-Hill, New York (1960).
2. Kernighan, B. and Lin, S. "An efficient heuristic procedure for partitioning graphs", The Bell System Technical Journal, (February 1970).
3. Holland, J. "Adaptation in Natural and Artificial Systems," The University of Michigan Press, Ann Arbor, Michigan (1975).
4. De Jong, Kenneth A. "An Analysis of the Behaviour of a Class of Genetic Adaptive Systems", Doctoral Thesis, Department of Computer and Communication Sciences, University of Michigan, Ann Arbor (1975).
5. Gelenbe, E. and Mitrani, I. "Analysis and Synthesis of Computer Systems", Academic Press, New York and London (1980).

6. Gelenbe, E. and Pujolle, G. "Introduction to Networks of Queues", J. Wiley and Sons, New York and London (1988), 2nd Edition (1998).

7. Muehlenbein, H., Schleuter, G. and Kramm, D. "Evolution algorithms in combinatorial optimization", Parallel Computing, Vol 7, No 2, (1988).

8. Goldberg, D. "Genetic Algorithms in Search, Optimization and Machine Learning", Addison Wesley, New York (1989).

9. Gelenbe, E. "Multiprocessor Performance". J. Wiley and Sons, New York and London (1989).

10. Talbi, E. and Bessière, P. "Un algorithme génétique massivement parallèle pour le problème de partitionement de graphes". Rapport de recherche. Laboratoire de Génie Informatique de Grenoble (1991).

11. Vose, M.D. and Liepins, G.E. "Punctuated equilibria in genetic search", Complex Systems, Vol. 5, (1991) 31-44.

12. Vose, M.D. "Formalizing genetic algorithms", Technical Report (CS-91-127), Department of Computer Science, The University of Tennessee (1991).

13. Vose, M.D. "Modeling simple genetic algorithms", in Whitley, L. Darrell ed.), "Foundations of Genetic Algorithms 2", Morgan Kaufmann, San Mateo (1993).

14. Gelenbe, E. "Learning in the recurrent random neural network", Neural Computation, Vol. 5, No. 1, (1993) 154-164.

15. Medhi, J. "Stochastic Processes," 2nd Edition, Wiley Eastern Ltd, New Delhi (1994).

16. Gunther, R. "Convergence analysis of canonical genetic algorithms", IEEE Transactions on Neural Networks, Vol. 5, No. 1, (1994) 96-101.

17. Xiaofeng Qui, Palmieri, F. "Theoretical analysis of evolutionary algorithms with infinite population size, Parts I and II", IEEE Transactions on Neural Networks, Vol. 5, No. 1, 102-129, 1994.

18. Gelenbe, E. "A class of genetic algorithms with analytical solution", Robotics and Autonomous Systems, Vol. 22, (1997) 59-64.

Back–Propagation Learning of Autonomous Behavior:
A Mobile Robot Khepera Took a Lesson from the Future Consequences

Kazuyuki Murase, Takaharu Wakida, Ryoichi Odagiri,
Wei Yu, Hirotaka Akita, and Tatsuya Asai

Department of Information Science, Fukui University,
3-9-1 Bunkyo, Fukui 910-8507, Japan.
{murase,wacky,ryo,wei,akita,asai}@synapse.fuis.fukui-u.ac.jp

Abstract. A modified back-propagation (BP) algorithm for the development of autonomous robots was proposed, and applied to a real mobile robot Khepera. Coefficients of a multi-layered neural network (NN), that determined the sensor-motor reflex of the robot, were first set randomly, and the robot was allowed to behave in an environment for some time. Sets of the sensor-motor values were continuously sampled during the free-moving period, and each set was evaluated by the behavior that occurred after the sampling by using an evaluation function. The set obtained the highest score was selected for each sensor pattern, and used to train the NN with BP. By repeating the above procedures, the robot obtained the adaptive behavior for the given environment in accordance with the evaluation function. The time needed for Khepera to acquire the ability of navigation was approximately one tenth of the conventional genetic evolution.

1 Introduction

Multi-layered neural networks (NN) have been widely used as a control scheme that generated motor outputs from sensor values in autonomous robots. Evolution with the genetic algorithm (GA) or genetic programming (GP) was a standard, and probably only one practical method that has been used to find best coefficients of the NN. However, a considerably long period of time was usually needed for the evolution, even for the acquisition of simple behavior. Therefore, some people have been suspicious for its practical and/or industrial application, where a much larger number of sensors were expected to generate outputs for a variety of motors and actuators.

The Back-propagation (BP) algorithm has been the most powerful tool to quickly obtain a NN that gives an input-output relation close to the training data sets. In case of constructing a robot autonomously behaving in environment, however, it has been difficult to think of getting the "training data", because autonomous behavior should have no template and only be obtained by gradual improvement in the environment.

In this study, we attempted to obtain the sets of training data through the behavior of robots in environment, and the NN was trained with the BP. As the matter of course, the sets were renewed through the development of autonomous behavior. We applied the method to Khepera, a small mobile robot, and let it learn (rather than evolve) to navigate in an environment with obstacles. The procedure and results of experiments were described below.

2   Evaluation Beyond the Causality

Let's think that a navigation robot was driven with a NN that had reasonably good sensor-motor reflexes so that it could navigate in environment. If one sampled the sensor values and motor movements at every second, for example, each sampled data set of the sensor and motor values represented a reflex behavior of the robot at the time of sampling. Whether or not the reflex behavior caused a good consequence would be found later on in the future. That is, whether a current behavior of turning-to-right, for example, might or might not cause a good consequence such as collision-to-wall would become apparent after a few seconds later. Therefore, the reflex should be evaluated after the particular reflex behavior was taken.

A reflex that occurred at a moment could be evaluated with the same method as the standard GA process by using an evaluation function. If the robot could navigate without collision for a certain period of time immediately after the reflex, the reflex should be given a good score. But, if the robot got into a trouble after the reflex, for example, touching a wall so that the navigation speed became slower, the reflex should be scored lower. After evaluating all reflexes, the reflexes could be sorted in accordance with the scores, and a set of good reflexes could be selected from the top of the list.

3   Back-Prop Learning

The selected set of good reflexes could be used to train the NN further so that the robot would navigate better in the environment. Since the set consisted of sensor and motor values, the standard BP algorithm could be used for the training. The newly trained NN would then replace the old one. The robot with the new NN could be again tested in the environment according to the above procedure. The reflexes during the behavior were sampled, evaluated, and selected for a new set of training data. By repeating this procedure, one could expect that the robot would acquire better behavior in accordance with the evaluation function.

Two problems could be expected to arise, namely the over-learning problem and the lack-of-diversity problem, both of which would correlate each other. The former was, we meant here, the problem that the NN learned some patterns but not others. Certain reflex patterns might appear too frequently in the training data so that the other necessary and important reflexes would be forgotten. The latter was that the new innovative behavior would not appear because sensory patterns might converge within a certain limit.

We took three strategies in order to overcome these problems. Firstly, we selected one best sample for each sensory pattern occurred so that the training data contained a variety of reflexes evenly. Secondly, the BP learning for a group of sensor-motor data sets was terminated before the error became minimum. Finally, we considered to add noise to the training data.

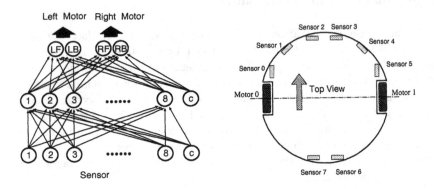

Fig. 1. The neural network

4   Implementation into Khepera

A three-layered NN, having 8 proximal sensor inputs, 2 motor outputs and 8 association units, was constructed in the CPU of the Khepera (Fig.1). The coefficients, that initially set randomly, were down loaded from the external workstation, and the robot was allowed to move in a test field with obstacles for 50s (Fig.2). The input and output values of the NN, as well as the actual number of rotation of the motors, were sampled at every 100ms. After the period of data collection, the data was transmitted to the host workstation.

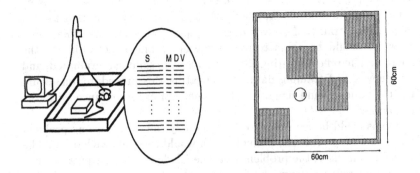

Fig. 2. The experimental setup

Each set of sensor-motor values was evaluated by the following evaluation function.

$$E = \sum D \times (1 - S) \times (1 - V)$$

where, D, S and V were the measures for the mileage of both motors, the average of proximity sensor values, and the difference of speeds between the motors, respectively. The motor mileage and speeds were the counts of rotary encoders attached to the wheels, not the outputs of the NN. The summation was performed for the period of 5s (50 samples) after each sampled reflex.

The data was sorted with the evaluation score. We selected one best-scored reflex for each sensor pattern appeared. Usually, the total of approximately 30–100 patterns occurred among $2^8 = 256$ possible sensory patterns. After 2000 times of training with the data sets by the BP, the new NN was down-loaded to the Khepera for the second round. We here called this one round of process as one iteration. We performed 50 iterations.

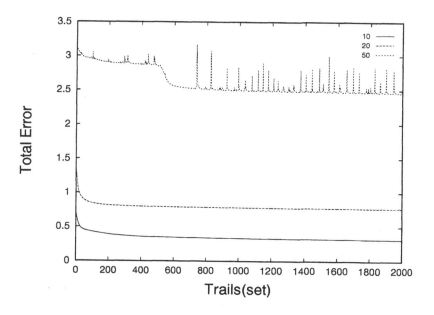

Fig. 3. The learning process

5 Learning Process

Learning processes at three different iterations (10, 20, and 50) were shown in Fig.3, where the errors steadily decreased by learning, but at 2000 where the training was terminated, the errors were not minimum yet. Since the number of patterns increased depending upon the iteration, 56, 76 and 111 for iteration of 10, 20 and 50 ,respectively, in this case, the errors at 2000 increased by iteration. Fig.4 showed the maximum and average values of the evaluation function in the course of free movement. The evaluation value was improved by iteration. Accordingly, the behavior of the robot was gradually improved, and finally achieved a good navigation as illustrated in Fig.5.

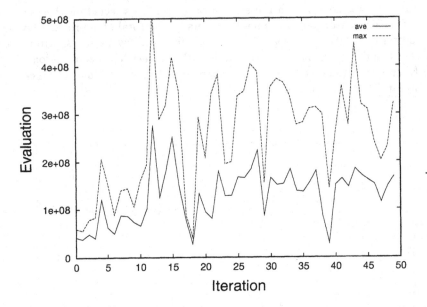

Fig. 4. The evaluation

The time needed for developing good behavior was 10–30 iterations, i.e., $(20 iterations) \times (50s\,free - movement) = 1000s$ in total. This was approximately one tenth of the time necessary for the conventional evolution, where it took usually 100 generations of 10 individuals with the life time of 10s, i.e., $100 \times 10 \times 10 = 10000s$.

As seen in Fig.4, the evaluation reached its maximum at the early period, at around 12th iteration in this case, and fluctuated afterward. This was largely due to the short evaluation period (50s free-movement period). The data selected for the period might not be good for the new environment appeared in the next iteration period. In our preliminary study, this could be overcome by adding the inertia component. That is, the training data could be selected from the sensor-motor patterns appeared in the past several iterations, in addition to the present ones.

Another problem sometimes seen in the experiments was that the robot was trapped in a local minimum and could not jump out from it. We here call it the lack-of-diversity problem. This was apparent in Fig.5, where the robot exhibited the same sequence of reflexes, that turned him too much backwards when he hit a wall. This is the essential problem similar to the one associated with the BP learning that goes for the largest gradient. In our experiments, it was observed that two out of five experiments started with identical initial weights, but with different initial positions of the robot, acquired the identical pattern of behavior which was not quite good, and never escaped from it.

In order to avoid the problem of the "lack-of-diversity", we did some experiments where noise was added in sensor data or also in coefficients. However, it did not speed up the learning, or rather disturbed it in some cases. This might be due to the fact that the sensor data and/or motor-encoder output data already contained environmental noise due to the fluctuation of light intensity and sticking and/or slipping of wheels on the running field, for example. Therefore, the further addition of noise might have exceeded the limit for the stable convergence. We are currently searching for the way to disturb the weights without loosing the ability for convergence.

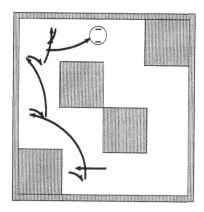

Fig. 5. The behavior

6 Remarks

In this study, we proposed to apply the BP learning for the development of autonomous robots, and described results of its application to a real mobile robot Khepera. The training data set was obtained in free movement in an environment, and selected in accordance with an evaluation function which represented the consequence of behavior in the future. Here the data needed to be further selected in order to avoid "over-training" of particular reflexes. With this method,

Khepera learned to navigate in an environment with obstacles faster than the conventional evolution.

Another advantage of this method was the easy access to the future per se. In evolutionary methods, individuals were evaluated by the behavior of their entire life. In contrast, in this learning method, the duration of the evaluation period was adjustable, just by changing the summation period for the evaluation. The duration may or may not include the past and the future (cause and consequence).

Although further analysis on the learning process would be needed to clarify the applicability of this method, it seemed to be successful at least in the case of the navigation task for Khepera with a dramatic improvement of learning speed.

References

1. Floreano, D. and Mondada, F. (1996) Evolution of homing navigation in a real mobile robot. IEEE Trans. Syst. Man, and Cybernet. Part B, 26:396–407.
2. Koza, J.R. (1992) Genetic programming: On the programming of computers by means of natural selection. MIT Press, Cambridge, MA.
3. Rumelhart, D.E., Hinton, G.E., and Williams, R.J. (1986) Learning representations by back-propagating errors. Nature (Lond.) 323:533–536.

SPIKE_4096: A Neural Integrated Circuit for Image Segmentation

Jean-Luc Rebourg [1], Jean-Denis Muller [1], and Manuel Samuelides [2]

[1] Commissariat à l'Energie Atomique - DAM
B.P. 12, F-91680 Bruyères-le-Châtel, France

[2] Office National d'Etudes et Recherches Aérospatiales - CERT
2 avenue Edouard Belin, B.P. 4025, F-31055 Toulouse Cedex, France

Abstract. An image segmentation algorithm, based on Pulse-Coupled Neural Networks, was implemented in silicon. We aimed at simplifying neuron hardware implementation while maintaining segmentation efficiency. Some algorithmic tricks have then been added, improving the results. The main components of the underlying neuron architecture are a single 8 bits register, a simple incrementer, and some glue logic. A prototype, using a data flow architecture, implementing a 64×64 neuron array, and based on a 0.2 μm CMOS SOI technology, will be released in 1998. A 64×64 segmentation is expected in less than 50 μs.

1 Introduction

In this paper, we describe the design of SPIKE_4096, a neural integrated circuit dedicated to very fast image segmentation. The main interest of this work lies in the specific neural algorithm we considered; we have simplified and optimized significantly this algorithm towards a specific hardware implementation, using a new 0.2 μm SOI (Silicon On Insulator) radiation-hardened technology. This leads us to a 1 cm² chip able to perform 20,000 64×64 image segmentations per second.

2 A particular image segmentation neural algorithm

Segmentation consists in dividing up the whole image into separate regions, for instance by grey level equalization. It is an important step in image processing. This first stage can be followed by other higher level algorithms such as image matching between various data bases. Image compression can also be carried out more readily after adequate segmentation. Since we aimed at real time operation, we decided to design an application-specific integrated circuit (ASIC), allowing a very high segmentation rate. The neural model that inspired our work belongs to the *Pulse-Coupled Neural Network* (PCNN) family; it is based on neurons of the *Integrate & Fire* type, described by Eckhorn et al. in [5]. The PCNN approach describes explicitly information exchange between neurons through pulse generation as outputs. Prominent image features are first obtained, imitating biological systems more closely than static classical neural networks do [8]. First we describe the algorithm main steps. A detailed presentation is given in [3].

The neuron includes the following elements :

(i) an input integrating soma (cell body) which induces output pulses generation. To this soma are associated an internal potential U(t), a discharge threshold θ(t), and a pulse generator producing an output signal Y(t).

(ii) dendrites connected to two kinds of synapses: *feeding synapses* for direct signal propagation, and *linking synapses* receiving auxiliary synchronizing signals (*linking inputs*) that modulate direct signals (*feeding inputs*).

Incoming signals into the ith neuron are:

(i) direct stimuli $E_{ij}(t)$ (feeding inputs) originating from image pixels or from preceding layer neurons. Feeding synapses convey weighted stimuli to the soma. Here neurons have but one direct input $E_i(t)$, the weighting coefficient being 1.

(ii) interacting signals $Y_{ik}(t)$ produced by neighbouring neurons. Those coupling stimuli act upon linking synapses which weight (through W^L_{ik} coefficients) and convey excitation to the neuron soma.

The neuron behaves as a leaky integrator, with respect to $E_i(t)$ and $Y_{ik}(t)$. $U_i(t)$ potential derives from two corresponding contributions, respectively $F_i(t)$ and $L_i(t)$. We use a discrete time scale, which gives the following equations:

$$L_i(t) = L_i(t-1).\exp(-1/\alpha_L) + \Sigma_{neighbours}W^L_{ik}.Y_k(t) \tag{1}$$

$$F_i(t) = F_i(t-1).\exp(-1/\alpha_F) + E_i(t) \tag{2}$$

When feeding input on a given neuron remains at a constant level, [4] proved that $F_i(t)$ tends to be proportional to $E_i(t)$. Our application takes advantage of this asymptotic property; in the sequel we use:

$$F_i(t) = E_i = constant \tag{3}$$

The Eckhorn model internal potential is given by $U_i(t) = F_i(t).(1 + \beta L_i(t))$. This multiplicative hypothesis for neural cross-influence is in agreement with cats visual cortex observed results. But, in order to obtain symmetric processing for all grey levels, additive inter-neural weighting is more convenient than a multiplicative one. Consequently we define $U_i(t)$ by:

$$U_i(t) = F_i(t) + \beta L_i(t) = E_i + \beta L_i(t) \tag{4}$$

When $U_i(t)$ is larger than a time-decreasing threshold $\theta(t)$, a spike is delivered as the corresponding neuron output, and is transmitted to neighbouring neurons. We consider here only binary output spikes, yielding the simplified expression:

$$\text{if } U_i(t) \geq \theta(t) \text{ then } Y_i(t) = 1, \text{ else } Y_i(t) = 0 \tag{5}$$

We replace the exponentially decreasing Eckhorn model for threshold $\theta(t)$ by a linear decreasing threshold, allowing for uniform segmentation over the whole grey level range. After neuron firing, $\theta(t)$ is re-increased in order to prevent any immediate second firing; this can be expressed as:

$$\text{if } Y_i(t) = 0 \text{ then } \theta(t+1) = \theta(t) - \Delta\theta, \text{ else } \theta(t+1) = \theta_0 \tag{6}$$

Johnson et al. [6] use a PCNN for image segmentation, but they must wait for transients to vanish: valid results are obtained only at network equilibrium. The neural model we present here converts pixel luminance into latency time. It gathers similar grey levels by neighbouring neurons synchronization. This process yields the desired segmentation in a progressive way, as opposed to Johnson's algorithm. One can note that in our case, a finite iteration number is required. We use a single layer network with one neuron per pixel. E_i is the ith pixel grey level, and $F_i(t)$ is made equal to E_i (3). $L_i(t)$ is a weighted sum of neighbouring neurons outputs. A synchronous avalanche mechanism [2] is introduced, involving two time scales: the first one related to network iterations, and a much faster one for neural interactions. When a neuron fires during a given clock cycle, its output spike is sent to neighbours, increasing their internal potential $U_i(t)$. In turn those neighbours are triggered, if $U_i(t) \geq \theta(t)$. Those triggered firings take place during the same main clock cycle, and propagate from neighbour to neighbour over the whole network. All those firings are synchronous with respect to network iterations clock: this constitutes a synchronizing avalanche mechanism.

3 General circuit specification

3.1 Manufacturing technology

This circuit will be issued using a new CMOS / SOI radiation hardened technology, with a 0.2 µm gate length, recently developed by CEA [7].

3.2 Digital circuit

Our first design is an entirely digital circuit, dedicated to 256 grey level image processing. We plan in the near future to design an analog version with this CMOS / SOI technology.

3.3 Parallel processing

The algorithm implies all neurons to cooperate during the segmentation process, and a one-to-one neuron association for proper avalanche propagation.

3.4 Neural area minimization

Neural area must be kept to the lowest possible value, in order to integrate a large neural array.

4 Algorithm adaptation to hardware

Circuit feasibility can only be reached through drastic algorithm simplification. We thus eliminated all costly hardware operators implied by initial algorithmic principles, while keeping global image segmentation efficiency.

4.1 Unique time scale

Two time scales were mentioned above: a relatively slow one for network iterations and one for neighbour neural interactions. But, as avalanche propagation can be implemented through combinatorial glue logic, we do not need this faster time scale; only the former and slower one remains in the final design.

4.2 Single firing technique

Initial neural model involves successive firings, separated by variable latency periods. Our modified algorithm for image segmentation is based on existing latency before neuron first firing. The segmentation process is over when all neurons have been fired once.

4.3 All-numerical approach

We use integer arithmetic exclusively, and manage to drastically reduce internal data word lengths and global register number. (3), (4), (5), (6) lead us to a straightforward but efficient algorithm that needs only a single 8 bit register and one memory point.

4.4 Linking signal influence

Neuron firing affects for a long time the internal potential of all neighbours to which it is connected (1). The $\exp(-1/\alpha_L)$ factor can be chosen to reduce multiplication complexity, such as $\exp(-1/\alpha_L) = 2^{-n}$. Even with that choice, $L_i(t)$ evaluation needs a two terms addition which involves word length extension. We decided to limit neuron firing effects on neighbours in time to the very cycle during which firing occurred. This influence time limiting scheme has the following side benefit: possible successive avalanches are strongly decorrelated. (1) reads now:

$$L_i(t) = \Sigma_{neighbours}W^L_{ik}.Y_k(t) \tag{7}$$

By incorporating β into all W^L_{ik}, (4) can be rewritten as:

$$U_i(t) = E_i + \Sigma_{neighbours}W^L_{ik}.Y_k(t) \tag{8}$$

4.5 Direct influence range

In the original algorithm we started from, every neuron could be influenced through its linking inputs by neighbours contained in a large neighbourhood (7×7, 9×9...). In order to reduce circuit complexity we decided to retain only the four nearest neurons (East, West, North and South). This choice simplifies markedly neuron interconnection. Drawbacks induced by such a short range connection have been evaluated in [1]. In this respect we mention that this neighbourhood limitation prevents avalanche from jumping over a neuron line. Moreover, this cross-shaped neighbourhood does not allow diagonal avalanche transmission. But this is not a serious drawback, since single pixel diagonal lines are rather scarce in real world images. On the other hand, such an iso-distance neighbourhood leads to four equal W^L_{ik} coefficients. If we define by $N_{FNi}(t)$ the number of ith neuron neighbours being fired at time t, with $0 \leq N_{FNi}(t) \leq 4$, (8) becomes:

$$U_i(t) = E_i + W_L.N_{FNi}(t) \tag{9}$$

4.6 Constant thresholds

Instead of linearly decreasing $\theta(t)$, and keeping $U_i(t) = E_i$ = constant when there is no firing neighbour, we decided to hold θ constant and to increase linearly $U_i(t)$, starting from $U_i(t=0) = E_i$. (9) and (6) become:

$$U_i(t+1) = U_i(t) + \Delta\theta + W_L.N_{FNi}(t), \ U_i(t=0) = E_i, \ \theta(t) = \theta_0 = \text{constant} \tag{10}$$

For simplicity we take $\Delta\theta = 1$, and rewrite (10) and (5) into (11) and (12):

$$U_i(t+1) = U_i(t) + 1, \ U_i(t=0) = E_i \tag{11}$$

$$\text{if } U_i(t) \geq \theta_0 - W_L.N_{FNi}(t) \text{ then } Y_i(t) = 1, \text{ else } Y_i(t) = 0 \tag{12}$$

We take θ_0 equal to 255. For a fixed W_L, the four other possible thresholds $\theta_1 = \theta_0 - W_L$, $\theta_2 = \theta_0 - 2W_L$, $\theta_3 = \theta_0 - 3W_L$, and $\theta_4 = \theta_0 - 4W_L$ can be implemented in hardware. θ_0 represents the auto-firing threshold. θ_1, θ_2, θ_3, and θ_4 are the avalanche mode firing thresholds. Equation (12) amounts to compare $U_i(t)$ incrementer output to one of the five thresholds defined above, according to the number of firing neighbouring neurons at time t. As a conclusion, it is much simpler for hardware implementation purposes to compare a variable quantity $U_i(t)$ to a fixed built-in number, than to compare two varying quantities such as $E_i + W_L.N_{FNi}(t)$ and $\theta_0 - t$. In one case we need a complete 8 bit comparator, while in the other case, comparison logic can be optimized and drastically reduced in size.

4.7 Inter zone boundaries

We defined $N_{FNi}(t)$ as the number of neurons surrounding the ith neuron and firing at time t. The higher $N_{FNi}(t)$, the more likely the ith neuron is to fire (12). Conversely, a still unfired neuron, surrounded by some neighbours that have been fired during previous clock cycles, receives a $N_{FNi}(t)$ value limited to the number of unfired neighbours. Consequently its firing stimulus cannot be at the highest possible level, when a new avalanche reaches it. When whole image segmentation has been completed, this phenomenon may result in isolated points between segmented zones, which were not included in any of the surrounding avalanches. In order to correct this undesired effect, we define $N_{OFFi}(t)$ as the number of ith neuron neighbours which have been fired during previous clock cycles. The new firing conditions are:

$$\text{if } U_i(t) = \theta_0 \text{ then auto-firing}$$
$$\text{else if } U_i(t) < \theta_0 \text{ and } N_{FNi}(t) = 0 \text{ then no firing}$$
$$\text{else if } (U_i(t) < \theta_0 \text{ and } N_{FNi}(t) \geq 1) \text{ then firing if } U_i(t) \geq \theta_0 - W_L.(N_{FNi}(t) + N_{OFFi}(t)) \tag{13}$$

4.8 Thresholds adjustment

Dedicated hardware for comparing $U_i(t)$ to θ_i can be more or less complicated according to θ_i numerical value. Taking $\theta_0 = 255$ leads to a very simple comparator for auto-firing, since only $U_i(t)$ incrementer carry-out flag needs to be checked. More generally, values such as $\theta_i = 256 - 2^p$ lead to more cost effective hardware than other values. The original avalanche mode thresholds θ_1, θ_2, θ_3 and θ_4 implied by (12) have therefore been replaced by specific values determined by extensive simulations. We think we reached a good compromise *comparator simplicity* versus *segmentation efficiency*. The first choice we made was $\theta_4 = 0$, which means forced firing for a given neuron, for example when its four neighbours fire at the same time t, or more generally when $N_{FNi}(t) \geq 1$ and $N_{FNi}(t) + N_{OFFi}(t) = 4$, as seen in (13). All dark isolated points inside homogeneous regions are thus eliminated through grey level equalization, since every central pixel takes its four neighbour grey level common value. This forced-firing mechanism results in reduced hardware, since one comparator is suppressed. A 4-input AND is ultimately needed to test the condition $N_{FNi}(t) + N_{OFFi}(t) = 4$. It can be shown that we must take $\theta_1 = \theta_2$ in order to avoid inter-region border distortion.

Another mechanism called "firing inhibition" will be introduced later on. It imposes $\theta_3 = \theta_4 = 0$, inducing single pixel concave gap elimination. Finally, all above algorithmic and hardware thresholding refinements boil down to a unique choice $\theta_1 = \theta_2 = \theta_{av}$, coupled with $\theta_0 = 255$ and $\theta_3 = \theta_4 = 0$. In connection with the above θ_i values, every neuron must be implemented with the following rules:

$$\text{if } N_{FNi}(t) = 0 \text{ then auto-firing if } U_i(t) = \theta_0 = 255$$
$$\text{else if } N_{FNi}(t) + N_{OFFi}(t) \geq 3 \text{ then avalanche forced firing}$$
$$\text{else if } U_i(t) \geq \theta_{av} \text{ then avalanche firing} \tag{14}$$

4.9 Firing inhibition

Any avalanche over some homogeneous image area is triggered by the neuron linked to the brightest pixel in that zone, prior to the segmentation process. Due to inter-neuron connections and to the synchronous avalanche mechanism described earlier, other neurons in that zone are bound to fire during the same cycle. The segmented zone common grey level is determined by the avalanche initializing neuron. If one looks for reasonable image segmentation, avalanche firing thresholds must be taken low enough, in such a way that all zone neurons get avalanched, even the less bright ones. A perfect segmentation would associate the average grey level of the original zone to the whole segmented zone, thus eliminating brightest points through delayed neuron firing. This brightest pixel time lag principle naturally introduces the above mentioned firing inhibition mechanism. This new device helps in decreasing higher level pixels down to average local grey level; isolated brightest points are eliminated too. The corresponding implementation is straightforward: when any $U_i(t)$ equals $\theta_0 = 255$, this i^{th} neuron does not fire as long as there are less than N_a ($1 \leq N_a \leq 4$) neighbours which have reached the θ_0 threshold too. One can note that every avalanche thus involves at least $N_a + 1$ neurons: the first one triggers the avalanche, and the N_a other ones, having allowed together this central neuron auto-firing, are themselves caught into the avalanche.

With segmented grey levels being closer to average local levels, avalanche thresholds $\theta_1 = \theta_2 = \theta_{av}$ can be increased, providing a more selective segmentation with less zone boundary overflow. This firing inhibition concept improves segmentation efficiency, with some drawbacks, however: the higher N_a, the higher the probability of finding permanently inhibited neurons. Fortunately, most of those permanently inhibited neurons will be reached by some future avalanche process. But segmentation can terminate, although some neurons have never been fired, neither in the auto mode, nor in the avalanche mode. Use of $\text{N}_{\text{OFFi}}(t)$, as shown in 4.7, decreases the number of those permanently excluded neurons. For $N_a = 1$, it eliminates this unwanted effect. For $N_a = 2$, the use of $\text{N}_{\text{OFFi}}(t)$ must be coupled with the $\theta_3 = 0$ value, in order to prevent any permanent exclusion situation. If we take $N_a = 3$, a new hardware device must be added to every neuron, thus increasing neuron complexity excessively. The case $N_a = 4$ cannot be solved practically. All in all, we chose $N_a = 2$, leading to an acceptable hardware complexity, with good segmentation results on many test images.

5 Circuit design

5.1 Neuron architecture

Neuron operating principles are as follow: the content of the 8 bit register $U_i(t)$ is initialized with the corresponding pixel value; it is incremented at every clock rising edge, and is compared to the avalanche mode and auto mode thresholds, taking $\text{N}_{\text{FNi}}(t)$, $\text{N}_{\text{OFFi}}(t)$ and $N_a = 2$ into account.

If $U_i(t) = \theta_0 = 255$, and if at least $N_a = 2$ neighbours are ready ($U_k(t) = \theta_0 = 255$), then the ith neuron is fired, triggering an avalanche process. This i^{th} neuron is then cut from its neighbours, and the $U_i(t)$ register is reset to 0. During remaining cycles, $U_i(t)$ will be incremented again: at the end of the 256 clock cycles segmentation process, the register will thus contain the adequate grey level.

If $U_i(t) = \theta_0 = 255$, but if there is some firing inhibition, the neuron waits either for auto-firing permission, or for an avalanche mode triggering.

If $U_i(t) < 255$, if at least one neighbour is being fired at time t, and if the avalanche mode threshold corresponding to $\text{N}_{\text{FNi}}(t) + \text{N}_{\text{OFFi}}(t)$ is reached, then the neuron fires. $U_i(t)$ register reset and neuron isolation are carried out as indicated above. In all other cases, $U_i(t)$ keeps increasing.

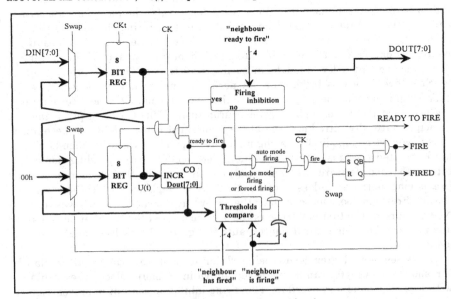

Figure 1: Neuron general organization

We recall that since our segmentation algorithm uses only every neuron first firing, the 8 bit register $U_i(t)$ can be used for segmented grey level evaluation, after i^{th} neuron firing. The internal potential is not needed any more.

5.2 Circuit architecture

Let us consider the image size to be NxN. We remind that every $U_i(t)$ incrementation is imposed simultaneously on all N^2 neurons; this parallel N^2 operation is in turn done 256 time successively, in order to carry out the complete segmentation process. The adequate time interval between two successive $U_i(t)$ ($1 \leq i \leq N^2$) incrementations is given by the longest occurring avalanche duration.

In order to optimize the total segmentation time, one could detect individual avalanche completion after each incrementation. But we deal here with a circuit involving large input/output data flows. Image loading and unloading times are $O(N^2)$ whereas maximum avalanche time is $O(2N)$, when one neuron is connected to its four closest neighbours. Consequently we aimed at globally optimizing the whole process *(loading + segmentation + unloading)*, rather than concentrating on segmentation time exclusively. We therefore chose a data flow architecture for our circuit.

This architecture includes two overlapping functional layers. The first one, named transfer layer, is an 8 bit shift register structure, which can progressively extract the segmented image I_n^*, while image I_{n+2} is progressively loaded into this very same layer. The time used for this progressive image substitution I_n^*/I_{n+2} is identically used at the second functional layer, named processing layer, in order to segment image I_{n+1}. Registers $U_i(t)$ ($1 \leq i \leq N^2$) and incrementors are located in this processing layer. At the very time when I_{n+2} loading is carried out into the transfer layer, I_{n+1} segmentation is over in the processing layer. An instant swap can then take place between transfer layer data (image I_{n+2}) and processing layer data (segmented image I_{n+1}^*): the circuit is now ready to start I_{n+2} segmentation.

This data flow architecture allows for an apparently non-stop continuous image segmentation: during one image segmentation time, next image is loaded. Finally we point out that, in the transfer layer, data are only moved from neighbour to neighbour, along natural image columns; consequently the circuit layout is basically local, with the only exception of clock signals.

6 Some results

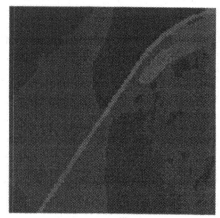

image SPOT_road (128×128)   *segmented image*

image Kitten (368×368)   *segmented image*

7 Conclusion

The development of those optimized neuron and global circuit architecture allows us to integrate a 64×64 neural array on a 1 cm² chip. Overall segmentation time for a 64×64 image would be about 50 μs (20,000 images/s); this processing time is closely related to the two 20 MHz, 32 bits external data buses for images loading and unloading. Moreover, in our circuit, the θ_{av} algorithmic constant can take three predefined values: $\theta_{av0} = 255$, $\theta_{av1} = 240$ (binary value: 11110000), and $\theta_{av2} = 224$ (binary value: 11100000). Each value leads to specific segmentation results: $\theta_{av0} = 255$ produces micro-segmentation, equivalent to local noise smoothing; $\theta_{av1} = 240$ is convenient for dark images (for instance: SPOT satellite images); $\theta_{av2} = 224$ is used for image with large grey level range. The circuit design is about complete, and the actual circuit should be tested in spring 1999.

The authors are grateful to Dr François Durbin for his contribution in the English adaptation of the original French material.

8 References

1. Ambiehl, Y., Pulse-Coupled Neural Networks for image segmentation (*in French*). Technical Report, Sup'Aéro-ONERA/CERT, Sept. 1997.
2. Bak, P., Tang, C., Wiesenfeld, K., Self organized criticality: An explanation of 1/f noise. Physical Review Letters, 59(4), 381-384, 1987.
3. Clastres, X., Freyss, L., Samuelides, M., Tarr, G. L., Dynamic segmentation of satellite images using Pulse-Coupled Neural Networks. ICANN'95, Paris, Oct. 1995.
4. Clastres, X., Samuelides, M., Pulse coupled neurons and image segmentation (*in French*). Technical Report ONERA n°1/35566.000/DERI, Dec. 1996.
5. Eckhorn, R., Reitboeck, H. J., Arndt, M., Dicke, P., Feature linking via synchronization among distributed assemblies: simulation and results from cat visual cortex. Neural Computation, 2, 293-307, 1990.
6. Johnson, J. L., Ritte, D., Observation of periodic waves in a pulse-coupled neural network. Optics Letters, 18 (15), 1253-1255, 1993.
7. Pelloie, J. L., Faynot, O., Raynaud, et al., A Scalable Technology for Three Successive Generations: 0.18, 0.13, and 0.1 μm for Low-Voltage and Low-Power Applications. Proceedings 1996 International SOI Conference, p.118, Oct. 1996.
8. Thorpe, S., Fize, D., Marlot, C., Speed of processing in the human visual system. Nature, 381, 520-522, 1996.

Analysis of the Scenery Perceived by a Real Mobile Robot Khepera

Ryoichi Odagiri, Wei Yu, Tatsuya Asai, and Kazuyuki Murase

Department of Information Science, Fukui University,
3–9–1 Bunkyo, Fukui 910–8507, Japan
{ryo, wei, asai, murase}@synapse.fuis.fukui-u.ac.jp

Abstract. In order to understand the dynamic interactions of autonomous robots with their environments, the flow of the sensory information perceived by a real mobile robot Khepera was analyzed with the return map. The plot of X_{n+1} v.s. X_n of chaotic time series X_n, called the return map, have been widely used to reveal the hidden structure. A real mobile robot Khepera was evolved in three different environments with various levels of complexity. The fitness function of GA operations included the complexity measure of the control structure, i.e., individuals with a simpler structure obtained a higher score. The evolution lead to develop the robot with the minimal structure sufficient to live and perform tasks in the given environment. The return maps of these robots differed each other considerably. However, when the robot evolved in the most complex environment was asked to navigate in the other two environments, the return maps obtained there were similar to (or a substructure of) the one in the most complex environment. These results indicated that the autonomous robot behaved such a way that the flow of sensory information did not depend much on the environment where he situated but largely on the one where he had evolved.

1 Introduction

For autonomous robot, the sensory information is essential for the action. The sensory information perceived by an robot was not simply the "static" real world, but rather the reflection of the "dynamic" interactions between the robot and the environment in which the robot was situated. The interactions include the sequence of the environmental scenery and the physical dynamics of the robot. The way how the dynamic interactions are perceived could be expected to vary among individuals because their physical dynamics differ from each other. The plot of X_{n+1} v.s. X_n of time series, called return map, has been widely used to reveal the chaotic dynamics [1, 2]. In the 1st order chaos represented by the 1st order difference equation $X_{n+1} = F(X_n)$, the plot of X_{n+1} v.s. X_n should visualize the function F. In this study, we attempted to see the differences of the perceived world among robots through the analysis of the sensory information with the return map as the first step to understand the hidden structure in the perception.

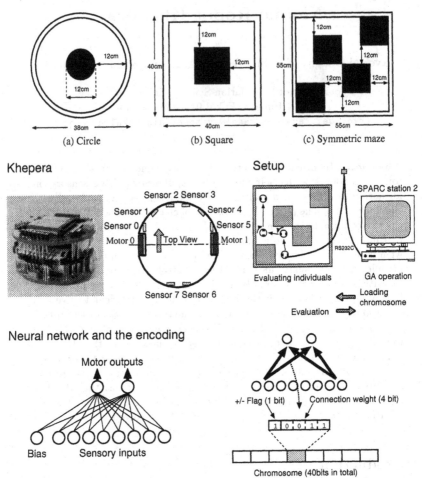

Fig. 1. Three environments (top) where a khepera robot with eight sensors and two motors was evolved by using the GA operation in a workstation (middle). A neural network was used to drive Khepera, where the weights were encoded in a chromosome (bottom).

2 Evolution in Various Environment

The real mobile robot Khepera[4] was evolved in three different environments; circle, square, and symmetric maze as shown in Fig.1. The robot was operated with a simple two-layered neural network. The network produced two outputs from the eight proximity sensor inputs and one bias unit. The weights from left sensors were set to equal with the ones from right sensors, i.e., the network

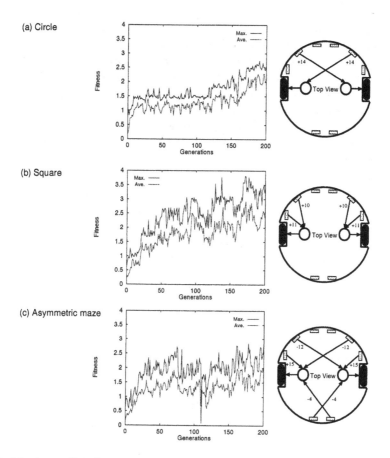

Fig. 2. Maximum (Max) and mean (Ave) fitness at each generation in three environments, and the final connections of the best individuals after 200 generations.

was arranged to have symmetrical weights. Each weight was represented with 5 binary bits, and therefore the total length of gene became 40 bits.

The task given to the robot was to navigate in a field by avoiding obstacles. The fitness function was designed to select the individuals that (1) moved a long distance (2) as straight as possible (3) without getting close to obstacles, and that had (4) a smaller number of connections. The number of connections was determined by the number of non-zero weights. If the value of a weight was zero, it was considered that there was no connection, and thus the structure was considered simpler.

The fitness function was therefore expressed as follows,

$$E = D\,(1 - V)\,(1 - S)\,Z \tag{1}$$

This function without Z has been a standard function used in the evolution of many navigation robots (e.g, [5]). Here, D, V and S represented the total mileage of both left and right motors, the difference of mileage between the motors, and the total of all the sensor outputs. The state of the robot was observed at every 100ms. V and S were detected at every observation, and D was obtained as a distance from the last observation made at 100ms before. Z was the number of zero weights, and it was set to 0.5 when there was no zero weight.

We employed the standard genetic algorithm with one–point crossover, bit–reverse mutation, fitness scaling, and elite preservation. The parameters used in the experiments were: number of individuals 10, crossover rate 0.7, mutation rate 0.02, and elite preservation rate 0.5.

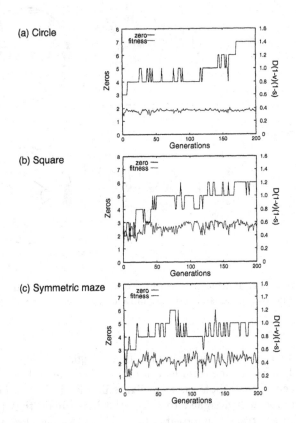

Fig. 3. The Z and $D(1-v)(1-s)$ components of the fitness values shown in Fig.2.

3 Fitness and Best Individuals

Results of evolution in the three test fields; circle, square, and symmetric maze were illustrated in Fig.2 and 3. In Fig.2, the fitness values of the best individual

(Max) and the mean of all individuals (Ave) in each generation and the final connection of the best individuals after 200 generations of evolution were shown.

The Z and $D(1 - v)(1 - s)$ values of the fitness were separately plotted in Fig.3. After evolution for the test fields of circle, square, and symmetric maze, the number of non-zero weights were 1, 2 and 3, respectively (see also inserts in Fig.2). For circle, the robot used only one contra–lateral sensor for each motor, which was the simplest we could think of for running around the field of circle. For square, the robot used two more sensors from the very lateral side. This seemed reasonable because the robot had to run straight between the walls. For symmetric maze, the robot used two more sensors from the back.

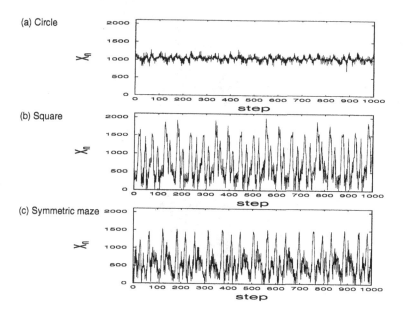

Fig. 4. Root square of all sensor values

4 Perceived Scenery of the Best Individuals

The root square (RS) of all the sensor values was sampled along the course of navigation in the given environment with a sampling rate of 0.1ms. Examples were illustrated in Fig.4, The RS values X_n of the best individuals obtained in the three different environments were plotted against time. The time series X_n should reflect the dynamics of environment and the robot. The return maps were constructed as shown in Fig.5. The difference of the dynamic structures became apparent from the figures. For example, in comparison to the one for symmetric

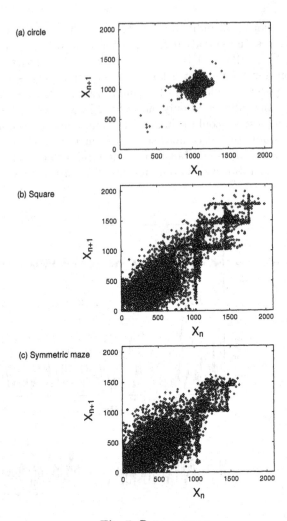

Fig. 5. Return maps

maze, an additional sensory–motor reflex seemed to appear in the one in square, because an additional organized structure appeared in the return map of square at the range of 1700–2000.

5 Self-Selection of Sensory Information

The best robot evolved in symmetric maze (shown in the bottom of Fig.2) navigated smoothly in symmetric maze, and also in square, because the pattern of the field of square was a part of the field of symmetric maze, only containing pointed angles. He could also navigated well in the circle by following the wall. Fig.6 showed the return maps obtained by the best robot of the symmetric

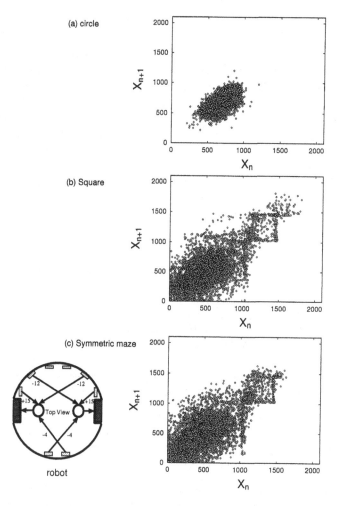

Fig. 6. Return maps of RS values which were obtained in three different environments by the best robot for symmetric maze shown in the insert at the bottom.

maze in three different environments. Although the environment was different, the return maps resembled each other (compare with the ones in Fig.5). That is, the robot "saw" three environments similarly. In other words, the flow of sensory information did not depend much on the environment where he situated but depend much on the environment where he had evolved, and the robot self-organized his action in such a way that the sensory information he gathered remained identical.

6 Conclusion

Here we tried to understand how the robot perceives the environment, and found that the return map could be used to visualize it. Robots obtained by evolutionary methods could behave similarly in a certain environment. For example, mobile robots with any control scheme could look alike when one watched their behavior. However, their way of looking at the world might be different. This study has shown that the perception of the world could become apparent by the analysis of sensory information with the return map. The sensory information should be a chaos that reflects the nonlinear dynamics of the environment and the robot. The analysis of the chaos would further reveal the hidden structure of the perception.

References

1. Degan, H., Holden, A. V., Olsen, L. F.: Chaos in Biological Systems, Plenum, New York. (1987)
2. Holden, A.V. : Chaos, Manchesterand Princeton Univ. Press., Princeton. (1986)
3. Langton, C.: Life at the Edge of Chaos. In Langton, C., Tayler, C., Farmer, J.D., Rasmussen, S.(eds.): Artificial Life II. Santa Fe Institute Studies in the Science of Complexity, Proceedings Volume X. Addison-Wesley, Redwood City, CA, (1992) 41–91
4. Mondada, F., Franzi, E., Ienne, P.: Mobile robot miniaturization: A tool for investigation in control algorithms, Proceedings of the Third International Symposium on Experiment Robotics. Kyoto, Japan, (1993)
5. Nolfi, S., Floreano, D., Miglino, O., Mondada, F.: How to evolve autonomous robots: Different approaches in evolutionary robotics, In Brooks, R., Maes, P.(eds.): Artificial Life IV Proceedings of the Fourth International Workshop on the Synthesis and Simulation of Living Systems. MIT Press, Cambridge, MA, (1994)
6. Odagiri, R., Wei, Y., Asai T., Murase, K.: Measuring the complexity of the real environment with evolutionary robot: Evolution of a real mobile robot Khepera to have a minimal structure, Proceedings of the IEEE International Conference on Evolutionary Computation, Anchorage, AK, (1998 in press)

Evolution of a Control Architecture
for a Mobile Robot

Marc Ebner

Eberhard-Karls-Universität Tübingen, Wilhelm-Schickard-Institut für Informatik
Arbeitsbereich Rechnerarchitektur, Köstlinstraße 6, 72074 Tübingen, Germany
ebner@informatik.uni-tuebingen.de

Abstract. Most work in evolutionary robotics used a neural net approach for control of a mobile robot. Genetic programming has mostly been used for computer simulations. We wanted to see if genetic programming is capable to evolve a hierarchical control architecture for simple reactive navigation on a large physical mobile robot. First, we evolved hierarchical control algorithms for a mobile robot using computer simulations. Then we repeated one of the experiments with a large physical mobile robot. The results achieved are summarized in this paper.

1 Motivation

Currently mobile robots are usually programmed by hand. For many real-world applications this introduces considerable difficulties due to the complexity of the task [10]. Instead of programming the robots directly Darwinian evolution may be used to automate this process [2]. Thus one would reduce the problem to finding an appropriate fitness function which describes how well a particular individual solves the task. The different approaches in evolutionary robotics are described by Nolfi et al. [21]. A number of researchers have already evolved control architectures for mobile robots using computer simulations. Few have worked with physical mobile robots. Brooks [4] emphasized the importance of using physical mobile robots as opposed to computer simulations. To our knowledge, those who used physical mobile robots only experimented with miniature robots such as the Khepera or a gantry-robot. Most work in evolutionary robotics was done using neural net control architectures [5, 11, 12, 9, 7, 19, 8, 18]. In contrast to this work, we wanted to see if genetic programming [13, 15] can be used to evolve a hierarchical control architecture for a simple reactive navigation task on a large physical mobile robot. If evolution is carried out on a large physical mobile robot safety measures have to be taken such that possible damage to the environment or to the robot is avoided.

As Matarić and Cliff noted [17], most research in evolutionary robotics focused on the evolution of simple tasks with long evaluation times. Recently, Nolfi [20] evolved a garbage collecting behavior for a mobile robot using accurate computer simulations of the Khepera and a neural net control architecture which also executed successfully on the real robot. Evaluation times may be reduced using computer simulations especially if more difficult tasks are addressed. Therefore it is also important to see if results from computer simulations can be transferred into the real-world.

2 Background

In this section we briefly review related approaches using genetic programming to evolve control architectures for a mobile robot.

Koza [16] used computer simulations to evolve a control program for a grid based mobile robot to mop an 8×8 area containing obstacles on some of the grids. In a more realistic setting, but also with discrete movements, Koza [14] evolved subsumption architectures [3] for wall-following for a mobile robot. Koza's representation consisted of the terminals S00, ..., S11 which return distance information from twelve sonar sensors, SS which returns the minimum distance of all sensors, MSD which specifies the minimum safe distance to the wall, EDG which specifies the desired edging distance and the four terminals for control of the robot. The control terminal (MF) moves the robot 30cm forward, (MB) moves the robot 40cm backwards, (TR) turns the robot 30° to the right, and (TL) turns the robot 30° to the left. The evaluation was always done from exactly the same position. As primitive functions Koza used PROGN2 which evaluates its two arguments in sequence and the conditional IFLTE. IFLTE is a four argument function. If the first argument is less than or equal to the second argument the third argument is evaluated, otherwise the fourth argument is evaluated.

Reynolds [26] used genetic programming to evolve obstacle avoidance behavior with a simulated critter. The critter moves with constant forward velocity one half of its body length per simulation step. The primitive function turn is used to specify the amount the critter should turn. In addition to the function turn the standard arithmetic functions +, -, *, % (protected division), abs, and the conditional iflte are used. The robot can perceive its environment by using the function look-for-obstacle, which returns a measure of the distance to an obstacle in the direction specified by its argument. As terminals he used the set of constants 0, 0.01, 0.1, 0.5 and 2. Reynolds also experimented with noise to evolve more robust controllers [27, 29] and investigated the influence the representation has on the difficulty of the problem [28]. Reynolds experimented with fixed sensors and with sensors that could be pointed dynamically. Using fixed sensors simplified the problem considerably.

Nordin and Banzhaf [22, 23, 24] used a miniature mobile robot to evolve a linear control program. Their control algorithms consisted of a variable length genome which represents an assembly program. The programs could use the following arithmetic, logical and shift operations: ADD, SUB, MUL, XOR, OR, AND, and SHL, SHR. Thus their programs are executed linearly one statement after the other. Olmer et al. [25] extended the approach by evolving control algorithms for lower level tasks and a strategy for selection of the different tasks. An overview about the research is given by Banzhaf et al. [1].

Wilson et al. [30] evolved hierarchical behaviors to locate a goal object in a maze for a mobile robot constructed from Lego Technic bricks. The primitive operations were MoveForward, MoveBackward, TurnLeft and TurnRight. Hierarchical structures are produced by chunking simple behaviors after each loop through the evolutionary process. This loop included the evolution of behaviors in simulation and then evaluating successful individuals on the real mobile robot.

In contrast to the work of Koza and Reynolds, we are working with a physical autonomous mobile robot, a Real World Interface B21. Wilson et al. and Nordin and Banzhaf also used a physical mobile robot. However our representation also includes a conditional statement. Thus a hierarchical structure emerges during the evolution. To our knowledge, so far, no one has tried to evolved a hierarchical control algorithm on a large physical mobile robot using Koza's genetic programming.

3 Representation

The representation used during the experiments described below was determined by performing a series of experiments in computer simulations. We had to chose a representation which could be used with a small population size with a reasonable number of generations [6]. Using computer simulations we found a representation suitable for the evolution of a control architecture during the allotted time.

3.1 Terminals

The terminals are used to provide sensor readings and to control the movement of the robot. All 24 sonar sensors are used to obtain information about the environment of the robot. We combined the sensors into blocks of four producing six virtual sensors as shown in figure 1. Each of the virtual sensors returns the minimum distance of the four physical sensors. The terminals for the virtual sensors are named: FL, FM, FR, BL, BM, BR.

The robot moves with a constant translational velocity of $0.1\frac{m}{s}$. Three terminals are available to the robot to control the direction of movement. The terminal TL starts a rotation with velocity $40\frac{\circ}{s}$ to the left, TR starts a rotation with velocity $40\frac{\circ}{s}$ to the right, and RHALT stops the rotational movement of the

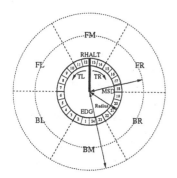

Fig. 1. Robot model seen from above with 24 sonar sensors combined to form six virtual sensors FL (front left), FM (front middle), FR (front right), BL (back left), BM (back middle) and BR (back right) which return the minimum distance of the corresponding 4 physical sonar sensors. The radius of the robot together with the minimum safe distance (MSD) and the desired edging distance (EDG) is also shown. The commands TL (turn left), RHALT (halt rotation) and TR (turn right) control the heading of the robot. The robot, a RWI B21, in its environment is shown on the right.

robot. In addition to executing the control command each of these three terminals also returns sensor information. The terminal TL returns the average value of the sensors 11 and 12, RHALT returns the average value of the sensors 12 and 13, and TR returns the average value of the sensors 13 and 14.

3.2 Primitive functions

As primitive functions we used the connective function PROGN2 and the conditional IFLTE. The function PROGN2 takes two arguments which are evaluated in sequence. The value of the last argument is returned. The function IFLTE takes four arguments. The first two arguments are evaluated. If the first argument is less than or equal to the second argument, then the third argument is evaluated. Otherwise the fourth argument is evaluated.

3.3 Fitness measure

In a small desktop evolution with a miniature mobile robot one may continue the evaluation of a mobile robot even if the robot hits a wall. The trajectory of the robot is simply changed due to the force exerted by the wall. With a large mobile robot such as the B21 this is no longer possible. The robot is capable of exerting a force which might cause considerable damage if the robot crashes into an obstacle. The information from the sonar sensors is monitored at all times. If an object is reported by the sonar sensors at a distance closer than 40cm from the center of the robot the motors are turned off and the evaluation is aborted. The bumpers of the robot are also monitored in case an object is not perceived by the sonar sensors. To avoid damage the next translatory movement of the robot has to be done in a direction away from the wall.

Therefore, we used a fitness measure which tries to maximize the time the robot survives until it bumps into an obstacle or its allocated time runs out. At the same time the sum of rotations should be kept at a minimum. After each evaluation the robot is repositioned using a repel operation. This operation evaluated the sonar data and the bumpers to move the robot 15cm away from the nearest obstacle. By using this operation we try to give all individuals an appropriate chance to survive. It makes no sense to start the evaluation of an individual if the robot is facing a wall in close distance.

Let T be the time available to the robot per fitness case. Let t be the time until the robot bumps into an obstacle. Let r_s be the sum of all signed rotations performed during the run and let ω be the rotational velocity of the robot. Then the raw fitness measure to be minimized is calculated according to: $\text{fitness}_{\text{raw}} = \frac{1}{n} \sum_{i=1}^{n} \left\{ 1 - D(1 - \sqrt{R_s}) \right\}$ where n is the number of fitness cases and $D = \frac{t}{T}$ and $R_s = \frac{|r_s|}{\omega t}$. Best possible raw fitness of zero is achieved if the robot avoids obstacles while performing a balanced number of turns to the right and left. A similar fitness function was previously used by Floreano and Mondada [7, 19, 8]. The adjusted fitness to be maximized is calculated according to $\text{fitness}_{\text{adj}} = \frac{1}{1+\text{fitness}_{\text{raw}}}$. This fitness measure tries to maximize survival time while penalizing unbalanced turning.

4 Experiments

First we performed a number of experiments using computer simulations and an environment very similar to the real environment in our lab. The environment used to evolve the hierarchical control architecture can be seen in figure 1. The environment we used is rather simple because of space limitations in our lab. Occasionally people had to move through the environment but this did not prevent evolution from making progress. In other computer simulations a larger and more complex environment was used [6].

For the simulation experiments we used two different processes. One process simulates the mobile robot in its environment. The other process evaluates the control structure. Sensor information is sent from the robot process to the evaluation process and commands are sent back to the robot process. 20 experiments were performed with 1,2 and 3 fitness cases. The experiment was run for 50 generations with a population size of 75. Tournament selection with size 7 was used and the crossover, reproduction and mutation probabilities were set to 85%, 10% and 5% respectively. As maximum time per fitness case we used $300s$.

In the following text we say that an individual is successful if it follows the environment without bumping into a wall. 2 successful individuals evolved with 1 fitness case, 6 with 2 fitness cases and 8 with 3 fitness cases. The performance of the individuals which evolved using 2 fitness cases is shown together with the fitness statistics in figure 2. In 50 generations with a population size of 75 a total of 3750 individuals are evaluated. We performed a control experiment with 3750 and 75000 random individuals. No successful individual was found among 3750 random individuals. The best individual found among 75000 random individuals together with a plot of the fitness values is shown in figure 2.

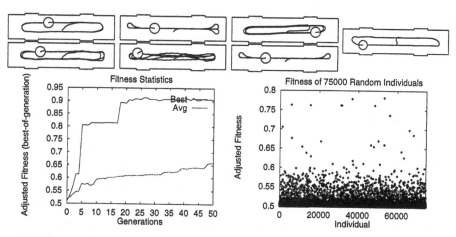

Fig. 2. 6 successful individuals evolved with 2 fitness cases. Of these the one shown on the top middle achieved an adjusted fitness of 0.9135. Fitness statistics for 2 fitness cases averaged over 20 runs and for the best run are shown on the left. The individual shown on the top right was the best individual found among 75000 random individuals. The fitness of 75000 random individuals is shown on the right for 3 fitness cases.

Next we performed the experiment using only 2 fitness cases (to limit the time needed for the evolution) on the physical mobile robot. The experiment was run for 50 generations. It took 2 months to perform the experiment on the real mobile robot. Actual time needed for the evolution was 197 hours. Initially batteries were exchanged after a generation was completely evaluated. During later generations batteries were also exchanged in between generations (just before a new individual was about to be evaluated). While the batteries were exchanged the evolutionary process was temporarily halted.

During this experiment an individual with 457 nodes reached the highest fitness. Due to the large size of the individual it is not shown here. In a series of 10 experiments where this individual was finally tested for 300s it managed to survive 8 of the 10 runs. The performance of this individual is shown in figure 3. The path was recorded using the robot's odometry and also with a standard camera using long term exposure. The average survival time of the individuals and the adjusted fitness of the best individual for each generation is shown in figure 3.

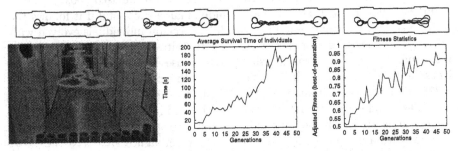

Fig. 3. Behavior of the best individual which evolved on the physical mobile robot. The paths were recorded using the robot's odometry. The photo on the left was taken at the same time the last path was recorded. The photo was taken with long term exposure while the robot was moving with a small light bulb mounted on its top. Average survival time of individuals is shown in the graph in the middle. The adjusted fitness (best-of-generation) for each generation is shown on the right.

5 Conclusion

Due to the length of the experiment we were only able to perform a single run with the real mobile robot. Although this experiment does not provide enough data to calculate reliable statistics, it shows that one is able to evolve a hierarchical control architecture using genetic programming on a large physical mobile robot. If evolution is carried out on a large physical mobile robot safety measures must be taken to avoid damage to the environment and to the robot. Although the time required for the experiment described here is prohibitive for most practical applications, one is able to use high speed computer simulations instead provided that reality is accurately modelled. On our case computer simulations were used to find a suitable representation and the main parameters for the run on the physical mobile robot.

6 Acknowledgements

The author is currently supported by a scholarship according to the Landes-graduiertenförderungsgesetz. For our experiments we used the lil-gp Programming System [31].

References

1. W. Banzhaf, P. Nordin, and M. Olmer. Generating adaptive behavior using function regression within genetic programming and a real robot. In *2nd International Conference on Genetic Programming, Stanford*, 1997.
2. V. Braitenberg. *Vehicles: Experiments in Synthetic Psychology*. The MIT Press, Cambridge, Massachusetts, 1984.
3. R. A. Brooks. A robust layered control system for a mobile robot. *IEEE Journal of Robotics and Automation*, RA-2(1):14–23, March 1986.
4. R. A. Brooks. Artifical life and real robots. In F. J. Varela and P. Bourgine, editors, *Toward a practice of autonomous systems: Proceedings of the First European Conference on Artificial Life*, pages 3–10, Cambridge, MA, 1992. The MIT Press.
5. D. Cliff, P. Husbands, and I. Harvey. Evolving visually guided robots. In J.-A. Meyer, H. L. Roitblat, and S. W. Wilson, editors, *From animals to animats 2: Proceedings of the Second International Conference on Simulation of Adaptive Behavior, Honolulu, Hawaii, 1992*, pages 374–383. The MIT Press, 1993.
6. M. Ebner and A. Zell. Evolution of a control architecture under real world constraints using genetic programming, unpublished manuscript, 1997.
7. D. Floreano and F. Mondada. Automatic creation of an autonomous agent: Genetic evolution of a neural network driven robot. In D. Cliff, P. Husbands, J.-A. Meyer, and S. W. Wilson, editors, *From animals to animats 3: Proceedings of the Third International Conference on Simulation of Adaptive Behavior, Brighton, England, 1994*, pages 421–430. The MIT Press, 1994.
8. D. Floreano and F. Mondada. Evolution of homing navigation in a real mobile robot. *IEEE Trans. Syst., Man, and Cybern. B*, 26(3):396–407, 1996.
9. I. Harvey. Artificial evolution and real robots. *Artificial Life and Robotics*, 1(1):35–38, 1997.
10. I. Harvey, P. Husbands, and D. Cliff. Issues in evolutionary robotics. In J.-A. Meyer, H. L. Roitblat, and S. W. Wilson, editors, *From animals to animats 2: Proceedings of the Second International Conference on Simulation of Adaptive Behavior, Honolulu, Hawaii, 1992*, pages 364–373. The MIT Press, 1993.
11. I. Harvey, P. Husbands, and D. Cliff. Seeing the light: Artificial evolution, real vision. In D. Cliff, P. Husbands, J.-A. Meyer, and S. W. Wilson, editors, *From animals to animats 3: Proc. of the Third International Conference on Simulation of Adaptive Behavior, Brighton, England, 1994*, pages 392–401. The MIT Press, 1994.
12. I. Harvey, P. Husbands, D. Cliff, A. Thompson, and N. Jakobi. Evolutionary robotics: the sussex approach. *Robotics and Autonomous Systems*, 20:205–224, 1997.
13. J. R. Koza. *Genetic Programming, On the Programming of Computers by Means of Natural Selection*. The MIT Press, Cambridge, Massachusetts, 1992.
14. J. R. Koza. Evolution of Subsumption. In J. R. Koza. *Genetic Programming I: On the Programming of Computers by Means of Natural Selection*, pages 357–393. The MIT Press, Cambridge, Massachusetts, 1992.

15. J. R. Koza. *Genetic Programming II, Automatic Discovery of Reusable Programs.* The MIT Press, Cambridge, Massachusetts, 1994.

16. John R. Koza. *Obstacle-Avoiding Robot.* In J. R. Koza. *Genetic Programming II: Automatic Discovery of Reusable Programs*, pages 365–376. The MIT Press, Cambridge, Massachusetts, 1994.

17. M. Matarić and D. Cliff. Challenges in evolving controllers for physical robots. *Robotics and Autonomous Systems*, 19:67–83, 1996.

18. L. A. Meeden. An incremental approach to developing intelligent neural network controllers for robots. *IEEE Trans. Syst., Man, and Cybern.B*, 26(3):474–485,1996.

19. F. Mondada and D. Floreano. Evolution of neural control structures: some experiments on mobile robots. *Robotics and Autonomous Systems*, 16:183–195, 1995.

20. S. Nolfi. Evolving non-trivial behaviors on real robots: A garbage collecting robot. *Robotics and Autonomous Systems*, 22:187–198, 1997.

21. S. Nolfi, D. Floreano, O. Miglino, and F. Mondada. How to evolve autonomous robots: different approaches in evolutionary robotics. In R. A. Brooks and P. Maes, editors, *Artificial Life IV: Proceedings of the Fourth International Workshop on the Synthesis and Simulation of Living Systems*, pages 190–197. The MIT Press, 1994.

22. P. Nordin and W. Banzhaf. Genetic programming controlling a miniature robot. In *Working Notes of the AAAI-95 Fall Symposium Series, Symposium on Genetic Programming, MIT, Cambridge, MA, 10-12 November 1995*, pages 61–67, 1995.

23. P. Nordin and W. Banzhaf. A genetic programming system learning ostacle avoiding behavior and controlling a miniature robot in real time. Technical Report Sys-Report 4/95, Dept. of Computer Science, University of Dortmund, Germany, 1995.

24. P. Nordin and W. Banzhaf. Real time evolution of behavior and a world model for a miniature robot using genetic programming. Technical Report SysReport 5/95, Dept. of Computer Science, University of Dortmund, Germany, 1995.

25. M. Olmer, P. Nordin, and W. Banzhaf. Evolving real-time behavioral modules for a robot with gp. In M. Jamshidi, F. Pin, and P. Danchez, editors, *Robotics and Manufacturing, Proc. 6th International Symposium on Robotics and Manufactoring (ISRAM-96), Montpellier, France*, pages 675–680, New York, 1996. Asme Press.

26. C. W. Reynolds. An evolved, vision-based model of obstacle avoidance behavior. In C. G. Langton, editor, *Proceedings of the Workshop of Artificial Life III SFI Studies in the Sciences of Complexity*, pages 327–346. Addison-Wesley, 1994.

27. C. W. Reynolds. Evolution of corridor following behavior in a noisy world. In D. Cliff, P. Husbands, J.-A. Meyer, and S. W. Wilson, editors, *From animals to animats 3: Proceedings of the Third International Conference on Simulation of Adaptive Behavior, Brighton, England, 1994*, pages 402–410. The MIT Press, 1994.

28. C. W. Reynolds. The difficulty of roving eyes. In *Proceedings of the First IEEE Conference on Evolutionary Computation*, pages 262–267. IEEE, 1994.

29. C. W. Reynolds. Evolution of obstacle avoidance bahavior: Using noise to promote robust solutions. In K. E. Kinnear, Jr., editor, *Advances in Genetic Programming*, pages 221–241, Cambridge, Massachusetts, 1994. The MIT Press.

30. M. S. Wilson, C. M. King, and J. E. Hunt. Evolving hierarchical robot behaviours. *Robotics and Autonomous Systems*, 22:215–230, 1997.

31. D. Zongker and B. Punch. *lil-gp 1.01 User's Manual (support and enhancements Bill Rand).* Michigan State University, March 1996.

Field Programmable Processor Arrays

Pascal Nussbaum, Bernard Girau, Arnaud Tisserand

Centre Suisse d'Electronique et de Microtechnique (CSEM)
Jaquet-Droz 1. CH-2007 Neuchâtel. Switzerland
Email : pascal.nussbaum@csemne.ch, bernard.girau@csemne.ch,
arnaud.tisserand@csemne.ch

Abstract. The FPPA concept has been developed in 1996, to address two main goals. The first one is to develop massively parallel processing unit arrays that require large chip areas, and the second one is to obtain a good substratum for evolutive algorithms. In such algorithms, phenotype evaluation is the most computing intensive task, so that if such kind of architecture is scaled beyond the conventional limits, the efficiency is significantly improved. However, several problems occur when chip size increases, as clock-skew and fabrication defects. The proposed solution is an array of cells, in which each cell contains a small low-power processor and a hierarchy of memories. The array tolerates defects by self-test and autonomous reconfiguration. Clock-skews are compensated by interconnecting the cells through an asynchronous protocol communication network. The FPPA is a single-chip MIMD machine that should be considered as a very coarse grain FPGA.

1 Introduction

The FPPA concept has been developed as an extension of the Biologiciel project [MGM95, MPMS94, PPC96] born in 1994 at the Logic Systems Laboratory (LSL), a laboratory of the Ecole Polytechnique Fédérale de Lausanne (EPFL). Our main goals are to provide a very large single-chip MIMD (Multiple Data Multiple Instruction) architecture, as well as a substratum enhancing experiments for genetic programming/configuring. The proposed chip architecture is interesting for hardware genetic programming since it allows both processor-level programming and interconnection-level configuring. The parallelism grain size is the one of a (small) processor, and the network bandwidth and configurability is similar to FPGAs (Field-Programmable Gate Arrays) capabilities. This approach implies the *FPGA implementation philosophy* to be applied to a processor network. Therefore, the main goal is no more to balance the computation load over the network like in conventional parallel computers, but rather to reserve exactly the needed computing, memory and communication resources for an application.

1.1 Applications

Standard applications may take advantage of the intrinsic parallelism of the target architecture, as digital signal and image processing or digital control do.

But maybe the most interesting ones are those being discovered in the field of evolving hardware/software, where the needs of parallel evaluation of phenotypes are critical.

- In genetic programming: a set of procedures of a program could be evolved according to their specifications (fitness), while the program is running on a separate part of the substratum. For instance, if a procedure is dedicated to data transformation before computation, a cluster of processors (one procedure phenotype per processor) could evolve it to obtain the fastest code. Another procedure dedicated to computation could be evolved to obtain a good precision/time ratio.
- In a higher level of hierarchy, if clusters of cells are considered, the configuration of the communication network may be included in the phenotype. This possibility enables fast treatment of datas by pipelining operations over multiple cells, as in adaptive ATM cell scheduling [LMH96]. The reduced sensitivity to timing variations allows complex systems to be evolved, while keeping phenotype reproducibility. Embedding a genetic farm into portable or transportable machines is the keypoint.

1.2 Massive parallelism

Nowadays, a very wide range of architectures are provided for parallel programmers. However communications are often the critical bottleneck. When several chips have to be connected together (SLiM [CLH97], C-NAPS [Sol95], ...), linear arrays are chosen to simplify inter-chip connexions, so that an efficient system may be obtained. This network topology does not provide efficient data propagation. If grids or hypercubes topologies are really needed, proposed solutions are often assigning one processing node to a chip, and interconnection is performed at board/system level as in CM1, Cray or Parsytec systems. The Transputer concept [Lim91] is the closest to ours, even if it requires one chip per node, this node contains the communication routing system and may be directly connected to neighbors to realize a grid. In general, all proposed solutions have a high node processing power compared to their communication network bandwidth and **latency**, which is supressed in the FPPA.

1.3 Single chip solution

Being able to design a very large IC without paying very expensive production costs is interesting for conventional design, but particularly interesting for genetic programming. Distributing phenotypes evaluations over several independent processors or processor clusters is a major advantage. Large arrays allow important populations of phenotypes to be evaluated. A system based on one chip lowers hardware costs and gives an opportunity to embed evolutive systems in a wider range of applications. However, large silicon areas force us to deal with two major problems:

1. Yield collapse: as size increases, the probability to have a fabrication defect in the circuit increases too.
2. Clock-skew: long metal lines imply long delays for signals propagation.

For each of them a particular solution has been proposed that takes advantage of the architecture based on an array of cells.

– The fault tolerance concept used in FPPAs has been defined within the Embryonic project developed in close collaboration with EPFL. The concept is based on an isotopological reconfiguration of the cell array. This means that the grid topology is not affected by the reconfiguration that avoids the defect. Therefore, the application configuration doesn't need to be replaced and re-routed [MPMS94], [MGM95]. All cells are numbered by X- and Y-coordinates. When a defect is detected (by self-test), the faulty cell is disconnected from the network by rendering the cells of its line and column transparent for message propagation. The remaining active cells are re-numbered, without considering the transparent cells (fig. 1). When renumbering is done, the new functionality of each cell can be again determined from the global system description. The description resides entirely in each cell, or best, may be sent again over a communication network to save memory resources [PPC96]. Isotopological reconfiguration is not performant for

$$
\begin{pmatrix}
S_{nn} & S_{nn} & S_{nn} & S_{nn} & S_{nn} \\
C_{04} & C_{14} & C_{24} & C_{34} & S_{nn} \\
C_{03} & C_{13} & C_{23} & C_{33} & S_{nn} \\
C_{02} & C_{12} & C_{22} & C_{32} & S_{nn} \\
C_{01} & C_{11} & C_{21} & C_{31} & S_{nn} \\
C_{00} & C_{10} & C_{20} & C_{30} & S_{nn}
\end{pmatrix}
\xRightarrow{Self-Test}
\begin{pmatrix}
C_{04} \leftrightarrow C_{14} & C_{24} & C_{34} \\
C_{03} \leftrightarrow C_{13} & C_{23} & C_{33} \\
C_{02} \leftrightarrow C_{12} & C_{22} & C_{32} \\
\updownarrow \quad \otimes & \updownarrow \quad \updownarrow \quad \updownarrow \\
C_{01} \leftrightarrow C_{11} & C_{21} & C_{31} \\
C_{00} \leftrightarrow C_{10} & C_{20} & C_{30}
\end{pmatrix}
$$

Fig. 1. Isotopological reconfiguration. C_{xy} are the numbered and configured cells. S_{nn} are the inactive spare cells used when reconfiguring

any size of cell, array and yield. Lines and columns of cells are lost when reconfiguration occurs, that is a maximum of $2n - 1$ cells in an n^2 cells array per defect. Foundries often give the yield for a specific chip size. For FPPAs, we must consider the yield concerning cell size (not FPPA size) to compute the probability of faulty cells in the array.

– The clock-skew over long metal lines in circuits makes synchronous designs slow or very power-consuming. Each cell of the FPPA being of reasonable size, it can be locally synchronous, and communicate with neighbors by using an asynchronous protocol. This concept allows the designer to distribute clocks without taking care of skews, which are compensated by the asynchronous network. Note that if an application covers more than one cell, it can't be programmed as if the system was synchronous over the corresponding cluster. A new principle of design synchronization (*macrosynchronization*, section 2.4)has been developed to answer this problem.

2 FPPA architecture

We designed the *Field Programmable Processor Array* chip as an array of function programmable cells with communication resources, like FPGAs. The system-level programmation that we address (2.5) has determined the parallelism grain size (a processor), which must be able to execute one or more of the application processes, as well as testing and configuring the hardware of the cell.

2.1 Array of cell architecture

To achieve testability and reconfiguration, cells are only connected to their nearest neighbors (fig. 2). This precaution ensures that no link over an arbitrary number of cells should be checked, and would make the test more complex. The communication channels are 8bit parallel buses for high bandwidth, with asynchronous handshake signals. This disposition allows the chip designer to scale

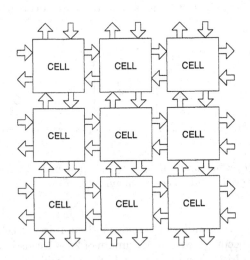

Fig. 2. Array overall architecture, based on a cell that connects nearest neighbors (N,E,W,S) with communication channels. The communication grid is totally homogenous.

up the system beyond the conventional limitations. The only global signal is the clock, which is not *skew-sensitive* anymore. Power distribution might be a problem however, because of the power lines which must scale with the number of cells in a column. The first array to be integrated will contain about one hundred cells. The cell (fig. 3) contains a small 8bit processor, three different memories, and an interface to the communication network. The processor is based on a Harvard model. Two independent static RAM memories are devoted to user's data and programs. Libraries containing configuration, self-test and math functions are located into a ROM memory. The cell's architecture and inner modules

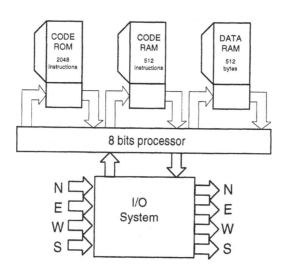

Fig. 3. Cell architecture, based on a processor and a communication interface (IO System) connected to nearest neighbors (N,E,W,S)

design have been evaluated by *emulating* different solutions on an industrial emulator (MetaSystem). An array of 3x5 cells has been realized and emulated. This array has been configured and programmed for different applications in order topoint out both advantages and weaknesses of the architectural choices. The applications were:

- Kohonen self-organizing map
- Character Recognition
- Industrial Process Control

Different testings induced changes of the processor core (register number and size, handling of special signals, ...), of the communication interface properties and of the memory sizes.

2.2 Interconnection resources: the communication network

The communication network is a grid. Each cell is connected to its nearest neighbors. Byte messages are exchanged in order to simplify the interface with the processor. When a message has to be communicated towards a cell over several intermediate cells, it is propagated from cell to cell by achieving the asynchronous handshake at each time. The testability of such an architecture is easier than if connections over multiple cells were allowed, but the delays induced by a same path length are longer.

The communication interface *(I/O System)* is independent from the cell processor. It is dedicated to route messages between the processor and the neighbours, and between the neighbors themselves. The interface relieves the processor of the big load due to message routing and asynchronous protocol achievement. The concept is close to FPGA, but it adds flexibility. A Look-Up Table *(LUT)* is programmed with the routing schemes, as in an FPGA. In order to reduce connection reservations for signals, resulting in a fast overload of resources when implementing an application, the communication interface emulates *virtual channels* to allow different signals to share the same hardware resources.

The asynchronous handshake implemented in our solution is a standard 4-phases request/acknowledge protocol. Two versions have been realized: A high bandwidth self-timed version demonstrated the feasability of a completely asynchronous interface, and a more compact synchronous design where the handshake signals are resampled by cells local clocks. The last one has been chosen for the first test integration because it is easier to debug.

Sharing hardware resources for different signal channels is performed by adding an identification label to each message. We call it *index*. If we consider only one signal (a particular channel for message propagation) propagated through the cell, the folowing routing constraints can be determined:

1. One message has one and only one source: North cell, East cell, West cell, South cell or local Processor *(N,E,W,S,P)*
2. One message can be propagated to any destination but the source:

$$\begin{pmatrix} N \to E\ W\ S\ P \\ E \to N\ W\ S\ P \\ W \to N\ E\ S\ P \\ S \to N\ E\ W\ P \\ P \to N\ E\ W\ S \end{pmatrix}$$

Multicast can be performed by routing a signal to several destinations. This possibility is similar to FPGAs, where an output signal may be connected to several gates. Such a simple message multiplication does not exist in conventional parallel platforms.

The index used in the FPPA design is 3 bits wide, giving the possibility to manage 8 virtual channels over one hardware channel. Each input of the interface (N,E,W,S,P) has a routing LUT where the destinations of each virtual channel can be programmed. 7 channels are available to the user, since channel 0 is reserved for configuration and special functions.

2.3 The processor

The choice of an 8bit processor could appear to be a weakness. However, a large number of applications only require 8bit coding. On the other hand, one may compensate this weakness by configuring clusters of cells to handle longer words if required: it is particularly interesting to be able to choose, for instance, to implement a 64-bits multiplier with only 8 cells in a small area, or with 64 cells for better timing performances.

The chosen processor is a CoolRISC 816, developed at CSEM [PMA97] for low-power applications. The main features of the core are: 16 registers, one-cycle 8x8 multiplier, 4 data pointers, automatic post-increment of data pointers, halt mode, programmable 1 to 16 times frequency divider. The memory model is a Harvard one. The Clock Per Instruction ratio is 1. The frequency targeted for the first integration is 10MHz (10MIPS).

As shown in the inner architecture of the cell, 3 memories have been added to the chosen processor core. A data static RAM of 512 bytes, a code ROM of 2048 instructions containing the self-test and math functions, and a code static RAM of 512 instructions, where the application program is stored.

2.4 FPPA particularities

Embedded applications require to deal with **power management**. The chosen processor is a low-power one. It can be halted when the local process is terminated, and awaken when an event (incoming message) occurs. The memories have been generated using CSEM know-how in low-power memories to ensure the lowest possible consumption. The estimated consumption for an array of 10x10 cells achieving 1GIPS of peak performances is about 1W @ 10MHz.

Defect tolerance imply their detection. The FPPA cell has been designed to be able to self-test after power-on reset. If the test fails, the cell sends a reconfiguration message to the neighbors. When the cell is too damaged to be able to send a message, it is detected by the neighbors. The test concerns the different parts of the cell:

1. The clock distribution. Redundancy is used, by doubling the clock signal (column and row signal distribution).
2. The processor core. A self-test program scans the core parts through different instruction sequences. If this test completes in less clock cycles than a hardwired prediction, the processor core is bug-free.
3. The memories. Data RAM is tested by a program. Code RAM and ROM content is not accessible as data (Harvard model). These memories have been made fail-safe by Hamming encoding of instructions. The address decoders (not covered by the Hamming detection) are tested by a program which stores and calls small subroutines at different addresses of the RAM code, and jumps between different parts of the ROM code.
4. The communication interfaces and channels. This test is performed in collaboration with neighboring cells in two phases. In the first phase, the chess-

board black cells are *message reflectors* (fig. 4), and white cells are testing their own I/O channels. In the second phase, the roles are exchanged.

In this last part of the test, the neighbors of totally inoperant cells are able to perform the diagnostic and to send the reconfiguration messages (fig. 4).

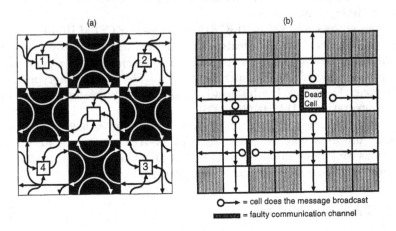

Fig. 4. (a) I/O testing with the help of the neighbors in the *chessboard* mode. The 4 channels to be tested in one phase are represented in the corners. (b) the reconfiguration messages are broadcast by the remaining healthy cells.

The user cannot considerate the cell clock as a synchronization signal, but only as a local working frequency. Another signal is distributed for synchronization at the system level, called *macrosynchronization*. As described in fig. 5, the macrosynchronization can be achieved over large silicon areas, because its frequency is 10^2 to 10^3 slower than the fast cell clock. When using this feature, applications need to be decomposed in a set of small processes that perform a time-limited task in each cell. The macro-synchronization cycle must be longer than the longest task. This principle may be used in order to perform run-time self-test of the application: when receiving macrosynchronization, each cell can check its neighbours to ensure task completion. If a neighbour does not answer, reconfiguration may be performed, so that the application is again ready to run. As local data are lost, the algorithm cannot be considered as a fail-safe mechanism, but rather as a run-time self-repair one.

2.5 FPPA programmation

An industrial point of view requires to provide a programmation tool to simplify standard and less standard programming of such a platform. The system design specification that is the closest to our architecture is the data-flow/state-machine

Conventional:

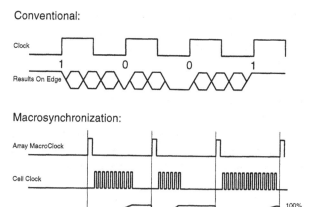

Macrosynchronization:

Fig. 5. System macro-synchronization. In standard systems, data are available every clock cycle. In FPPA, each cell contains a process requiring several local clock cycles to complete

one. Data-flows can describe how cells exchange data, and state-machines can describe the inner behavior of the cell. The tool automatically maps the system description into FPPA configuration files. A generic diagram capture tool (Mentor Graphics Renoir) is used to generate a VHDL description of the system. The block mapping and communication routing is performed by a tool (Kandinsky) developed at CSEM. This software automatically maps the different blocks of the VHDL data-flow description, according to the link densities between blocs. After mapping, it routes the communication channels according a set of different heuristics, and computes the routing tables (LUTs) to connect the blocks according to the most efficient heuristics. A software being designed at the LSL will convert VHDL description of state-machines in a CoolRISC assembly program achieving the necessary data exchanges and state computations in two successive steps. The state machine update is clocked by means of a macrosynchronization signal. In a second phase, this software will handle state machine spliting and merging, in order to optimize the distribution over the cells.

3 Evolvable hardware

After having described the main properties of the proposed substratum, it is time to prospect the field of evolvable hardware experimentations.

3.1 Genetic Algorithms

Genetic Algorithms *GAs* have been introduced by John Holland [Hol75]. They are a valuable tool for high-dimensional parameter space exploration. GAs have been used for a wide range of applications, but particularly focusing on function optimization or on the design of structures (programs, ICs, mechanics, ...) where the possible variations of the phenotype cover a too large domain to be explored by conventional means.

3.2 Two cases study

Two examples are demonstrating the high interest to involve software **and hardware** of a machine in a continuous evolution process which gives the possibility to cover simply defined tasks with means that have not to be clearly defined by the designer.

- The works of Thompson [Tho96] about gate-level configuration issued by GAs. Thompson uses a new family of fastly reprogrammable FPGAs (Xilinx 6200 Family) to perform the design of a tone-discrimination device. The phenotype evaluation lasts 5 seconds, the population is composed of 50 individuals, and the 5000 generations of the experiment last about 14 days. The design is dependent on the intrinsic physical properties of the hardware.
- The works of Salami [Sal96] about function-level configuration issued by GAs. The addressed problem is image compression using a Predictive Coding algorithm. The solution proposed by Salami is based on an array of floating point units (FPUs) that are pre-configured. Each cluster can achieve an operation like multiplication, division, if-then-else, sin, cos and so on. Fitness is based on signal-to-noise ratio and compression rate.

3.3 FPPA as a substratum for evolvable hardware

Considering both previous examples, one might appreciate the necessity of a configurable substratum well adapted to the complexity level of the addressed problem.

- For applications of high complexity level, it seems that the functional level is a better solution. The FPUs used by Salami offer very high bandwidth, but local control is costly (one FPU per if-then-else). Wider local memories are also a requirement for algorithms that need more intermediate storage. These two weaknesses may be compensated by FPPAs.
- For the reproducibility of the obtained codes (sharing configuration between different machines) as well as for high complexity levels, it seems that only well controled properties of the substratum should be used. The intrinsic architecture of the FPPA makes it insensitive to mapping changes or bias of fabrication parameters (only the speed could be altered).
- Though the MIMD-like arrangement of the processors may generate programming hardships (as for any MIMD platform), the FPPA chip will be a low-cost solution for massively-parallel genetic programming that is currently performed on expensive and bulky machines.

4 Conclusion

We have developed and implemented a new concept of very coarse grain FPGA leading to a single-chip MIMD grid. However, FPPAs must be programmed with the same approach as FPGAs, by reserving only the resource needed by the target application The architecture has been validated by testing a set of representative applications. The possibility is given to program such a platform with high-level tools for non-specialized users. Depending on further results given by experiments, new generations of FPPAs could be released with more node power, and chip sizes beyond the standard limits. We are convinced to have launched a new computing paradigm that might be of great help to open the era of *single chip genetic farm*.

References

[CLH97]   Hyunman Chang, Changhee Lee, and Myung H.Sunwoo. Slim-ii: A linear array simd processor for real-time image processing. In *Proceedings of the International Conference on Parallel and Distributed Systems*. IEEE Computer Society, 1997.

[Hol75]   John H. Holland. Adaptation in natural and artificial systems. *The University of Michigan Press*, 1975.

[Lim91]   INMOS Limited. *The T9000 Transputer products overview manual*. INMOS Limited, 1991.

[LMH96]   Weixin Liu, Masahiro Murakawa, and Tetsuya Higuchi. Atm cell scheduling by function level evolvable hardware. In *Lecture Notes in Computer Science*, volume 1259. International Conference on Evolvable Systems (ICES96), Springer, 1996.

[MGM95]   Daniel Mange, Maxim Goeke, and Dominik Madon. Embryonics: A new family of coarse-grained field-programmable gate arrays with self-repair and self-reproducing properties. In E.Sanchez and M.Tomassini, editors, *Towards Evolvable Hardware*. volume 1062 of Lecture Notes in Computer Science, Springer-Verlag, 1995.

[MPMS94] Pierre Marchal, Christian Piguet, Daniel Mange, and Andre Stauffer. Embriological development on silicon. In *Proceedings of Artificial Life IV*. Rodney A.Brooks and Pattie Maes, 1994.

[PMA97]   Christian Piguet, Jean-Marc Masgonty, and Claude Arm. Low-power design of 8-b embedded coolrisc microcontroller cores. *IEEE Journal of Solid-state Circuits*, 32(7):1067–1077, July 1997.

[PPC96]   P.Nussbaum, P.Marchal, and C.Piguet. Functionnal organisms growing on silicon. In *Lecture Notes in Computer Science*, volume 1259. International Conference on Evolvable Systems (ICES96), Springer, 1996.

[Sal96]   Mehrdad Salami. Data compression based on evolvable hardware. In *Lecture Notes in Computer Science*, volume 1259. International Conference on Evolvable Systems (ICES96), Springer, 1996.

[Sol95]   Adaptive Solutions. *CNAPS Data Book 2.0*. Adaptive Solutions, Inc., 1400 N.W. Compton Drive, Suite 340, Beaverton, OR 97006, 1995.

[Tho96]   Adrian Thompson. An evolved circuit, intrinsic in silicon, entwined with physics. In *Lecture Notes in Computer Science*, volume 1259. International Conference on Evolvable Systems (ICES96), Springer, 1996.

General Purpose Computer Architecture Based on Fully Programmable Logic

Kiyoshi Oguri, Norbert Imlig, Hideyuki Ito, Kouichi Nagami,
Ryusuke Konishi, and Tsunemichi Shiozawa

NTT Optical Network Systems Laboratories,
1-1 Hikarinooka Yokosuka-shi Kanagawa-ken 239-0847 Japan
{oguri, imlig, hi, nagami, ryusuke, shiozawa}@exa.onlab.ntt.co.jp

Abstract. We propose a new general-purpose computer architecture based on programmable logic. It consists of a dual-structured array of cells accommodating a fixed "built-in part" and a freely programmable "plastic part". We call this composition the "Plastic Cell Architecture" (PCA). The built-in part with its fully connective two- dimensional mesh structure serves as a communication platform based on the cellular automata model. It is responsible for the configuration of the plastic part which implements a sea of logic gates similar to programmable devices (FPGA). The key point of our architecture is dynamic, distributed object instantiation during runtime. An object can encapsulate data and/or behavior and communicates with other objects through a unique type of message passing implemented in hardware. Thus, PCA combines the merits of fine grained, high performance hardware implementation and the dynamic memory allocation capabilities of software.

1   Introduction

A wide range of methods can be adopted to implement functionality. One of the easiest and most generic ways is programming a CPU. The main advantage of this approach is its flexibility and cost-effectiveness since there is no need for device production. With the introduction of field programmable devices the same has come true for wired logic, too. Because of performance concerns "computing in the space domain" (hardware) is often more attractive than "computing in the time domain" (software) [2]. However, despite the development of high- level hardware description languages and sophisticated synthesis tools, programmable logic is still playing a minor role. One reason for this might be that the performance gap between hardware and software is often marginal given the torrid increase in CPU performance seen in the last few years. Another reason why co-processing systems usually are not able to boost overall performance lies in the von Neumann bus bottleneck indicated in Fig. 1. While the CPU is capable of hiding this effect by employing a sophisticated memory hierarchy, the co- processor engine implemented as a slave in the FPGA performs rather poorly. In other words: as long as memory is employed, hardware will lose out over software. Another way to realize high performance computing systems is to use multi CPU configurations. Except for some special applications, the overheads caused by CPU-CPU communication, cache coherency protocols, and high cost make such systems not significantly more attractive than single CPU configurations.

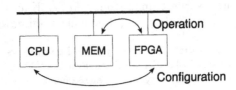

Fig. 1. Von Neumann bottleneck slowing down configuration and operation

In order to overcome the above dilemma, we must think of a composition that implements functionality not as a central block but in a more distributed manner. In order to fully exploit the advantages of fine- grained parallelism unique to wired logic we have already developed a hardware description language SFL (Structured Function description Language) and its dedicated synthesis and CAD tools PARTHENON [1]. SFL is an object-oriented register-transfer-level language that allows describing functionality at a high level of abstraction. PARTHENON takes an SFL description as an input and generates highly optimized circuits for various target implementations (ASIC, FPGA).

In order to maximize the merits of wired logic we first thought of a language that excludes constructs for describing memory arrays. However, some designers of the PARTHENON user community did not stop utilizing memory. This is why their designs lost the cost-performance race against highly integrated CPU-memory configurations. But why has this dualistic architecture been so successful? One reason lies in the fact that VLSI fabrication processes for memory and random logic differ. Due to the regular structure the density of the memory process is higher. However this is not the major reason that led to the dominance of software.

Figure 2 summarizes the common and different points when implementing functionality in software and hardware. While software is compiled into a memory image, hardware is synthesized into a static circuit image. Starting from this netlist presentation a mask layout is generated for ASIC production in a fab or a configuration bit-stream for downloading into an FPGA. The major difference lies in the runtime behavior. Software supports dynamic memory allocation. On the other hand dynamic reconfiguration is a basic omission of hardware. Since many algorithms operate on dynamically generated complex data structures, software is the only way to implement such functionality. Algorithms implemented as wired logic can also be realized by memory. Until the appearance of FPGA about ten years ago, nobody thought about exploiting this dynamic nature of wired logic. Even today, FPGAs are mainly used as dumb static devices for prototyping. One reason for this is that the overhead of the runtime configuration is still too high. Again the master-slave relationship between CPU and FPGA leads to the bus bottleneck familiar with von Neumann configurations.

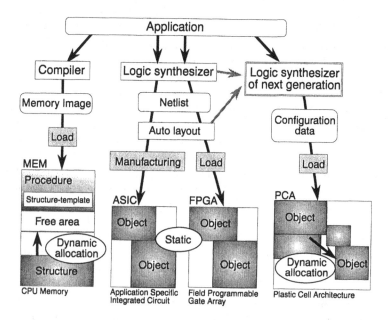

Fig. 2. The common and different points when implementing functionality in software and hardware

Apart from the dynamic allocation of data, programmable logic devices allow the dynamic allocation of fine-grained functionality according to the processing load. Such a feature cannot be supported by software even with a multi CPU configuration. In order to implement a general purpose computing system, which utilizes only wired logic, one must create a dynamic allocation mechanism for data and functions.

The rest of the paper is organized as follows. In Sect. 2 we analyze the basic properties of a dual-structured general-purpose computer architecture. We then describe the elements of the plastic cell architecture. Section 3 discusses objects, dynamic module instantiation, and message passing. After presenting a short example in Sect. 4 we conclude our paper by pointing out some related work and considerations for VLSI implementation.

2 Dual-Structured General Purpose Computer Architecture

In order to allow dynamic runtime reconfiguration after the initial state has been set, a system must be composed of two distinct elements: a variable part as the place holder of the new configuration and a fixed part responsible for setting the former. This dual-structured composition is a must for any general-purpose computer architecture. In the example of a CPU-memory configuration, the memory corresponds to the variable part and the CPU to the fixed part. The same is true for FPGAs. Their configuration SRAM memory corresponds to the variable part while the attached CPU can be regarded as the fixed part.

Figure 3-a shows the basic cell of our architecture. We call the variable part the "plastic part" and the fixed part the "built-in part". In order to exploit parallelism, the cells are arranged in an orthogonal array as indicated in Fig. 3-b. There are various connection methods such as one-, two-, three-dimensional or hyper-cube meshes. For straightforward communication between plastic and built-in parts and easy VLSI implementation we choose a two-dimensional mesh. The array of built- in parts forms a network of *cellular automata*. A cellular automaton is a processing model in which a collection of identical state machines exchange information with their neighbors by a simple set of rules. A similar principle is applied in the *systolic array* configuration for implementing inherently parallel algorithms [12]. Cellular programming sets the functions of the state machines. The cellular automata array is responsible for configuring the plastic part and message passing between encapsulated hardware objects during operations. A cluster of configured plastic parts forms a processing element which itself is able to configure other plastic parts through the built-in part array.

Due to its dual-structured property, this architecture allows dynamic, distributed runtime configuration of data and functions. Thus we call this composition the "Plastic Cell Architecture" (PCA) [3] [4]. In addition, the regular structure simplifies the development of efficient CAD tools and guarantees high density VLSI implementation. Since functionality is configured "in the field", yield problems caused by manufacturing defects can be reduced. Moreover, if LSI processes become more sophisticated, the two-dimensional cell array can be folded into a compact three-dimensional representation. Because of yield problems such technology can not be used for memories or gate-arrays.

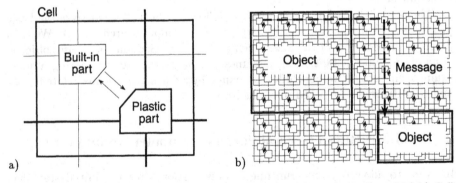

Fig. 3. a) Basic cell with built-in part (fixed) and plastic part (reconfigurable) b) Cell array with two hardware-objects communicating by message passing

3 Design of Plastic Cell Architecture

3.1 Objects and message passing

In order to allow dynamic instantiation of circuits, objects are formed as shown in Fig. 3-b. An object encapsulates data and/or behavior and has no input/output terminals. The only way to communicate with other objects is by message passing implemented in the built-in part.

During runtime there are no constraints and objects can freely grow or merge. If message passing by the built-in part is too slow it is possible to statically connect two objects using the routing resources of the plastic part.

3.2 Dynamic allocation mechanism

In order to dynamically allocate new objects the following steps are necessary:

1. Free area is searched and reserved
2. A new object is generated by injecting its behavior into the plastic part
3. The newly formed object is enabled for runtime operation
4. The communication path between built-in part and plastic part is opened to allow message passing
5. Operation of hardware object, communication with other objects
6. After receiving a release message, the area is freed for new allocation

Objects and generated objects have a mother-child relationship. Thus, only the mother object can send a release message and free the resources of an allocated child.

3.3 Message routing

In order to enable proper configuration, the route of each message should be fixed. In fact, we employ exact routing rather than adaptive message routing. All information is packed into the following 10 commands:

- Routing commands (SET_WEST, SET_NORTH, SET_EAST, SET_SOUTH, SET_PLASTIC)
- Configuration commands (CONFIG_IN, CONFIG_OUT, OPEN, CLOSE)
- Clear command (CLEAR)

The commands are processed by the built-in part. They consist of simple orders like "set the route west, north" etc. After being set once the cellular state machine holds its state until a CLEAR command is delivered. This guarantees that the route of a message is constant. Every built-in part that is in the initial state consumes one command of the message. For routed cells, all commands are passed from cell to cell.

Each built-in part has 5 input- and 5 output ports as shown in Fig. 4-a. Except in the case of an output port conflict, all input ports behave independently. This enables a wide range of message crossing as depicted in Fig. 4- b.

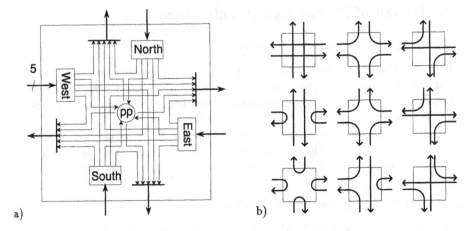

Fig. 4. a) Built-in part b) Possible routing configurations

3.4 Configuration message

There are two types of interconnection paths between the built-in part and the plastic part. The first type is a message sending- and receiving path, the other one is a configuration path as shown in Fig. 5-a. A gate in the message path can be opened/closed by the OPEN/CLOSE commands. In the initial state, all gates are closed. If a gate is closed the corresponding object cannot send/receive any message. This is because we do not want any hazardous effects while an object is being constructed. Furthermore, the resources of an active object are freed with the CLOSE command.

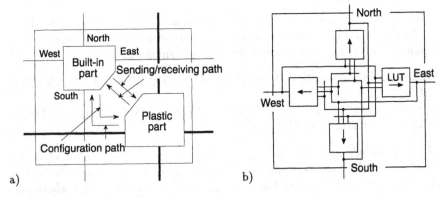

Fig. 5. a) Interconnection paths between built-in part and plastic part b) Orthogonal LUT structure of plastic part

Figure 6 shows a configuration message. It consists of 5 parts. The first part specifies the route to the cell that will be configured first. The second part starts with the CONFIG_IN command. The following fixed sized data packet is downloaded into the plastic part's configuration memory. The following route command is related to the cell that has just been configured. This sequence is repeated until all cells of a hardware object are configured. At this point the hardware object is in a stand-by mode ready for operation. To activate the object, its "task register" must be set. The "task concept" is a part of the semantics of our hardware description language SFL. The behavior of an object is only enabled for execution if the corresponding task is activated. Again, to prevent unpredictable operation before configuration is completed, the task register is set in the third part of the configuration message. The fourth part opens the gate mentioned above. Finally, the CLEAR command at the end of the message cleans up the fixed routing path set by the first part.

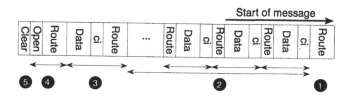

Fig. 6. The 5 parts of a configuration message

3.5 Read out message (memory object)

If the gate of the message path is closed, objects cannot affect anything outside. Because of this we can use the plastic part's configuration memory for general purpose storage. We call such an object a memory object. In order to read out data from a memory object the read out message uses the CONFIG_OUT command. After receiving this command, the cell switches to the configure-out mode. This mode operates in the same manner as the normal mode with one exception: after receiving a CLEAR command in the configure-out mode, the cell outputs the contents of its configuration memory via the routed path to the target object. After the reading out process has finished the memory object forwards the CLEAR command in order to release the routing states of the built-in parts. This sophisticated mechanism allows multiple copies of a hardware template to be freely routed to various places.

3.6 State machine of built-in part

The state machine of the built-in part handles message passing. Using above 10 commands we can construct the following 4 messages:

1. Configuration message (Fig. 6)
2. Release message

3. Read out message
4. Message to object

Messages 1. and 3. were explained above while messages 2. and 4. are simple and need not be discussed in detail. The state machine exists at every input port of the two dimensional built-in part array. After logic synthesis with the PARTHENON CAD tool, the size of one built-in part becomes 1038 gates (register= 5 gates, 3-input logic= 1 gate). There is room for further optimization.

3.7 Plastic part

Figure 5-b shows the basic structure of the plastic part. The most important feature is its regular orthogonal structure. While the configurable part of conventional FPGAs consists of several building blocks like look-up tables (LUT), flip-flops, and routing switches, our plastic part is very simple. Every cell consists of four identical LUT memories with output tristate buffers connected in a symmetric way. A cell can implement any logic function with up to 3 input variables, register, or routing resources. A *Muller C latch* is a principal element of an asynchronous circuit and can be constructed with two feedback coupled LUTs.

An important issue is the balance between built-in part and plastic part elements. It depends on the size of the two circuits and is not fixed yet.

4 Discussion

In the following section we discuss the properties that make applications highly suitable for PCA mapping. As an example, a dynamic pattern matching engine was implemented and simulated. Here are the factors that play key roles:

- Simple, regular, and modular structure for efficient PCA mapping and dynamic template allocation
- Locality of communication for extensive pipelining (stream processing with little data- and control dependencies)
- Large support of parallelism at the bit-, object- and system levels

In fact, the pattern matching engine described below satisfies almost all of these conditions. Compared to software implementations, pattern matching algorithms realized in hardware achieve large speed increases [11]. Due to the dynamic allocation properties of the PCA architecture, our pattern matcher is very flexible. It combines the functionality of a *content addressable memory* (CAM) and dedicated processing agents. Thus we call this composition plastic CAM (P-CAM). Its basic structure is depicted in Fig. 7-a. It consists of 3 parts:

1. Hardware-OS for dynamic allocation services and template library with hard-wired object bit-slices
2. Heap for search pattern allocation
3. Processing agents that perform dedicated action after a hit event

4.1 Hardware-OS and template library

The Hardware-OS is the core of the system. It is responsible for resource management and provides dynamic allocation services such as 'new( )' and 'delete( )'.
After a user request for new pattern installation is received, the OS generates
a configuration message. This stream passes through the template library and
copies the data stored in the memory objects to the designated area in the pattern heap. This process works in parallel because of the decentralized message
passing control scheme located in every cell automaton of the built-in part. Furthermore, no expensive pins are wasted for memory data- and address buses
because all intelligence is located on chip. If the Hardware-OS must be partitioned into several chips, only the injection port pins are needed.

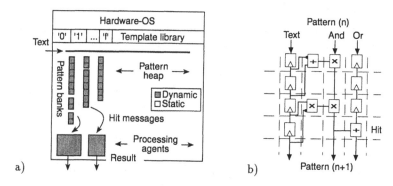

Fig. 7. a) The basic structure of the pattern matcher. b) The layout of an atom of a
pattern.

Figure 7-b depicts the layout of a 4-bit pattern library template. The advantage of bit-serial implementation lies in the fact that communication is highly
localized [13]. Thus mapping and routing of bit-slices is a straight forward task.
Since the PCA architecture uses the LUT resources for routing within objects,
mapping and routing tasks can be optimized in a single CAD tool. Powerful
languages such as LOLA [14] for efficient mapping of bit- sliced, parameterized
hardware modules can be found in the public domain.

4.2 Pattern heap

After all users have installed their desired patterns in dedicated pattern banks,
the text is streamed through the pattern heap. Since latency is not important
while throughput is, the whole datapath is heavily pipelined. The amount of text
in a database is in the range of Mega- to Gigabytes. It can be processed without
any interruption. Thus we do not have to care about expensive hazard processing
and pipeline flushing. Retiming the registers at the pattern boundaries keeps the
overall delay constant. The hit results of all patterns in a bank are fed into the

shift register or-queue that generates trigger signals for the agent processor units. If the user wants, they can alter or allocate new patterns during the streaming process. This makes P-CAM an ideal platform for concurrent processing.

Exploiting parallelism is one of the major reasons for hardware implementation. The PCA architecture allows a designer to take advantage of various types of parallelisms. Because of the fine grain architecture, bit-level parallelism can be fully exploited. This makes the hard-wired patterns in the P-CAM application very compact. Parallelism at the object level is also possible since all pattern banks work independently. Due to the scalability of the PCA cell array even parallelism at the system level can be accomplished if the patterns are partitioned among multiple chips.

4.3 Processing agents

The communication between pattern banks and dedicated processing agents is fully asynchronous. As soon as a pattern in the bank matches the input text a hit message is sent to its processing agent. In our P-CAM application these objects work as simple counters, one per user. Every user accesses his result count after the whole stream has passed through the system. Of course, the counter themselves are allocated dynamically. The user can further specify the desired counter resolution.

Above P-CAM was implemented and tested on our *pcasim* software simulator. Due to the complexity of the Hardware-OS and the fact that the synthesis tools are still under development, the allocation task was implemented on a CPU. It generates the configuration messages and injects the streams via a dedicated port. All other functions are available as bit-sliced hardware objects in the template library. The simplicity, regularity, and modularity of the P-CAM make it an ideal candidate for PCA implementation. It is easy to balance resource constraints (size of template library, inter-object hardware communication) and performance constraints (logic and routing delay, dynamic allocation overhead).

Many digital signal processing operators (addition, multiplication, shifter, filter) can be realized in a bit-sliced systolic manner resulting in designs with very attractive performance-area products. In addition, the dynamic allocation properties of the PCA architecture make it possible for an application to adapt to the computation needs (resolution, algorithms) in a changing environment.

We are now investigating the feasibility of an adaptive, high performance, optical link transmission system with intelligent, bufferless routing. In order to further explore the semantics of dynamic module instantiation the "Constraint Satisfaction Problem" (CSP) [5] and Genetic Algorithm (GA) implementations are being examined.

5   Related Research

Our plastic cell architecture with its built-in and plastic facilities is a combination of a reconfigurable logic device (FPGA) and a cellular automaton. A von

Neumann architecture is obtained if only one cell part is extracted. The memory corresponds to the plastic part and the built-in part can be compared with a mini CPU engine. On the other hand, if we take out the plastic part of the cell, the array of built-in parts form a regular cellular automaton array. Further, if the plastic part can function only as data memory, the configuration can be regarded as a multiple CPU network. A *connection machine* utilizes the plastic part only as wired logic that is configured simultaneously by a central control unit.

Some runtime reconfigurable SRAM based FPGAs have recently entered the market. However, a central control unit, i.e. a CPU, is required for configuration. A more decentralized configuration approach is proposed in the *wormhole system* [6] [7] [8] [9]. The difference between their approach and ours is the fine granularity of the Plastic Cell Architecture. The authors in [10] propose a highly fault-tolerant cellular architecture that mimics nature and biological concepts. It is widely known that many algorithms can be speeded up if memory can realize data storage and useful functions. This is the case with the plastic part of our cell.

6 Conclusion

We have introduced a new general-purpose computer architecture based on programmable logic. We are now simulating the architecture in order to balance performance, resource, and VLSI implementation issues. With a 0.5 μm process technology, about 4 Mbit of static SRAM memory or 400 Kgate of random logic (gate array) can be fabricated. On the other hand, only 8000 FPGA LUTs (including configuration memory) or so can be placed on today's state of the art chips. Supposing that about 10 gates can be realized per LUT (after intensive synthesis and optimization), a design with up to 80 Kgates can be implemented on one chip. Therefore, the gate density of an FPGA LSI is about 5 times smaller than that of a gate array. If we assume that about 100 memory bits are consumed per LUT (16 bits for memory, the rest for configuration information) we get 800 Kbit of static memory as a whole. Again this is only one fifth the density achieved by regular SRAM chips. Therefore, in terms of density, FPGAs lie in an unfavorable position between memory chips and gate arrays.

That's one reason why some people in the VLSI industry still take a rather pessimistic stance against programmable logic devices. They would rather stick to the familiar von Neumann paradigm. In order to boost the density of programmable logic devices, more research should be performed in mixed memory/logic process technology. By proposing this new general- purpose computer paradigm, we hope that a breakthrough in this area will soon be achieved. In our opinion, programmable devices will become a core technology and make obsolete today's dualistic memory-processor computing architectures.

References

1. Y. Nakamura, K. Oguri and A. Nagoya: "Synthesis From Pure Behavioral Descriptions," High-Level VLSI Synthesis, Edited by R. Camposano and W. Wolf, Kluwer Academic Publishers, pp.205-229, June, 1991
   http://www.kecl.ntt.co.jp/car/parthe/
2. R. Hartenstein, J. Becker, M. Herz and U. Nageldinger: "Parallelization in Co-Compilation for Configurable Accelerators," in Proc. of Asia and South Pacific Design Automation Conf. (ASP-DAC'98), pp. 23-33, February, 1998
3. K. Nagami, K. Oguri, T. Shiozawa, H. Ito and R. Konishi: "Plastic Cell Architecture: Towards Reconfigurable Computing for General-Purpose," in Preliminary Proc. of IEEE Symposium on Field-Programmable Custom Computing Machines (FCCM'98), April, 1998
4. H. Ito, K. Oguri, K. Nagami, R. Konishi and T. Shiozawa: "Plastic Cell Architecture for General Purpose Reconfigurable Computing," to appear in Proc. of IEEE International Workshop on Rapid System Prototyping (RSP'98), June, 1998
5. T. Shiozawa, K. Oguri, K. Nagami, H. Ito, R. Konishi and N. Imlig: " A Hardware Implementation of Constraint Satisfaction Problem Based on New Reconfigurable LSI Architecture," to appear in Proc. of International Workshop on Field Programmable Logic and Applications (FPL'98), August, 1998
6. H. Schmit: "Incremental Reconfiguration for Pipelined Applications," in Proc. IEEE Workshop on FPGAs for Custom Computing Machines, pp. 47-55, April, 1997
7. W. Luk, N. Shirazi and P. Y.K. Cheung: "Compilation Tools for Run-Time Reconfigurable Designs," in Proc. IEEE Workshop on FPGAs for Custom Computing Machines, pp. 56-65, April, 1997
8. J. Burns, A. Donlin, J. Hogg, S. Singh and M. Wit: "A Dynamic Reconfiguration Run-Time System," in Proc. IEEE Workshop on FPGAs for Custom Computing Machines, pp. 66-75, April, 1997
9. R. A. Bittner, Jr and P. M. Athanas: "Computing Kernels Implemented with a Wormhole RTR CCM," in Proc. IEEE Workshop on FPGAs for Custom Computing Machines, pp. 98-105, April, 1997
10. P. Nussbaum, P. Marchal and Ch. Piguet: "Functional Organisms Growing on Silicon," in Proc. of The First International Conference on Evolvable Systems: From Biology to Hardware (ICES96), vol. 1259 of Lecture Notes in Computer Science, Springer Verlag, pp. 139-151, October, 1996
11. E. Lemoine and D. Merceron: "Run Time Reconfiguration of FPGAs for Scanning Geomic DataBases," in Proc. FCCM95, P. Athanas and K.L. Pocek (eds.), IEEE Computer Society Press, pp. 85-89, 1995
12. S. Y. Kung: "VLSI Array Processors," Prentice Hall, ISBN 0-13-942749-X, 1988
13. P. Denyer and D. Renshaw: "VLSI Signal Processing: A Bit-Serial Approach," Addison-Wesley Publishing Company, 1985
14. H. Eberle, S. Gehring, S. Ludwig, and N. Wirth: "Tools for Digital Circuit Design using FPGAs," Departement Informatik, Institut fuer Computersysteme, ETH Zurich, 1994
    http://www.cs.inf.ethz.ch/cs/group/wirth/projects/cad_tools/

Palmo : Field Programmable Analogue and Mixed-Signal VLSI for Evolvable Hardware

Alister Hamilton, Kostis Papathanasiou, Morgan R. Tamplin and Thomas Brandtner

Department of Electrical Engineering, University of
Edinburgh, King's Buildings, Mayfield Road,
Edinburgh EH9 3JL, Scotland.
Email : Alister.Hamilton@ee.ed.ac.uk

Abstract. This paper presents novel pulse based techniques initially intended to implement signal processing functions such as analogue and mixed-signal filters, data converters and amplitude modulators. Field programmable devices using these techniques have been implemented and used on a demonstration board to implement analogue and mixed-signal arrays. The rich mix of analogue and digital functionality provided by Palmo systems combined with the fact that they may accept random configuration bit streams makes them most attractive as platforms for evolvable hardware.

1   Introduction

Research in Edinburgh on Field Programmable Devices has produced a novel range of analogue and mixed-signal techniques, implementations and results. Evolvable Hardware [1] (EHW) can extend the application base for our techniques by exploiting novel, complex or adaptable circuitry without the need for explicit programming. In this paper we demonstrate the inherent computational flexibility of our techniques and their suitability to evolvable hardware platforms.

Until recently, the lack of suitable programmable devices has limited EHW work to simulation. While this extrinsic evolution has produced interesting results [2], the physical peculiarities of the actual material from which the components are made, have not been exploited (for better or worse) during the evolution process. Evolution takes place at a higher abstract level using models of the components involved. This approach is therefore incompatible with real time adaptable hardware.

Some [3] believe intrinsic (on-chip, non-simulated) evolution is necessary to take full advantage of EHW for circuit design. With the introduction of the XILINX XC6216 and other Field Programmable Gate Arrays (FPGA), intrinsic evolution became possible. Each member configuration of an evolving population could be tested on-chip during evolution rather than in simulation. Intrinsic evolution has produced configurations which are compact but, unlike good hand-designed examples, less tolerant to environmental variations like temperature, and virtually impossible to analyse [4, 5, 6].

FPGA devices have increased the possibilities of EHW. Yet, they continue to limit the work to a digital domain. Current examples of *analogue* EHW demonstrate circuit design tasks of a "pick and place" nature [7, 8, 9, 10], used to choose component values and decide their location in a circuit [7, 2]. These are extrinsically rather than intrinsically produced circuits.

Desirable target applications for EHW, such as signal processing and agent control or autonomous robotics, are often well-suited to analogue processing. These applications could take particular advantage of the potential compactness, lower energy consumption, and adaptability of a reconfigurable evolved circuit. While it may be desirable to broaden EHW to analogue and mixed signal circuitry, accurate circuit simulation is not always practical for EHW [3, 2].

Commercially available programmable analogue devices such as the Motorola Field Programmable Analogue Array (MPAA020) [11] and the Zetex Totally Reconfigurable Analog Circuit (TRAC020) [12] are emerging in the marketplace, but neither is ideally suited to evolvable hardware. The Motorola device is a switched capacitor sampled analogue data device that may be damaged by illegal configurations. Proprietary information released under non-disclosure agreements is required to avoid damaging the chip. The Zetex device is interesting as it is continuous time but requires support switches and external components to make it useful in this context.

Here we present novel programmable analogue and mixed signal techniques that provide suitable platforms for both intrinsic analogue and mixed-signal hardware evolution without the restrictions of the Motorola or Zetex devices.

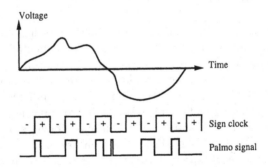

Fig. 1. *Palmo* signal representation.

2   Signal Coding in Programmable Analogue Arrays

In most programmable analogue systems signals are represented by voltages or currents. In the ideal case the output of each programmable analogue cell must be able to be connected to the input of any other programmable analogue cell in the array. This ideal requirement is not achievable using voltage mode or current mode circuit techniques. In voltage mode systems there is a limit to the drive capability of the cell that is generating the signal and noise may be introduced into the signal from a variety of sources. Therefore voltage mode programmable analogue cells tend only to be connected to others in a local region of the chip. Currents on the other hand may be distributed across large chip areas. However, each current output from the cell requires a separate circuit so that an arbitrarily high fan out is not possible.

Considerations such as these led us to consider a novel signal representation for the implementation of programmable analogue arrays. Instead of using a voltage or current to represent a signal value, a digital pulse is used to represent discrete analogue signals where the magnitude of the signal is encoded in the width of the pulse. The sign of the signal is determined by whether the pulse occurs in the positive or negative cycle of a global *sign* clock [13]. Figure 1 shows an input voltage waveform converted into a Palmo signal. The novelty and computational flexibility of the systems reported here arise directly from this method of representing analogue signal values. Due to the significance of the signal representation we have named these systems Palmo systems, where Palmo is Hellenic for pulse beat, pulse palpitation or series of pulses.

3 Palmo Programmable Analogue Cells

In choosing the functionality of the Palmo analogue cell the strategy was to implement a basic analogue function that could be used in a variety of different signal processing tasks [14]. The most common analogue signal processing tasks are filtering, analogue to digital conversion and digital to analogue conversion. These signal processing tasks may be implemented using only three simple analogue functions; an integrator, a scaler (i.e. a circuit that multiplies a signal by a constant value) and a comparator. An integrator may be turned into a scaler if the integration is reset between analogue samples and the gain of the integrator is programmable. It turns out that if we design an integrator that accepts Palmo signals as inputs and generates Palmo signals at the output, then the integrator will contain a comparator. We therefore chose to make the programmable analogue Palmo cell a programmable integrator.

In the design of analogue circuits great care should be taken to make the operation of the circuit insensitive to variations in the materials that the circuits are made from and variations in the environment in which the circuit operates. It is not possible to make exact capacitor values and generate exact current values on chip, for example, but it is possible to get highly accurate *ratios* of capacitors and currents. Palmo integrators have been designed with these considerations in mind [13, 14, 15, 16] and the gain of the integrator is defined by

$$K = \frac{I_{int}}{I_{ramp}} \cdot \frac{C_{ramp}}{C_{int}} \tag{1}$$

where I_{int} and C_{int} are the integrator current and capacitor and I_{ramp} and C_{ramp} generate an internal ramp waveform that is normally used to derive the Palmo output signal from the cell. In order to make the gain, K, of the integrator programmable, we can make the currents and capacitors programmable. It is trivial to design current digital to analogue converters and if we make each capacitor out of an array of smaller capacitors, then we can easily switch in different capacitor ratios depending on the gain required. Thus the gain, K, for each Palmo integrator may be defined by a few digital bits stored locally on chip within each integrator. Typically K has 9 bits of resolution.

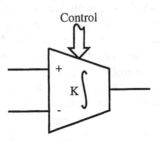

Fig. 2. *Palmo* integrator cell symbol.

Figure 2 shows the Palmo integrator cell symbol. The signals that control the cell comprise a bus so that the scaling factor, K, may be loaded into the cell, and some simple clocked digital signals that control the flow of data through the cell.

4 Palmo Systems

At present our Palmo chips contain an array of programmable analogue Palmo cells. The pulsed signals for each Palmo cell and all the digital control signals are fed straight off chip and into a standard Xilinx FPGA. The FPGA may be programmed with digital logic circuits that control the functionality of the Palmo cells and that control the interconnection of the Palmo cells. Thus Palmo cells may be connected together and programmed to perform various analogue signal processing functions. A photograph of a Palmo development system is shown in Figure 3. This development system contains two Palmo chips each containing 8 programmable analogue Palmo cells, a Xilinx FPGA, a microcontroller and some analogue electronics for power supply management and reference signal generation for the Palmo chips. The host PC downloads the programming bitstream for the FPGA via the microcontroller. This sets up the control circuits and interconnect for the Palmo array. The host PC then downloads the K factors for the Palmo cells and the circuit is ready to operate.

We have implemented a number of analogue filters and analogue to digital converters using this system.

Fig. 3. *Palmo* development system.

5 Mixed-Signal Palmo Systems

Since Palmo signals are digital pulses it is not necessary to go through an analogue to digital conversion process before manipulating these analogue quantities
in the digital domain [14, 16]. Due to the temporal nature of the pulse representing the analogue quantity, efficient digital circuits may be devised to operate
on these signals. For example, consider the digital circuit of Figure 4 which multiplies an incoming pulse signal by $a0 = 5$. Assume that the latch is cleared

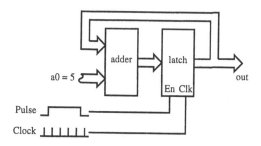

Fig. 4. *Palmo* signal multiplied by a constant, $a0 = 5$.

initially. While the pulse signal is high, $a0 = 5$ is added to the contents of the
latch at every clock cycle. Since the width of the pulse is proportional to its
magnitude, at the end of the pulse the output of the latch will contain a digital
word that represents the result of the multiplication of the pulsed signal by the
constant. Thus processing of Palmo signals may occur in the digital domain.

A digital word may be converted to a Palmo signal and fed into the programmable analogue cells for processing. This might be achieved by parallel loading a down counter with the digital word and using the time the counter takes to count down to zero to generate the Palmo signal.

The digital domain may be used to generate non-linear functions for example, by using a digital look-up table or adaptive systems where digital circuits monitor the signal processing as it occurs and adapt the parameters as required.

6 Performance and Implementation Issues

Our early Palmo devices are linear integrators implemented in CMOS technology that require a good analogue fabrication process and a relatively high power supply voltage (10V) [14]. The sampling frequency of this system is below 1MHz, largely due to the relatively slow operation of the voltage mode comparator used in the design.

Recent Palmo devices being fabricated at the time of writing use log domain integrators implemented in BiCMOS technology and may be operated at 5V [15]. The maximum sampling frequency of this system is 5MHz achievable by the use of fast current mode comparators. We hope to be able to operate BiCMOS circuits at much higher sampling frequencies (up to 20MHz) or much lower voltages (down to 1V) once we have proved the concept with working VLSI devices.

As sampling frequencies increase we will experiment with Palmo systems that integrate the programmable analogue cells and digital logic on the same device. This will allow us to evaluate the Palmo technique as a solution to mixed-signal circuit problems.

In all these systems the bit pattern that defines the gain of the cell and the control signals that operate the circuits may be set in any random pattern. There are some bit patterns that will prevent the circuit from working, and others that will make the circuit operate in a way that was not originally intended, but none that will fatally damage the cells. It is not difficult to allow only bit patterns that give operational cells to be programmed either by subtly changing the design of the cells or by providing a mask in software.

7 Palmo Application Results

Here we present results obtained from working Palmo devices for four different applications. The circuit solutions in each case were arrived at using traditional engineering techniques rather than evolutionary ones. In the first application the Palmo system has been configured as a 1st, 2nd and a 3rd order low pass filter [14, 16]. The results show the frequency response of the Palmo filter compared to the ideal response of the mathematical transfer function.

The second application demonstrates the mixed signal capability of the Palmo system where it is used to implement a z-domain finite impulse response filter

(FIR) with 24 coefficients [14, 16]. In this application, the analogue cells perform the function of a short-term analogue memory. The Palmo outputs Out_i gate the coefficients a_i, and the sum of those outputs is integrated in time. The waveform diagram in Figure 6 demonstrates the operation of this circuit for two pulsed inputs Out_0 and Out_1. The coefficients associated with these inputs are $a_0 = 3$ and $a_1 = 5$. At each integrating clock epoch, the sum of the coefficients of active inputs is added to the previous accumulated value. Thus $3 + 5 = 8$ is added for the first two epochs, while for the third epoch 5 is added. The overall output appears at the end of the sample period (in this example the output would have a value of 21). The results of a 24 tap FIR filter implemented using these techniques is shown in Figure 7. The VLSI results show an excellent match to the theoretical filter characteristic.

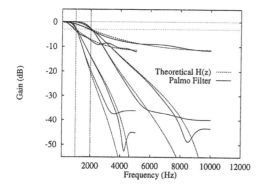

Fig. 5. *Palmo* 1st, 2nd and 3rd order analogue low pass filter results for cut-off frequencies of 1kHz and 2kHz. Theoretical results and results from Palmo VLSI devices.

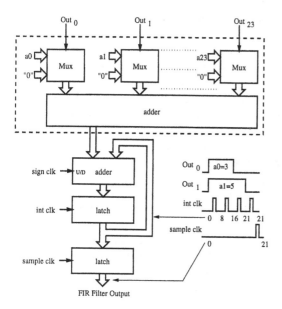

Fig. 6. Digital processing of *Palmo* signals for a 24 tap FIR filter implementation.

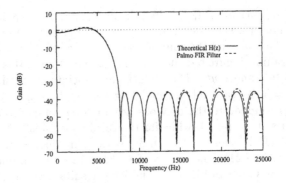

Fig. 7. *Palmo* mixed-signal 24 tap FIR filter implementation results.

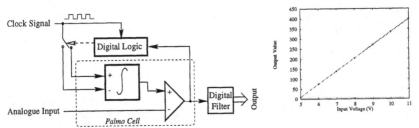

Fig. 8. *Palmo* $\Sigma - \Delta$ implementation.

Fig. 9. *Palmo* $\Sigma - \Delta$ implementation results

The third application is a Palmo $\Sigma - \Delta$ modulator (figure 8) which was implemented using the analogue Palmo cell with a clock signal fed to the *positive* input [16]. This is integrated in time, until the level of the integrated value reaches the value of the *analogue input*, which forces the comparator to change state. The *digital logic* block consequently redirects the *clock signal* to the negative input of the *Palmo* cell changing the direction of the integration. Thus, the integrated value oscillates about the true value of the input. A digital filter similar to the one used by a typical $\Sigma - \Delta$ modulator, can be used to average the output of the comparator. Results from this implementation verify the functionality of the circuit: we observed almost 9 bits of accuracy, and concluded that the conversion is sufficiently accurate.

The fourth application which has been demonstrated in simulation is continuous time signal multiplication. This extends the application area of Palmo circuits so that two signals may be multiplied together in continuous time rather than multiplying an analogue sample with a constant number (scaling). One possible application is amplitude modulation (AM). Since the BiCMOS circuits operate at MHz frequencies, it is conceivable that this technique can be used to modulate radio waves. Demodulation of AM signals is easily achieved using rectification and filtering - trivial tasks for Palmo cells. Thus radio communication may be a task that Palmo devices could perform. Figure 10 shows results from the multiplication of two sine wave signals; one derived from a Palmo in-

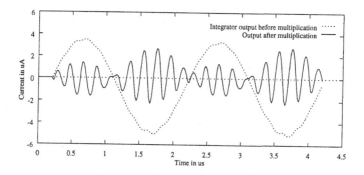

Fig. 10. *Palmo* amplitude modulation simulation results.

tegrator representing the signal and the other, which may be derived using 3 Palmo integrators, representing the carrier signal.

8 Discussion

While the circuit solutions to the applications presented in this paper have been engineered, there is no reason why circuit solutions may not be evolved either intrinsically or extrinsically.

Since the interface to Palmo cells is entirely digital, digital hardware description languages such as Verilog and VHDL may be used to simulate mixed-signal Palmo systems where the Palmo cell is modelled by a simple, but complete behavioural model of a programmable integrator. Extrinsic evolution may be used to determine a circuit solution to a problem in simulation before being downloaded to the Palmo system.

Since the Palmo devices can accept random bit patterns without damage, intrinsic evolution may be used to evolve circuits directly on-chip.

We have demonstrated analogue applications of the Palmo device, which may also be implemented using the Motorola device, for example. We have simulated continuous time multiplication which could be achieved using the Zetex device. However, due to a mixed-signal capability and the underlying signal representation, the Palmo device is also uniquely capable of realising algorithms normally associated with Digital Signal Processors.

These capabilities arose from human ingenuity; an evolvable system would be able exploit the rich mix of analogue and digital resources available in Palmo systems in ways that are currently unforeseen. Add to that the wireless communication inherent in Palmo systems and one could realise autonomous evolvable agents establishing their own unique communication protocols and behaviours.

Acknowledgements

This work is supported by the UK EPSRC grant reference GR/L36031.

References

1. X. Yao and T. Higuchi, "Promises and challenges of evolvable hardware," in *Proceedings of The First International Conference on Evolvable Systems: from Biology to Hardware (ICES96)* (T. Higuchi and M. Iwata, eds.), pp. 55–78, Springer Verlag LNCS 1259, 1997.

2. J. R. Koza, F. H. B. III, D. Andre, M. A. Keane, and F. Dunlap, "Automated synthesis of analog electrical circuits by means of genetic programming," *IEEE Transactions on Evolutionary Computation*, vol. 1, pp. 109–128, July 1997.

3. A. Thompson, *Automatic design of electronic circuits in reconfigurable hardware by artificial evolution.* Dphil thesis, University of Sussex, 1996.

4. A. Thompson, "Evolving electronic robot controllers that exploit hardware resources," in *Proceedings of the 3rd European Conference on Artificial Life (ECAL95)*, no. CSRP 368 in Lecture notes in Atificial Intelligence, pp. 640–656, 1995.

5. A. Thompson, "An evolved circuit, intrinsic in silicon, entwined with physics," in *Proceedings of The First International Conference on Evolvable Systems: from Biology to Hardware (ICES96)* (T. Higuchi and M. Iwata, eds.), pp. 390–405, Springer Verlag LNCS 1259, 1997.

6. A. Thompson, "Silicon evolution," in *Proceedings of Genetic Programming 1996 (GP96)* (J. K. et al, ed.), pp. 444–452, MIT Press, 1996.

7. R. Zebulum, M. Vellasco, and M. Pacheco, "Comparison of different evolutionary methodologies applied to electronic filter design," in *IEEE International Conference on Evolutionary Computation (ICEC98), Anchorage, Alaska*, in press.

8. D. H. Horrocks and Y. M. A. Khalifa, "Genetically derived filter circuits using preferred value components," in *Proc. IEE Colloq. on Analogue Signal Processing*, (Oxford, UK), pp. 4/1–4/5, 1995.

9. D. H. Horrocks and M. C. Spittle, "Component value selection for active filters using genetic algorithms," in *Proc. IEE/IEEE Workshop on Natural Algorithms in Signal Processing*, vol. 1, (Chelmsford, UK), pp. 13/1–13/6, 1993.

10. D. H. Horrocks and Y. M. A. Khalifa, "Genetic algorithm design of electronic analogue circuits including parasitic effects," in *Proc. 1st Online Workshop on Soft Computing (WSC1)*, 1996. See: http://www.bioele.nuee.nagoya-uac.jp/wsc1.

11. A. Bratt and I. Macbeth, "Design and implementation of a Field Programmable Analogue Array," in *FPGA '96*, Monterey California: ACM, February 1996.

12. D. L. Grundy, "A computational approach to VLSI analog design," *Journal of VLSI Signal Processing*, vol. 8, no. 1, pp. 53–60, 1994.

13. K. Papathanasiou and A. Hamilton, "Pulse based signal processing: VLSI implementation of a palmo filter," in *International Symposium on Circuits and Systems (ISCAS)*, vol. I, (Atlanta), pp. 270–273, IEEE, May 1996.

14. A. Hamilton and K. Papathanasiou, "Preprocessing for pulsed neural VLSI systems," in *Pulsed Neural Networks* (W. Maass and C. Bishop, eds.), MIT Press, in press.

15. T. Brandtner, K. Papathanasiou, and A. Hamilton, "A palmo cell using sampled data log–domain integrators.," *IEE Electronics Letters*, Vol. 34, No. 6, March 1998.

16. K. Papathanasiou and A. Hamilton, "Novel palmo techniques for electronically programmable mixed signal arrays," in *to appear in the International Symposium on Circuits and Systems (ISCAS)*, (California), IEEE, May 1998.

Feasible Evolutionary and Self-Repairing Hardware by Means of the Dynamic Reconfiguration Capabilities of the FIPSOC Devices

J.M. Moreno[1], J. Madrenas[1], J. Faura[2], E. Cantó[1], J. Cabestany[1], J.M. Insenser[2]

[1]Universitat Politècnica de Catalunya, Dept. of Electronic Engineering, Building C4,
Campus Nord, c/Gran Capità s/n, 08034 – Barcelona, Spain
moreno@eel.upc.es
[2]SIDSA, PTM, c/Isaac Newton 1, Tres Cantos, 28760 – Madrid, Spain
faura@sidsa.es

Abstract. In this paper we shall address the paradigms of evolutionary and self-repairing hardware using a new family of programmable devices, called FIPSOC (Field Programmable System On a Chip). The most salient feature of these devices is the integration on a single chip of a programmable digital section, a programmable analog section and a general-purpose microcontroller. Furthermore, the programmable digital section has been designed including a flexible and fast dynamic reconfiguration scheme. These properties provide an efficient framework for tackling the specific features posed by the emerging field of evolutionary computation. We shall demonstrate this fact by means of two different case studies: a self-repairing strategy for digital systems, suitable for applications in environments exposed to radiation, and an efficient implementation scheme for evolving parallel cellular machines.

1 Introduction

During the last years, the field of programmable logic has evidenced an increasing interest in the addition of dynamic reconfiguration capabilities to conventional FPGA architectures [1]. This is due to the necessity to provide resources which should allow for handling efficiently such applications as programmable/customisable processors or co-processors, adaptive signal processing architectures or custom computing machines, among others. This has resulted in a variety of proposals, coming from both the academic [2] and industrial [3], [4], [5] communities.

In this paper we shall concentrate our attention on the new family of devices introduced in [5], which is called FIPSOC (Field Programmable System On a Chip). The distinguishing characteristic of FIPSOC is the integration on a single device of an FPGA, a set of analog programmable cells and a standard microcontroller. Since there exists an efficient interface between the three sections of the device, it constitutes a natural platform for the prototyping and integration of mixed signal applications. Furthermore, a fast and efficient dynamic reconfiguration scheme has been included in the programmable digital section. It permits to work with two independent (in terms of functionality) hardware contexts, being possible to switch between them in just one memory access cycle (if the microcontroller handles the reconfiguration process) or immediately through a hardware signal. Additionally, the physical implementation of the configuration scheme for the programmable digital part permits to modify the configuration of one context while the other is being used. As we shall demonstrate by means of two case studies, these features make this device a good candidate for handling the specific requirements associated with evolutionary computation.

The organisation of this paper is the following: In the next section we shall provide an overview of the general organisation and capabilities of the FIPSOC device, focusing on the dynamic reconfiguration properties of its programmable digital part. Then we shall present the first case study, which consists of the implementation of a self-repairing scheme for digital systems, especially suitable for applications operating in environments exposed to radiation. Afterwards, an efficient implementation for parallel cellular machines will be introduced. Finally, the conclusions and future work will be outlined.

2   Architectural Overview of the FIPSOC Device

Figure 1 depicts the global organisation of the FIPSOC device.

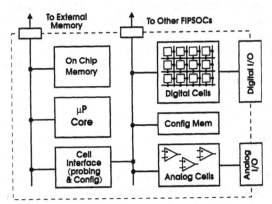

Fig. 1. Global organisation of the FIPSOC device.

As it can be seen, the internal architecture of the FIPSOC device is divided in five main sections: the microcontroller, the programmable digital section, the configurable analog part, the internal memory and the interface between the different functional blocks.

Because the initial goal of the FIPSOC family is to target general purpose mixed signal applications, the microcontroller included in the first version of the device is a full compliant 8051 core, including also some peripherals like a serial port, timers, parallel ports, etc. Apart from running general-purpose user programs, it is in charge of handling the initial setup of the device, as well as the interface and configuration of the remaining sections.

The main function of the analog section, whose structure is depicted in figure 2, is to provide a front-end able to perform some basic conditioning, pre-processing and acquisition functions on external analog signals.

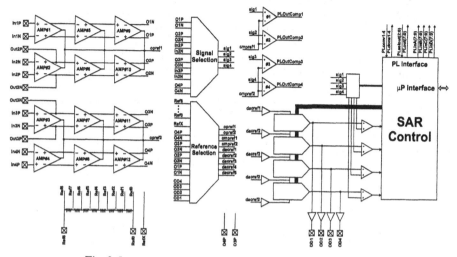

Fig. 2. Internal structure of the programmable analog section.

This section is composed of four major sections: the gain block, the data conversion block, the comparators block and the reference block. The gain block consists of twelve differential, fully balanced, programmable gain stages, organized as four independent channels. Furthermore, it is possible to have access to every input and output of the first amplification stage in two channels. This feature permits to construct additional analog functions, like filters, by using external passive components. The comparators block is composed of four comparators, each one at the output of an amplification channel. Each two comparators share a reference signal which is the threshold voltage to which the input signal is to be compared. The reference block is constructed around a resistor divider, providing nine internal voltage references. Finally, the data conversion block is configurable, so that it is possible to provide a 10-bit DAC or ADC, two 9-bit DAC/ADCs, four 8-bit DAC/ADCs, or one 9-bit and two 8-bit DAC/ADCs. Since nearly any internal point of the analog block can be routed to this data conversion block, the microprocessor can use the ADC to probe in real time any internal signal by dynamically reconfiguring the analog routing resources.

Regarding the programmable digital section, it is composed of a two-dimensional array of programmable cells, called DMCs (Digital Macro Cell). The organisation of these cells is shown in figure 3.

As it can be deduced from this figure, it is a large-granularity, 4-bit wide programmable cell. The sequential block is composed of four registers, whose functionality can be independently configured as a mux-, E- or D-type flipflop or latch. Furthermore, it is also possible to define the polarity of the clock (rising/falling edge) as well as the set/reset configuration (synchronous/asynchronous). Finally, two main macro modes (counter and shift register) have been provided in order to allow for compact and fast realisations.

The combinational block of the DMC has been implemented by means of four 16x1-bit dual port memory blocks (Look Up Tables – LUTs – in figure 3). These ports are connected to the microprocessor interface (permitting a flexible management of the LUTs contents) and to the DMC inputs and outputs (allowing for their use as either RAM or combinational functions). Furthermore, an adder/subtractor macro mode has been included in this combinational block, so as to permit the efficient implementation of arithmetic functions.

A distinguishing feature of this block is that its implementation permits its use either with a fixed (*static* mode) or with two independently selectable (*dynamic reconfigurable* mode) functionalities. Each 16-bit LUT can be accessed as two independent 8-bit LUTs. Therefore it is possible to use four different 4-LUTs in static mode, sharing two inputs every two LUTs, as depicted in figure 3, or four independent 3-LUT in each context in dynamic reconfigurable mode. Table 1 summarises the operating modes attainable by the combinational block of the DMC in *static* mode and in each context in *dynamic reconfigurable* mode.

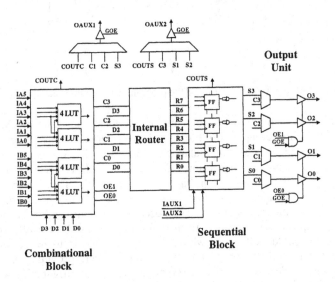

Fig. 3. Organisation of the basic cell (DMC) in the programmable digital section.

Table 1. Functionalities of the combinational block in *static* and *dynamic reconfigurable* modes.

Static mode	Dynamic reconfigurable mode
• 4 x 4-LUTs (sharing 2 inputs every two LUTs) • 2 x 5-LUTs • 1 x 6-LUT • 1 x 4-bit adder • 2 x 16x2-bit RAMs	• 4 x 3-LUTs • 2 x 4-LUTs • 1 x 5-LUT • 1 x 4-bit adder • 2 x 8x2-bit RAMs

Furthermore, since the operating modes indicated in table 1 are implemented in two independent 16x2-bit RAMs (8x2-bit RAMs in *dynamic reconfigurable* mode), it is possible to combine the functionalities depicted in this table. For instance, it is possible to configure the combinational block in order to provide one 6-LUT and one 16x2-bit RAM in *static* mode or two 3-LUTs and one 4-LUT in *dynamic reconfigurable* mode.

The multicontext dynamic reconfiguration properties have been provided also for the sequential block of the DMC. For this purpose, the data stored in each register has been physically duplicated. In addition, an extra configuration bit has been provided in order to include the possibility of saving the contents of the registers when the context is changed and recover the data when the context becomes active again.

In order to enhance the overall flexibility of the system, an isolation strategy has been followed when implementing the configuration scheme of the FIPSOC device. This strategy, depicted in figure 4(a), provides an effective separation between the actual configuration bit and the mapped memory through an NMOS switch. This switch can be used to load the information stored in the memory to the configuration cell, so that the microprocessor can only read and write the mapped memory. This implementation is said to have one *mapped* context (the one mapped in the microprocessor memory space) and one *buffered* context (the actual configuration memory which directly drives the configuration signals).

The benefits of this strategy are clear. First, the mapped memory can be used to store general-purpose user programs or data, once its contents have been transferred to the configuration cells. Furthermore, the memory cells are safer, since their output does not drive directly the other side of the configuration bit. Finally, at the expense of increasing the required silicon area, it is possible to provide more than one mapped context to be transferred to the buffered context, as depicted in figure 4(b). This is the actual configuration scheme which has been implemented in the FIPSOC device, and it permits to change the configuration of the system just by issuing a memory write command.

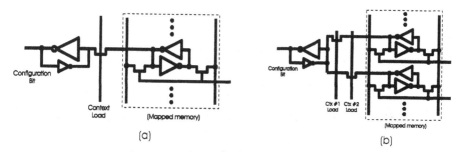

Fig. 4. (a) Configuration scheme. (b) Multicontext configuration.

In addition to this configuration scheme, an efficient interface between the microcontroller and the configuration memory has been included in the FIPSOC device, as depicted in figure 5.

As it can be seen, the microcontroller can select any section in the array of DMCs (the shaded rectangle depicted in figure 5), and, while the rest of the array is in normal operation, modify its configuration just by issuing a memory write command. Therefore, the dynamic configuration strategy included in the FIPSOC device shows two main properties: it is *transparent* (i.e., it is not necessary to stop the system while it is being reconfigured) and *time-efficient* (since only two memory write cycles are required to complete the reconfiguration, one to select the *logical rectangle* of DMCs to be reconfigured and one to establish the actual configuration for these elements).

Regarding the routing resources, the array of DMCs which constitutes the programmable digital section of the FIPSOC device is provided with 24 vertical channels per column and 16 horizontal channels per row. The routing channels are not identical, and have different lengths and routing patterns. Switch matrices are also provided to connect vertical and horizontal routing channels. There are also special nets (two per row and column) which span the whole length or height of the array, and whose goal is to facilitate the clock distribution.

In the next sections we shall show through two case studies how these properties can be efficiently exploited so as to handle the specific features of evolutionary computing systems.

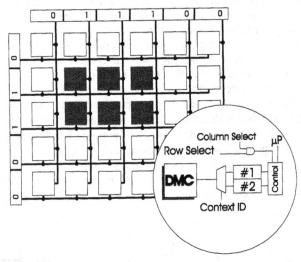

Fig. 5. Microcontroller interface for dynamic reconfiguration.

3 Case Study I: Self-Repairing Hardware for Space Applications

Digital circuits operating in space are subject to different types of radiation, whose effects can be permanent or transient [6]. The former are due to particles which get trapped in the silicon/oxide interface and appear only after long exposure to radiation, while the later are caused by the impact of a single charged particle (single event effects, SEE) in sensitive circuit zones.

An interesting type of SEEs corresponds to the so-called Single Event Upset (SEU), which is responsible for non-destructive changes (bit flips) of information bits stored by digital circuits [7]. Though there have been proposals to overcome this kind of transient errors by including self-test techniques in programmable logic systems

[8], the lack of efficient (in terms of time) reconfiguration mechanisms in commercial devices has prevented their use for real-time self-repairing applications.

The specific features included in the FIPSOC device which were explained in previous sections make it a natural candidate to overcome these limitations. In a first approach, it could be possible to map the entire digital application in the array of DMCs in *dynamic reconfigurable* mode. In this way, while one context is active and performing its functionality, the microprocessor can write in the other context the same configuration and then perform a context swap. Since the contents of the registers can be saved when the context is changed, a *clean* duplicate of the original system is thus obtained. If this process is performed periodically, this *hardware refresh* scheme could permit to correct automatically the errors induced by SEUs, with no need to stop the system.

However, the probability of an error induced by SEUs is proportional to the surface of the sensitive area in the system, which is related to the amount of internal memory. As a consequence, bearing in mind that the functionality of the DMCs is based on SRAM, this would impose very a short *hardware refresh* period to prevent errors, thus forcing the microcontroller to be used exclusively for testing purposes. Furthermore, an eventual transient error affecting a register could not be detected using this strategy, since it requires that the contents of the registers are held when the context swap is performed.

To overcome these problems, the self-testing and repairing strategy depicted in figure 6 is proposed.

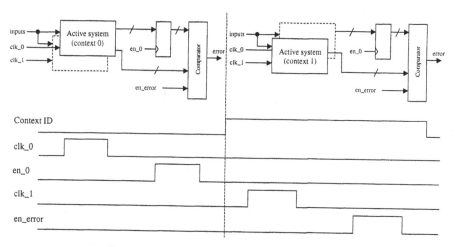

Fig. 6. Proposed self-testing and repairing strategy.

As it can be deduced from this figure, the basic principle of the self-repairing strategy consists in using the array of DMCs in *dynamic reconfigurable* mode and duplicating the functionality of the digital system in both hardware contexts. Both contexts are used with the same input signals in consecutive time windows (the signals clk_0 and clk_1 depicted in figure 6 are the global clock signals of the digital system mapped in the FIPSOC device), and their outputs are compared for equivalence, being the latch and comparator depicted in figure 6 implemented also in the array of DMCs. In this way, after a new set of input signals is presented to the system, they are processed by the copy of the digital system stored in context 0 and its internal state is advanced one step by activating the global clock signal (clk_0). Then, the outputs of the system are stored by means of a register activated by the clock signal en_0. Afterwards, a context swap is performed, but in this case the data stored in the registers of the DMC is not transferred between contexts. Once the input signals have been processed by the second context and its state advanced (by activating the clock signal clk_1), its outputs are compared (activating the en_error signal) against those produced by context 0, and if they are different an error signal is activated.

In the case the error signal is activated, the microcontroller fixes one of the two hardware contexts and, while it is still processing new inputs, refreshes the other one. Then the operation is inverted, so that the sequence depicted in figure 6 can be started again.

There is still a source for potential errors, constituted by the comparator stage of figure 6. Its influence can be however alleviated by implementing it in *dynamic reconfigurable* mode and forcing the microcontroller to refresh periodically the non-active context. Since the sensitive area is in this case much lower, the refresh period can be set long enough, so as to reduce the load over the microcontroller. Furthermore, since the microcontroller is able to access any input or output in the array of DMCs, the register and comparator stages can be actually implemented by the microcontroller, but at the expense of augmenting the time period required to process the input signals. As a consequence, if the programmable digital section is able to handle an operating frequency four times higher that the input frequency (since four steps are required between two successive new sets of input signals), it is possible to include a self-testing and repairing mechanism with almost no area overhead. In this way, the effect of SEUs caused by radiation is minimised, thus increasing the overall reliability of the application.

4   Case Study II: Evolution of Parallel Cellular Machines

Cellular Automata (CA) are dynamical systems whose behavior can be considered discrete both in terms of space and time [9]. It means that they consist of an array of cells, which are determined to be in one of a finite set of possible states, and whose update is performed synchronously in discrete time steps depending on local interaction rules.

As it was pointed out in [9], *non-uniform* CA (i.e., those in which the local interaction rules are not necessarily the same for every cell in the array) are of particular interest because they exhibit universal computation properties. Furthermore, by introducing the cellular programming paradigm, it is possible to make the non-uniform CA coevolve locally in order to solve a specific task. The dynamics of such a system can be divided in three main stages:

- *Initialisation:* The rules of the cells included in the array are initialised randomly.
- *Execution:* The cells interact locally, updating their state depending on its previous state and on their neighbors' current state. During this stage no change is produced in the rule associated with each cell.
- *Evolution:* Each cell is evaluated against a specific fit function and its rule is modified depending on the number of fitter neighbors.

It has been possible even to provide a hardware implementation for this principle [9], [10], thus demonstrating the possibility of attaining *online evolution*. In this section we shall demonstrate that the features included in the FIPSOC device offer an interesting alternative to cope the realisation of the dynamics inherent to parallel cellular machines. Recalling the configuration possibilities of the combinational block included in the DMC, it is possible to provide simultaneously in *dynamic configurable* mode the functionality of some logic functions (two 3-LUTs or one 4-LUT) and a 8 x 2-bit SRAM block, as depicted in figure 7.

Since the functionality of the cells in the CA is quite simple (a 3-LUT is enough for the one-dimensional synchronisation task handled in [9]), the combinational section of a DMC configured in this way can perform the rule of each cell in the CA during both *initialisation* and *execution* phases. The section configured as a SRAM cell plays an important role during the *evolution* stage of the CA. During this phase, depending on the number of fitter neighbors, the rule of each cell has to be modified for the next *execution stage*. It can be accomplished by an *evaluation unit*, able to determine the fitness score of each cell and its immediate neighbors and then to write the SRAM section accordingly. Once this modification is completed, the hardware context is changed, so that now the SRAM section of the previous context is used as a combinational function and the combinational function of the previous context is now the SRAM section.

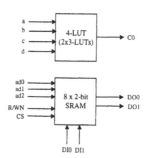

Fig. 7. Functionality of the DMC combinational block in *dynamic reconfigurable* mode.

Since the functionality of this *evaluation unit* is the same for every cell in the array, it could be possible to implement the *evolution stage* in a sequential fashion, one cell at a time. However, since very wide multiplexers have to be implemented, this solution results in a high area penalty, and usually, as in the realisation proposed in [9], this unit is included locally in each cell of the CA. In the FIPSOC device, these wide multiplexers can be *virtually* emulated by means of the dynamic reconfiguration scheme depicted in figure 4(b), which affects not only the configuration of the functional units, but also that of the routing resources. In this way, while one cell in the CA is evaluated and updated, the microcontroller can modify the configuration of the routing resources, so that in the next time step a context load is performed and the same evaluation/update process is performed for the next cell in the array. This strategy is depicted in figure 8, where we have represented with solid lines the actual connections between blocks and with dotted lines the *virtual* connections to be configured.

Using this *virtual multiplexing* strategy it is thus possible to obtain an reasonable trade-off between the cost of the system (since the overall complexity of the cells is lower) and its performance (the *evolution* phase is performed sequentially). It is therefore suitable for those applications where cost is the most limiting factor or where the system is able to cope with the additional timing penalty. As it can be easily deduced, this strategy gives an efficient solution to overcome the communication bottleneck inherent to massively parallel systems, like Artificial Neural Network models.

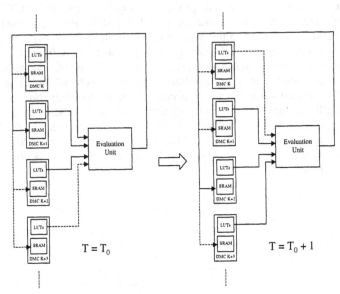

Fig. 8. *Virtual* multiplexing strategy by dynamic reconfiguration of the routing resources

5 Conclusions and Future Work

In this paper we have considered the hardware implementation of evolvable systems, taking as a reference a new programmable device, the FIPSOC chip, which includes a programmable digital section, a flexible analog section and a microcontroller.

After reviewing the main features of the device, especially those related with its dynamic reconfiguration properties, we have demonstrated its suitability for evolvable systems by means of two case studies: a self-repairing strategy for space applications and the implementation of the principles corresponding to evolving parallel cellular machines. Our current work is concentrated in finishing the last physical details of the device prior to fabrication, as well as in the development of tools able to aid the user in the management of the dynamic reconfiguration features of the system.

Acknowledgements

This work is being carried out under the ESPRIT project 21625 and spanish CICYT project TIC-96-2015-CE.

References

1. G. McGregor, P. Lysaght: Extending Dynamic Circuit Switching to Meet the Challenges of New FPGA Architectures. Field Programmable Logic and Applications, Proceedings of FPL'97. Springer-Verlag (1997) 31-40.
2. A. DeHon: Reconfigurable Architectures for General-Purpose Computing. A.I. Technical Report No. 1586. MIT Artificial Intelligence Laboratory (1996).
3. S. Churcher, T. Kean, B. Wilkie: The XC6200 FastMap Processor Interface. Field Programmable Logic and Applications, Proceedings of FPL'95. Springer-Verlag (1995) 36-43.
4. A. Hesener: Implementing Reconfigurable Datapaths in FPGAs for Adaptive Filter Design. Field Programmable Logic, Proceedings of FPL'96. Springer-Verlag (1996) 220-229.
5. J. Faura, C. Horton, P. van Duong, J. Madrenas, J.M. Insenser: A Novel Mixed Signal Programmable Device with On-Chip Microprocessor. Proceedings of the IEEE 1997 Custom Integrated Circuits Conference (1997) 103-106.
6. T. Ma, P. Dressendorfer: Ionizing Effects in MOS Devices and Circuits. Wiley Eds., New York (1989).
7. E.L. Petersen: Single event upset in space: Basic Concepts. Tutorial Short Course, IEEE Nuclear & Space Radiation Efects Conference (NSREC) (1983).
8. C. Stroud, S. Konala, P. Chen, M. Abramovici: Built-in self-test for programmable logic blocks in FPGAs. Proceedings of the IEEE VLSI Test Symposium (1996) 387-392.
9. M Sipper: Evolution of Parallel Cellular Machines. The Cellular Programming Approach. Springer-Verlag (1997).
10. M. Sipper: Designing evolware by cellular programming. Proc. of the first International Conference on Evolvable Systems: From Biology to Hardware (ICES96). Springer-Verlag (1996) 81-95.

Fault Tolerance of a Large-Scale MIMD Architecture Using a Genetic Algorithm

Philippe Millet[1], Jean-Claude Heudin[2]

[1] University of Paris-sud Orsay, Institute of Fundamental Electronic, FRANCE
Philippe.P.M.MILLET@AIRSYS.thomson.fr
[2] Pole Universitaire Léonard de Vinci, 92916 Paris La Défense CEDEX, FRANCE.

Abstract. This article presents a software routing algorithm used to increase fault tolerance of a large-scale MIMD computer based on an isochronous crossbar network. The routing algorithm is completed with a placement algorithm that uses an evolutionary approach that allows dynamic reconfigurations of the task-processor mapping when a fault is detected.

1 Introduction

In order to process Radar signals at a real time rate, Thomson AIRSYS has designed a modular parallel machine (CAMARO) based on a large-scale MIMD architecture [2]. The nodes of this computer can either be computational (C40 or 21K DSP based) or communication boards. All nodes are linked together within a two-dimensional network based on isochronous crossbars. Since each node has its own crossbar, when a board is plugged, it is connected to the rest of the network

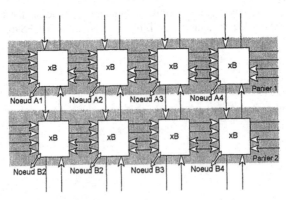

Fig. 1. The isochronous crossbar network.

A typical radar application includes approximately two hundred tasks. A typical machine include about 100 boards ($\approx$200 DSPs). Therefore, without considering any constraint on task mapping, there is about 200! possible mapping configurations.

Given an application, the computer needs to be configured in order to map the application's tasks onto the available processors and to route the messages through crossbars efficiently. This is done offline by a software environment called OCTANE [3].

However, it does not provide automatic reconfiguration at run-time when a board breaks down. Some works have already been done on fault-tolerant routing, but those works deal with dynamic routing in a dynamic crossbar network [5], or with faulty network links [1]. These problems are not similar to ours because in dynamic crossbar networks, each route has to be defined while the message is going from one node to another. When a path is no longer needed, the resources, used to route, are freed. In isochronous crossbar networks, the resources are allocated and paths defined at machine initialisation stage and cannot be modified at run-time. Other works have been done in VLSI-Layout optimisation, some of them with Genetic Algorithms (GA) [4]. Designing a router is very complex, so we decided first to keep the existing routing algorithm (LAURA) and to add a placement algorithm (GOLEM). GOLEM must map the tasks so that LAURA could find a route, through the machine crossbar network, for each data path. RADAR application being critical, GOLEM run-time must be as short as possible. It has been determined that two minutes delay is acceptable. A simple program that would examine all the cases by brute force does not fit the time constraint. Even if a computer is quick enough to determine whether each case is a solution or not in one nanosecond, the typical problem would take about 10^{365}s to be solved in the worst case. Due to this complexity, we have proposed to use heuristic approaches like GA or simulated annealing (SA) [7]. In a previous study [6], the two algorithms have been compared. Since SA is not intrinsically parallel, while GA is, and GOLEM has to run on a parallel computer, we decided to design and experiment a GA-based placement algorithm.

2   A GA-based placement Algorithm

2.1   Chromosome description

The basic objects of our study are tasks and processors. In this experiment we chose to encode the gtype using a task array, and allocating a processor to each task (Fig. 2).

Task (gene)	0	1	2	3	4	5
Mapped on pe	A1a	A1b	B1a	B1b	B2a	B2b

Fig. 2. A task array structure. The name of the processor (for example A1a) allocated to a task indicates the board (A1) and the processor used on the board (a).

2.2 Starting population

We used a randomly generated population that matched the rules defined by the structure of the boards, so that each chromosome exists in the search space area. As our algorithm has to be a solution to a fault problem, we introduce the old configuration in some chromosomes. However, as some boards are unavailable (because of the faulty context), the tasks that were on faulty boards are randomly moved to available ones to create new individuals that match the search space.

2.3 Fitness function

The higher the fitness, the better the chances for the chromosome to bring a successful routing. When a mapping has been done, a way for GOLEM to efficiently rate that mapping is to simulate LAURA. If the routing process terminates successfully, we have a solution; otherwise, we may consider the number of links that cannot be routed through the network to give a fitness value to the individual. LAURA uses the distance between two tasks and tries to find the shortest routes to minimise the number of links used. Thus, for each individual (i), the fitness function calculates the sum S_i of the length of each path. A weight can be associated to each physical link (link between two crossbars). To have a result between 0 and 1 we need to compare S_i with a maximum S_j and a minimum S_k. S_j and S_k can be calculated (1) for each generation within the population, or (2) once for all the generations. In the first case, we have to find, for each generation, the individual (k) that has the lowest sum S_k and the one (j) that has the highest sum S_j. In the second case, we have to find a maximum sum S_j that cannot be overstepped. So we consider the maximum distance that can be used within the network, and then multiply that distance by the number of data paths to route. Finding a S_k is the same problem and we considered the minimum distance that can be used in the machine. We will call (1) the local fitness and (2) the global fitness. The management of constraints like faulted boards or dedicated boards is integrated into crossover and mutation procedures rather than in the fitness computation.

2.4 Selection

The selection procedure used in this algorithm is the one described by Heudin [9]. In this procedure, at generation G, two individuals are randomly selected. In order to reduce the stochastic errors generally associated with roulette wheel selection, we used a remainder stochastic sampling without replacement as suggested by Booker [10]. Then one looks for the two worst individuals. One replaces the two worst by the two (expected) better ones. Then one applies crossover on those two offspring and mutation operators to each of them to reach generation G+1. This selection method is not the classical one where all the population is recombined between generation G and generation G+1. But this method allows the algorithm to keep partial solutions for a longer time than the standard one. Thus, the evolution between two generations is less important (only two individuals have changed) but the GA hardly stops in local sub-optimums.

2.5 Crossover

The crossover operator used in our algorithm comes from the Travelling Salesman Problem (TSP) resolution with a GA suggested by Goldberg [8]. A classic operator could not fit here because only one processor element (pe) can be associated to one task, and only one task can be mapped to one pe. We tested two crossing-over operators (1) a single point crossover and (2) a partially mixed crossover.

Single point crossover (SPC) : Consider the two following chromosomes (Fig. 3).

	Initial					basic crossover					cross gene 3					check				
offspring 1	A	B	C	D	E	A	B	**C**	**B**	**A**	A	B	C	**B**	E	A	**D**	C	**B**	E
offspring 2	E	D	C	B	A	E	D	**C**	**D**	**E**	E	D	C	**D**	A	E	D	C	**D**	A

Fig. 3. The bad effects of a single point crossover can be solved by adding a coherency check.

Using a basic crossover, the offspring cannot be solutions to the problem because (if examining the first offspring) the tasks 1 and 3 are mapped on node B while tasks 0 and 4 are mapped on node A. This configuration does not fit the previously given rule. A solution is to scan all the individuals each time two genes are crossed. In check phase, scanning the offspring 1, after crossing gene 3, the algorithm finds that task 1 is already on pe B, so it sets this task on pe D (previously task 3 pe). The same has to be done with the second offspring and so on for each gene.

Partially mixed crossover (PMC) : In this procedure, consider N the number of genes to swap between the offspring and chosen at random. Do the following N times. Select a gene by random, and proceed like in the single point crossover procedure to exchange the selected genes.

2.6 Mutations

We have designed four mutation operators. Two mutation operators introduce some disorder in the offspring and two other mutations help the algorithm find a better solution. Classically, mutation operators are used to prevent the algorithm from premature convergence. Our algorithm was too slow using only crossover and classical mutations. That is the reason why we used two mutation operators in addition to the crossover operator.

2.6.1 The first two functions (Exchange and change boards):

The first two functions used are inspired from mutation procedures used by Goldberg [8] to solve the TSP: (1) exchange boards, and (2) change board.

In *exchange boards* procedure, two tasks are chosen randomly in the string. Task T1 is running on the pe A1a, and task T2 is running on the pe B1a. The goal is to swap T1 and T2 so that T1 will run on B1a and T2 on A1a. We used to exchange all the tasks that are on each board because tasks that are grouped on one board need to be on the same board to keep running. For each exchange, the mutation procedure compares A1 and B1. If A1 is equal to B1, do nothing. Otherwise, find all the tasks that are on A1 and on B1, and, for each task, exchange the board reference. Starting with the chromosome described Fig. 2, and selecting tasks 0 and 2, this mutation would give chromosome A (Fig 4)

In *change board* procedure, one task T1 is chosen randomly. Consider that T1 is running on A1a. To each chromosome, there corresponds a table that contains the name of the free boards. That table is kept in date while the GA is going on. A free board (B1) is chosen randomly in the table. Then A1 is marked as free and each task previously on A1 is marked as running on B1. Starting with the same initial chromosome shown above (Fig 2) and considering { D1, E1, F1, G1 } free boards in the machine. Selecting task 2 and free board E, should give chromosome B (Fig 4) and { D1, B1, F1, G1 } the new list of free boards.

Chromosome A	**B1a**	**B1b**	**A1a**	**A1b**	B2a	B2b
Chromosome B	A1a	A1b	**E1a**	**E1b**	C1a	C1b

Fig. 4. Chromosomes A and B respectivly show the effects of exchange and change board.

2.6.2 The next two functions were (Special exchange and change board):

The principles used in those mutations are the same used in the two previous ones. But we added as a constraint that each time a task is moved, the sum of the lengths of the links held by the task must decrease.

Exchange board 2 procedure choose a task T1 at random. Construct a table A1 containing the index {T10, T11, T12, T13, ...} of the tasks linked directly to T1. For each task in A1, try to find an occupied board at distance 1 from T1. If not found, try at distance 2, and so on until the distance we are trying is higher that the present distance between the two tasks.

Change board 2 is exactly the same algorithm than exchange board 2, but we are looking for free boards instead of occupied ones.

3. First results

To show the effectiveness of our algorithm, we describe in this paper an experiment based on a real case (182 tasks) dealing 190 boards (349 processors). The number of boards is about two times higher than really needed, in order to see the coherency of the algorithm. To check its convergence we plotted in the same diagram the maximum fitness, the minimum fitness, and the average fitness at each generation and for 3000 iterations. We have also drawn the mapping of the tasks in the computer to have a qualitative analysis of the results.

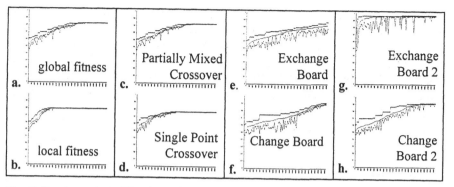

Fig. 5. Results analysis. We plotted in the same diagram the maximum fitness, the minimum fitness, and the average fitness at each generation and for 3000 iterations.

3.1 Fitness function analysis

These two diagrams are the results given by two different experiments: one using global fitness (fig 5.a) and the other using local fitness (fig 5.b). Local fitness is quicker to converge but the result is not as good as that of global fitness. The problem is that in local fitness, at each generation, the best individual has a fitness equal to 1. So the algorithm selects it a lot, conducting to a quick algorithm but also to a local optimum. We decided to use global fitness for further experiments.

3.2 Crossover function analysis

These two diagrams are the results given by two experiments: one using PMC (fig 5.c) and the other using SPC (fig 5.d). To plot those diagrams we did not use mutation procedures. We can see that in each case the crossover function leads to a maximum fitness that is higher than at the beginning of the process. The SPC leads to a lower maximum fitness than the PMC.

3.3 Mutations

The first two mutations are classical mutation procedures that introduce noise in the solutions found. The purpose is to slow down the convergence of the algorithm to prevent it from falling in local optimums. The results given by these two functions are disappointing because they do not bring better solutions in the population. So the maximum fitness given by the algorithm with these functions is not better than the maximum fitness given by the algorithm without these two functions. The two diagrams above show respectively the effect of the exchange (fig 5.e) and the change (fig 5.f) mutation. To allow fast convergence and better results, we tried two other procedures that we implemented as mutation processes in the algorithm. These two mutations do not disturb the crossing over process and do not really bring noise into the chromosome. They try to get a better individual, but if they cannot, they do nothing. So the mutation probability can be much higher than the one used for the previous two mutation procedures, say in range [0.1, 0.5]. Higher than 0.5 would lead the algorithm to converge quickly to a quasi-perfect individual, and thus that individual would become the only individual of the population. (Fig 5.h) shows the change2 mutation with a probability equal to 0.5 while (fig 5.g) shows exchange2 mutation with a probability equal to 0.01.

3.4 Qualitative results

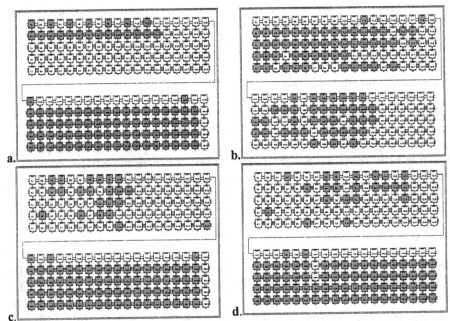

Fig. 6. Qualitative results, a. human solution, b. rounded blocs, c. cabinet structured, d. best result. Grey scares are used boards.

To configure the algorithm's constants (crossover probability, mutation probability, population size, ...), we ran our algorithm with each function separately. Then, we set all functions with the best parameter values found. In order to appreciate the quality of the results, we compared the repatriation found by the algorithm and the one implemented by human hand on a real cased composed of two cabinet linked with only one . The solution given by the user is drawn Fig. 6.a. The first solutions the algorithm found were composed of rounded blocs (Fig. 6.b). We considered that the links have the same weight and do not consider that only one physical link exists between the two cabinets. Then we applied a strong weight to that link (Fig. 6.c). This result could be great, but in a CAMARO computer, it is better (for routing reasons) to have the tasks horizontally arrangedThe best result given by this algorithm (Fig. 6.d) is much like the solution given by hand. We gave a different weight to horizontal links and vertical links.

4. Conclusion

In this article we have shown that a genetic algorithm can be used as a placement algorithm in a parallel computer to increase fault tolerance. We described the crossover, mutations, fitness functions, and the chromosome coding used. The results are encouraging. In future works, we will experiment with a two dimensional chromosome coding that should increase the crossover effectiveness. We will experiment to change our GA in an adaptive genetic algorithm. That will simplify the search for good parameters (crossover and mutation probabilities). We will also implement the algorithm on the parallel computer to give it the capability to work over failure.

5. Acknowledgements

This work is being supported by the Radar Development Technical Business Unit at Thomson AIRSYS (France). Special thanks to Mr. Ph. Ellaume and Mr. F. Giersch.

References

1. Olson, A., Shin, K.G.: Fault-tolerant Routing in Mesh Architecture, *IEEE Trans. on Parallel and Distributed Systems* Vol. 5 no. 11, University of Michigan, Ann Arbor (1994) 1225-1232.
2. Delacotte, V.: *Specification B1 CAMARO Produit 96*, Tech. Rep. no. 46 110 432-306. Thomson AIRSYS, Bagneux, France (1997).
3. Giersch, F.: *OCTANE User Manual*, Tech. Rep. no. 46 110 720-108. Thomson AIRSYS, Bagneux, France (1998)
4. Schnecke, V., Vornberger, O.: An Adaptive Parallel Genetic Algorithm for VLSI-Layout Optimization. In: *Proc. 4th Int. Conf. on Problem Solving from Nature (PPSN IV)*, Berlin, Germany (1996)
5. Su, C.-C., Shin, K.G.: Adaptive Fault-Tolerant Deadlock-Free Routing in Mesh and Hypercubes, IEEE Trans. on Computer Vol. 45 no. 6 (1996) 666-683.
6. Mahfoud, S.W., Goldberg, D.E.: Parallel Recombinative Simulated Annealing: a Genetic Algorithm. In: *Parallel Computing*, 21, (1995) 1-28
7. Schnecke, V., Vornberger, O.: Genetic Design of VLSI-Layouts. *IEE Conf. Publication* no. 414, Sheffield, U.K. (1995) 430-435
8. Goldberg, D.E.: *Algorithmes Genetiques Exploration, Optimisation et Apprentissage Automatique*. Addison Wesley (1991) 13-14, 186-194.
9. Heudin, J.-C.: *La Vie Artificielle*.Hermes, Paris (1994) 210-212.
10. Booker, L.B.: *Intelligent Behavior as an Adaptation to the Task Environment*. Tech. Rep. no. 243, University of Michigan, Ann Arbor (1982).

Hardware Evolution with a Massively Parallel Dynamicaly Reconfigurable Computer: POLYP

Uwe Tangen and John S. McCaskill

Institut für Molekulare Biotechnologie, Beutenbergstraße 11, D–07745 Jena

Abstract. POLYP is a second generation, massively parallel reconfigurable computer based on micro-reconfigurable Field Programmable Gate Arrays (Xilinx XC6000) with a high density of additional distributed memory under local control and broad-band dynamically reroutable optical interconnect technology. Inspired by and designed to study the dynamical self-organization of distributed molecular biological systems using the programmable matter paradigm (like its predecessor NGEN), the new hardware allows the study of large interacting evolving populations of functional design elements in hardware. POLYP includes 144 FPGAs and 400 MB of high speed distributed memory on twelve 18-layer extended VME boards each interconnected via 2 crossbars to 80 unidirectional optical fibers. It is extendable to 20 boards in a single chassis and further to asynchronous multiple host operation. Local reconfiguration of the hardware is mediated by an intermediate hierarchical level of distributed macro-reconfigurable FPGAs, so that the machine is capable of simultaneously evolving functional circuits and their binary representation under user-configurable local control. The process of hierarchical configuration reached the fine-grained level in November 1997, and this paper reports a first experiment in hardware evolution performed with the machine. In contrast with previous evolvable hardware examples, the example is designed to explore the evolution of interconnection structures. As a first step with the new hardware, it by no means yet exploits the powerful potential of the machine. Just as NGEN allowed the study of spatially distributed epigenetic effects in interacting populations of molecules in user-configurable hardware, POLYP allows the study of such effects with individuals dynamically reconfiguring the local hardware.

1 Introduction

Following the introduction of chemical kinetic models of sequence dependent evolution by Eigen in 1971 [1], there has been a steady increase in our understanding of adaptation and self-organization in molecular systems (see [2] and [3] for somewhat complementary reviews). Basic concepts of the role of populations extended in sequence space on the optimization of molecular properties have proven fruitful [4, 5]. Our primary interest has been in moving beyond the optimization of single molecular properties to look at functional interactions and their optimization in interacting populations. Spatial isolation plays a major role

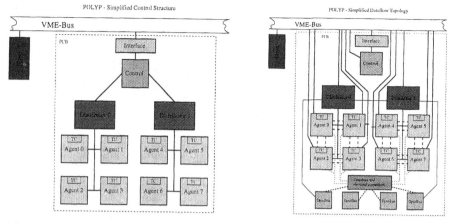

Fig. 1. The board is designed hierarchally. All FPGAs on the board are Xilinx chips. The VME bus interface is realized via a XC4006, the control FPGA and the distributors via XC4028 chips and the micro-reconfigurable agent FPGAs via the micro-reconfigurable XC6264 chips. Complete temperature control of the agent FPGAs is supported. LM75 circuits are attached via copper heat flow bridges through the board to the bottoms of the agent FPGAs. Several data buses are implemented on board to allow for a rich and flexible communication topology.

in evolution and allows the optimization of altruistic functions at the molecular level [6], and these developments prompted the design of a massively parallel hardware configurable computer NGEN which was constructed in FPGA technology [7, 8] to explore local computational paradigms for evolutionary processes.

The examples of hardware evolution demonstrated to date [9–11], have involved serial testing of digital circuits and population emulation in a host workstation. While this paper is not yet an exception, the hardware we have developed allows massively parallel hardware evolution across many chips and more significantly, the exploration of ecological interactions and the question of functional modularization in hardware. In the remainder of the paper we first describe the hardware and its potential and then turn to a proof of principle for our hardware by looking at a very simple first example of hardware evolution relating to hardware connections.

2 POLYP

POLYP stands for parallel online polymer processing in recognition of the remarkable proliferation of the largest biological organism. POLYP was used for the experiment presented below. This computer is on-line micro-reconfigurable and therefore ideally suited for investigations in the field of evolvable hardware (see fig. 3). On each board, eight agent FPGAs with six large SRAMs (4 MBits each) locally connected are available and up to 20 boards can be put into one

Connection design of POLYP

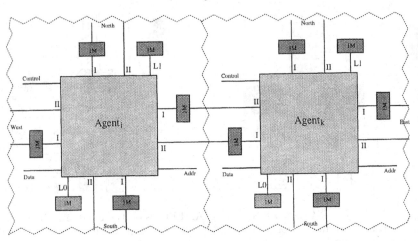

Fig. 2. With the connection topology shown, 2D, 3D and higher dimensional topologies can be realized. Every agent FPGA controls 4 inline SRAMs which are located, eight bit wide, on the connection lines between adjacent agent FPGAs. In addition, two further SRAMs with data and control lines only accessible to a single agent are provided. Each agent is able to communicate with the distributor via eight data lines. Sufficient control and clock distribution lines are also available.

standard 19" VME bus rack. So far 12 boards have already been integrated into POLYP and appear to be functional. For the online control of the eight agent FPGAs two standard user configurable distributor FPGAs are located on the board – the major part of these distributors currently dedicated to an enhanced micro-controller[12]. Each board is furnished with four high bandwith 20 fibre optical OptoBus (Motorola) tranceivers which are driven by two crossbars to allow for flexible routing schemes.

Its main purpose is to study evolutionary scenarios of a relevant size for the investigation of evolvable hardware in biology and electronics.

2.1 Control structure

The boards of POLYP are designed as slave cards in a VME64 bus system. A XC4006 from Xilinx does not only build up the interface to the VME bus but also realizes a temperature control for the agent FPGAs. The main control over the board is carried out by a second FPGA, a XC4028PQ240, called the control FPGA, fig. 1 left picture. This FPGA provides the interface to the Unix embedded operating system as in the first generation machine NGEN. The second level in the hierarchy is implemented by the two distributor FPGAs (XC4028PQ240) each controlling the hardware evolution in four agent FPGAs (XC6264PQ240). These agent FPGAs are micro-reconfigurable via the VME bus and the control FPGA or the distributors. The distributors autonomously have full control over

Top view   Bottom view

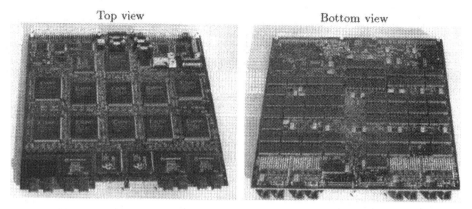

Fig. 3. The real board with both sides is shown. Whenever possible the parts are surface mounted. About 930 circuits were soldered, most of them bypass capacitors to minimize ground bounce.

the agent FPGAs. This structure was chosen to relieve pin-count and logic bottlenecks and to provide a separation of two types of design levels into separate chips, allowing different generalities of application.

2.2 Connection structure of the agent FPGAs

The principle connection topology stems from NGEN [8] which allows 2D, 3D and higher dimensional topologies to be implemented. The eight agent FPGAs on each board are grouped into constellations of four connected in a square, see also fig. 1. The connection topologies and the distributed inline memory differ from other massively parallel hardware configurable computers such as SPLASH [13]. Advanced optical communication allow a bandwidth of about 2 gigabits bidirectional I/O per second. Four optical devices have been placed in POLYP on one board of double height (6U). To allow for more flexibility, two crossbars IQ160 from ICube mediate the connection to the agent FPGAs. Thus, all agent FPGAs can have access to all optical fibres if configured appropriately.

3 Hardware evolution

One challenge in conjunction with evolvable hardware in optimization scenarios is to obtain a microscopic feedback between processed information and the reconfiguration of the hardware. The coupling of the evolving parts of the logic to suitable external selection pressures presents a further problem [10,9]. In what follows we present a brief account of the first experiment in POLYP.

Preliminary investigations with a simulation of the XC6200 function cells grouped in a square array forming a kind of evolution reactor showed a wide variety of evolved configuration sequences. These sequences in the large part expressed the functionality of constants – mostly glitch producing. The very simple selection scheme employed clearly points towards the problem of brittleness [14].

Eight functions allowed   Only buffers allowed

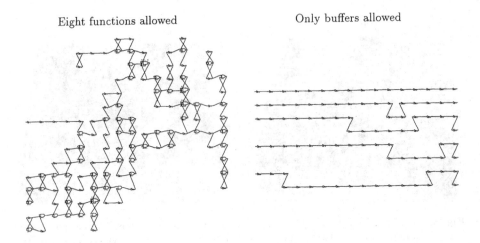

Fig. 4. Hardware evolution of a Single (8 bit wide) Bus. To increase the probability of getting at least one connection between the input of the 16x16 array and the output, wires of random shape are added. These wires are chains of simple buffer elements. In the left picture it is shown that evolution is trapped in a local minimum due to combinatorics. These X-shaped structures represent mostly constants or oscillators. On the right side only buffers have been allowed for the evolutionary process. After 10^5 processed individuals, five channels out of eight were correctly mapped.

Our first test design in POLYP, by no means optimized, employs a 16x16 array of function cells which are to be dynamically reconfigured for different individuals. The array is adjoined by an eight bit input register on the left side and an eight bit output register on the right side. The single example investigates the evolution of digital functions which implement a simple input output relation as a finite state machine via low level gates and flip-flops.

At the level of the distributors a micro-controller starts and stops the input/output test procedure, reads out the counter and mutates the individual actually located in the 16x16 array of the agent. The genetic algorithms[14, 15] used at the host level take into account that the access of agents is mediated by the micro-controller. In the current implementation, the population of individuals is kept in the host at the usual programming C-level. Typical population sizes are within the range of 1000 to 10000 individuals. Crossover and other evolution operators are easily applied. The genetic algorithm which was applied has been tested with the seven element sorter problem by Koza et al.[11].

The experiments carried out differed in the set of Boolean functions which can be addressed by evolution. In fig. 4 (left hand side), eight different Boolean functions were accessible by the evolutionary mechanism. Reverse routing from east to west was not allowed. This is due to a large basin of attraction which produces constants and oscillators. The evolved design shown in fig. 4 (left hand side, after 10000 processed individuals) exhibits a kind of X-like structure, mostly in connection with some OR-gates. These structures prevent connections between

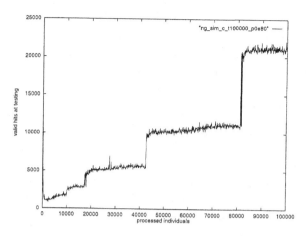

Fig. 5. Time Course of Hardware Fitness. In contrast to the simulation shown in fig. 4 (left hand side), the variety of functional elements has been reduced to simple buffers with three input directions. Given a set of eight Boolean functions the number of combinatorial solutions is currently too large to be handled by evolution. The mutation rate used in this run was about $7 \cdot 10^{-3}$ per Boolean function. The population size was 10000. After processing one individual, 10 crossovers from a collection of the 100 best individuals are introduced into the population. One of the designs evolved is shown in fig. 4 on the right hand side.

input and output and the output will no longer be disturbed by random mutation inside the 16x16 area. So far, no run has been able to escape from this attractor of the system.

To reduce the strong brittleness [14] the set of usable Boolean functions has been simplified. Now, only buffers can be used to connect input and output. Fig. 5 shows the time course of fitness in such a run. Even with this simple generating set of Boolean functions, large evolutionary steps have to be carried out. The system is able to acquire new information and in fig. 4 (right hand side) the corresponding hardware design is shown, which evolved after 100000 processed individuals.

4 Discussion

Logical networks can be accessed in software or in hardware, and by the connection to an external electronic device (forming the environment) or internally. The solution to the credit assignment problem, which would allow the selection of partial solutions within a complex system remains a thorny problem. The bucket brigade [14] or back propagation [16] type of solutions only appear to work within rather restricted contexts. Much recent work has been based on total selection in which the entire logical network is the unit of selection [9, 10, 17–21]. Of course, the use of recombination allows to some degree the benefits of functional unit selection (to the extent of which the schemata theorem [14] is valid). Spatial proximity is employed in nature for credit assignment, and can lead to the selection of systems composed of separate functional elements with selection on the individual level [22–24, 6]. Recognition schemes can in principle extend this type of credit assignment through locality into the functional domain. While POLYP has been constructed to provide a micro-reconfigurable parallel computer which allows high bandwidth dataflow applications based on our experience with NGEN, it would be premature to evaluate its computation potential at this stage. We anticipate both a speed up in design iteration as a result of the transparent configuration process and partial reconfiguration and also advantages from the increased flexibility of the overall topology. The first experiment reported here in no way exploits these high bandwith capabilites and will be further developed towards random Boolean networks [25, 3] with local interactions (length 4 and length 16 connections – high flyers – have been omitted in this experiment). In general, analog properties of the chips have to be taken into account as well [10, 9]. The focus of this work is on the digital approach as a first proof of principle for our powerful hardware. It in no way yet utilizes the unique power of POLYP. We anticipate that the hardware will be of special interest to those in the hardware evolution field needing large or interacting populations of evolving functional elements (i.e. interested in evolving systems). We hope to be able to report on further work with POLYP in this direction at the meeting.

The computer POLYP is a necessary step to bridge the gap between the thousands of individuals in workstation simulations and the 10^{14} individuals in biological *in vitro* systems. It allows spatially resolved and interacting populations of functional elements to evolve in hardware. It is certain that the evolvable hardware paradigm has now brought von Neumann's duality of construction and computation [26] into technological focus.

Acknowledgements   The authors wish to thank L. Schulte for his assistance in implementing the electronic design using the Board Station software package of Mentor Graphics. Routing was performed by a shape based router Specctra with final hand routing finishing required.

References

1. Eigen M. Selforganization of Matter and the Evolution of Biological Macromolecules *Naturwissenschaften*, 58:465–523, 1971.
2. Eigen M., McCaskill J., Schuster P. Molecular Quasi-Species *J. Phys. Chem.*, 92:6881–6891, 1988.
3. Kauffman S. A., Johnsen S. Coevolution to the Edge of Chaos: Coupled Fitness Landscapes, Poised States, and Coevolutionary Avalanches 1991.
4. McCaskill J. S. A Localization Threshold for Macromolecular Quasispecies from Continuously Distributed Replication Rates *J. Chem. Phys.*, 80:5194–5202, 1984.
5. Fontana W., Stadler P., Bornberg-Bauer E. G., Griesmacher T., Hofacker I. L., Tacker M., Tarazona P., Weinberger E. D., Schuster P. RNA Folding and Combinatory Landscapes *Phys. Rev. E*, 47:2083–2099, 1993.

6. Boerlijst M. C., Hogeweg P. Spiral Wave Structure in Pre-Biotic Evolution: Hypercycles Stable Against Parasites *Physica D*, 48:17–28, 1991.

7. M^c^Caskill J. S., Chorongiewski H., Mekelburg K., Tangen U., Gemm U. NGEN - Configurable Computer Hardware to Simulate Long-Time Self-Organization of Biopolymers (Abstract) *Physical Chemistry*, 98:1114–1114, 1994.

8. M^c^Caskill J. S., Maeke T., Gemm U., Schulte L., Tangen U. NGEN A Massively Parallel Reconfigurable Computer for Biological Simulation: towards a Self-Organizing Computer *Lec. Note Comp. Sci*, 1259:260–276, 1997.

9. Thompson A. An Evolved Circuit, Intrinsic in Silicon, Entwined with Physics *Lect. Not. Comp. Sci.*, 1259:390–405, 1996.

10. Harvey I., Thompson A. Through the Labyrinth Evolution Finds a Way: A Silicon Ridge *Lect. Not. Comp. Sci.*, 1259:406–422, 1996.

11. Koza J. R., Bennett III F. H., Hutchings J. L., Bade S. L, Keane M. A., Andre D. Rapidly Reconfigurable Field-Programmable Gate Arrays for Accelerating Fitness Evaluation in Genetic Programming In Koza J. R., editor, *Late Breaking Papers at the Genetic Programming 1997 Conference*, pages 121–131 Standford University Bookstore, Standford CA, 1997.

12. Chapman K. Dynamic Microcontroller in an XC4000 FPGA *Xilinx Application Note*, 1994.

13. Gokhale M., Holmes B., Kopser A., Kunze D., Lopresti D., Lucas S., Minnich R., Olsen R. SPLASH: A Reconfigurable Linear Logic Array *SRC-TR-90-012*, 1:1–16, 1992 Preprint.

14. Holland J. H. Adaptation in Natural and Artificial Systems University of Michigan Press, Ann Arbor, 1975.

15. Goldberg D. E. Genetic Algorithms in Search, Optimization & Machine Learning Addison Wesley, Reading, Massachusetts, 1985.

16. Rumelhart D. E., McClelland J. L. Parallel Distributed Processing, Vol. 1 The MIT Press, Cambridge, England, 1986.

17. Keymeulen D., Durantez M., Konaka K., Kuniyoshi Y., Higuchi T. An Evolutionary Robot Navigation System using a Gate-Level Evolvable Hardware *Lect. Not. Comp. Sci.*, 1259:159–209, 1997.

18. Armstrong W. W. Hardware Requirements for Fast Evaluation of Functions Learned by Adaptive Logic Networks *Lect. Not. Comp. Sci.*, 1259:17–22, 1997.

19. Nussbaum P., Marchal P., Piguet C. Functional Organisms Growing on Silicon *Lect. Not. Comp. Sci.*, 1259:139–151, 1997.

20. Bennett III F. H., Koza J. R., Andre D., Keane M. A. Evolution of a 60 Decibel Op Amp Using Genetic Programming *Lect. Not. Comp. Sci.*, 1259:312–326, 1997.

21. Koza J. R., Bennett III F. H., Andre D., Keane M. A. Reuse, Parametrized Reuse, and Hierarchical Reuse of Substructures in Evolving Electrical Circuits Using Genetic Programming *Lect. Not. Comp. Sci.*, 1259:312–326, 1997.

22. Lee, McCormick W. D., Pearson J. E., Swinney H. L. Experimental Observation of Self-Replicating Spots in a Reaction-Diffusion System *Nature*, 369:215–217, 1994.

23. Pearson J. E. Complex Patterns in a Simple System *Science*, 261:189–192, 1993.

24. Böddeker B., M^c^Caskill J. S. Do Self-Replicating Spots Provide a Platform For Heriditary Molecular Diversity *J. Theor. Biol.*, 1996 Submitted.

25. Kauffman S. A., Weinberger E. D. The NK Model of Rugged Fitness Landscapes And Its Application to Maturation of the Immune Response *J. Theor. Biol.*, 141:211–245, 1989.

26. von Neumann J. Theory of Self-Reproducing Automata Burks, A. W. University of Illinois Press, Urbana, 1966.

Molecular Inference via Unidirectional Chemical Reactions

Jan J. Mulawka, Magdalena J. Oćwieja

Warsaw University of Technology, Nowowiejska 15/19, 00-665 Warsaw, Poland
jml@ipe.pw.edu.pl

Abstract. Inference process plays an important role in the realisation of expert systems. In this paper it is shown that chemical reactions may by used to perform molecular inference according to the algorithm of forward chaining. This method is accomplished by an adequate interpretation of inorganic chemical compounds and unidirectional reactions. In our approach premise clauses are represented by the reactants while conclusion clauses are represented by the products of reaction. Different inorganic compounds and reactions have been discussed with respect to their utility for the molecular inference. Special attention is focused on qualitative chemistry and a number of reactions has been taken into account. Experimental results demonstrating application of these reactions in expert systems are provided.

1 Introduction

Ability to infer is a characteristic feature of intelligent beings. In artificial intelligence different paradigms of inference were developed a long time ago. From logical point of view the inference process involves some kind of search procedure. For example in expert systems [1 - 3] the inference process is performed by so called inference engine which is also known as some kind of knowledge processor. The other important part of any expert system is its knowledge base. In majority of such systems it consists of the set of heuristic rules that serve to determine the conclusion at which the expert system ultimately arrives. During the inference process the system is drawing a conclusion by means of a set of rules, for a specific set of facts, for a given situation. The inference process may be used to undertake different operations by the expert system.

Usually, the expert systems and also the inference engines are realised by adequate programs on classical electronic computers. Such computing systems suffer some restrictions which follow from disadvantages of traditional computers like small degree of parallelism as well as lack of associative memory and difficulty with scaling up to larger knowledge bases. Therefore, it is interesting to consider alternative implementations of computing systems.

Recently, there has been an interest in the field known as molecular computing. Deoxyribonucleic acid (DNA) as well as ribonucleic acid (RNA) constitute interesting groups of useful compounds [4]. The molecules which belong to these acids are strings on a four letter alphabet. These molecules are of great importance in genetics since they serve as heredity carriers in living matter. In [5] a concept of computational nucleic acid (CNA) is proposed. This hypothetical molecule would be a simultaneous generalisation of DNA, RNA, and protein although it can be considered as a string over a finite alphabet.

An interesting class of molecules which can represent states of computation based on modification of the electronic structure has been reported in [6,7]. These molecules are chains built of donor and acceptor subunits and are called mnemons. The other class of molecules which can be considered as information carriers are so called smart molecules, which are the subject of research in supramolecular chemistry [8,9].

Adleman who is a pioneer in this area has demonstrated that NP-complete problems can be solved by means of biochemical reactions using DNA [10]. The main advantages of the molecular computing are possibilities to obtain a massive parallelism of computation and an associative memory of high capacity. Many researchers reported concepts of general molecular computer [11,12]. It is also possible to implement an inference engine performing the backward chaining [13,14]. Good solutions to the inference problem have been given in [15,16]. All these implementations have been achieved by utilising self-assembly of DNA standards.

Molecular computing may be performed also on smaller and cheaper molecules than those used in biochemistry. The primary objective of this contribution is to check if simple chemical compounds may be useful in molecular inference. We try to implement simple inference engine using unidirectional reactions known in classical qualitative chemistry.

2 Interpreting Rules by Using Chemical Reactions

In our approach the knowledge base is created by a set of rules with either intermediate or final conclusions. We assume that premise clauses in rules are connected by the AND operators. Because the inference process is carried out by using unidirectional chemical reactions the knowledge base should be suitable for adequate chemical interpretation. The rules may be composed of the following statements:
premise clauses
    „In the probe there is reacting substance < name of reactant>"
conclusion clauses
    „In the probe is formed reaction product < name of product> "
The syntax of the rules may be written as:
If (premise clause A) AND (premise clause B) ....THEN (conclusion clause K) AND (conclusion clause L)...
or in the shorter form

$$A \wedge B....... \rightarrow K \wedge L ..... \tag{1}$$

From chemical point of view the premise clauses represent reactants while the conclusion clauses represent products of reaction. Since a chemical reaction can be performed for two or more reactants therefore in premise part of the formula (1) should be at least two statements . As an example consider the rule :

$$P \wedge Q \rightarrow S. \tag{2}$$

This rule may be represented by the reaction :

$$AgNO_3 + KOH \rightarrow AgOH\downarrow + KNO_3 \tag{3}$$

where $AgNO_3$ and KOH are reactants of the reaction; they represent premise clauses P and Q respectively while AgOH is the product of this reaction; it represents conclusion premise S. In our experiments sometimes happens that all reactants are in one probe. In such case a general problem is to identify individual products of reaction on the ground of characteristic properties occurrence.

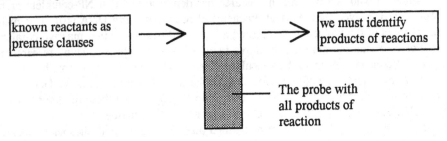

Fig. 1. The probe representing a rule

For simple identification of the reaction products it can be assumed that each reaction is performed in a separate probe, which means that particular rules may be fired separately.

3 A Method of the Forward Chaining Implemented Chemically

The algorithm of the forward chaining [1] can be implemented by means of chemical compounds and unidirectional reactions in the following manner. First, an adequate knowledge base for a given problem must be created. To this purpose let us consider a set of chemical reactions which products are characteristic and easy to distinguish. The premise clauses of these rules and in consequence the reactants of these reactions are known.

An initial step of the algorithm is equivalent to the preparation of a set of probes for the reactions. Each of them represents particular rule. Performance of the reaction is equivalent to the execution of two steps of the forward chaining algorithm - rule scanning and checking as well as rule firing. Because these reactions are unidirectional the products cannot react, and hence the products cannot be converted into the reactants.

Our approach may be applied to solve two kinds of the inference problems:

Problem 1. Derive all possible facts from a given set of rules and facts.

Problem 2. Prove a given hypothesis upon a give set of rules and facts.

If in our knowledge base there are rules with final conclusions then obtained products must be identified and final results of the inference process is reached.

However if we have the rules with intermediate conclusions, obtained products after identification are used in the further cycle of the algorithm.

4 Search for Adequate Reaction Sets

Identification of the reaction product is a crucial problem in molecular inferencing. Therefore the products should be easy to distinguish and not prone to an error. In devising a procedure for estimating any inorganic species, it is necessary to take account of all other constituents likely to be present in the sample being analysed. This is because reactions of reactants with inorganic species are usually non-specific.

To perform the inference process via chemical reactions it is important to have a set of reactions which react in a selective way. This requirement is necessary to have rules with a greater number of premise clauses. In the following we discuss our reactants to be used for this purpose. We consider the methods of analytical qualitative chemistry. The reactants in our selective unidirectional reactions should be in suitable mass concentration [mole/cm^3] while products of reactions should have unique symptoms as: colour of substance in the base-tube, value of pH, perceptible smell, precipitation of sediment etc.

Table 1. A set of reactions with easily identified products.

Rule notation	Chemical reaction and notation of respective clauses	Characteristic features of the reaction product
R1	$AgNO_3 + HCl \rightarrow AgCl_{(\downarrow)} + HNO_3$ $\quad P \quad\quad Q \quad\quad S$	$S \rightarrow$ black, caseous sediment
R2	$AgNO_3 + (NH_3*H_2O) \rightarrow AgOH_{(\downarrow)} + NH_4NO_3$ $\quad P \quad\quad A \quad\quad\quad T$	$T \rightarrow$ tawny sediment
R3	$AgCl_{(\downarrow)} + AgOH_{(\downarrow)} \rightarrow$ multicomponent mixture $\quad S \quad\quad T \quad\quad\quad\quad U$	$U \rightarrow$ little brown sediment
R4	$AgCl + K_2CrO_4 \rightarrow Ag_2CrO_4 + 2\ KCl$ $\quad S \quad\quad R \quad\quad V$	$V \rightarrow$ tawny - red sediment

As a result of our search for chemical unidirectional reactions we have chosen two sets of compounds suitable for molecular inferencing. They are listed in Tables 1 and 2 respectively. The chemical reactions listed in Table 1 represent the following set of rules which we denote as the knowledge base 1.

$$R1: P \wedge Q \rightarrow S$$
$$R2: P \wedge A \rightarrow T \qquad (4)$$
$$R3: S \wedge T \rightarrow U$$
$$R4: S \wedge R \rightarrow V$$

The rules R1 and R2 have intermediate conclusion clauses. They may be used as premise clauses for the rule R3.

Table 2. A set of reactions creating knowledge base 2

Rule notation	Chemical reaction and notation of respective clauses	Characteristic features of the reaction product
R5	$2AsO_3^{3-} + 5H_2S + 6H^+ \rightarrow As_2S_5 (\downarrow) + 8H_2O$ $\quad A \qquad B \quad C \qquad D$	D $\rightarrow$ yellow colour of the sediment
R6	$3AsO_3^{3-} + K_2Cr_2O_7 + 8H^+ \rightarrow 3AsO_4^{3-} + 4H_2O + 2Cr^{3+}$ $\quad A \qquad E \qquad C \qquad\qquad\qquad G$	G $\rightarrow$ green colour of the solution
R7	$3As_2S_3 + 28NO_3^- + 4H_2O + 4H^+ \rightarrow 6H_2AsO_4^- + 9SO_4^{2-} + 28NO_{(\uparrow)}$ $\quad D \qquad H \qquad\quad C \qquad\quad I \qquad\qquad K$	I $\rightarrow$ sediment dissolve K $\rightarrow$ visible airbubbles
R8	$2Cu^{2+} + 5OH^- + HAsO_3^{2-} \rightarrow Cu_2O_{(\downarrow)} + AsO_4^{3-} + 3H_2O$ $\quad L \qquad\qquad M \qquad\qquad N$	N $\rightarrow$ red solution

In Table 2 an assortment of more complicated reactions is provided. As is seen these reactions may be written in form of the knowledge base 2.

$$R5: A \wedge B \wedge C \rightarrow D$$
$$R6: A \wedge E \wedge C \rightarrow G \qquad (5)$$
$$R7: D \wedge H \wedge C \rightarrow I \wedge K$$
$$R8: L \wedge M \rightarrow N$$

The other reactions also have been considered. In case of simple analysis we deal with single cations and anions; however, in complex analysis we have multicomponent chemical mixtures. Generally, it is difficult to find greater number of specific reactions with many reactants.

5 Results of the Experiments

To demonstrate possibility of inference via simple unidirectional chemical reactions a number of experiments have been carried out in the laboratory. These experiments have been performed under the following assumptions. We have used intermicrochemical method. To perform unidirectional reaction the quantity of the reactants to be used have been chosen as $0.5 - 5$ cm^3 or $10 - 100$ mg. In such approach experimenting with a small amount of substance suppresses a time necessary to perform the reaction. During the experiments we have applied chemical qualitative techniques such as: heating the compounds, precipitation and centrifuging the sediments, dissolving the sediment.

Example 1

The knowledge base 1 consists of 4 rules which coprise 6 premise clauses: P, Q, A, S, T, R and 4 conclusion clauses: S, T, U, V. Suppose, the following interpretation for particular clauses has been assumed:

P - patient has a fit of shivers
Q - patient has scratchy throat
A- patient has swollen nose
S - patient has a cough
T- patient caught a cold
R - patient has a high temperature
U - patient has an influenza
V - patient has a pneumonia

Prove the hypothesis: patient caught a cold. Assume that two facts are known: patient has a fit of shivers, patient has swollen nose. According to Table 1 premise clauses are represented by the compounds

$$P \rightarrow AgNO_3$$
$$A \rightarrow (NH_3 {}^* H_2 0)$$

while conclusion clause is expected as

$$T \rightarrow \text{tawny sediment}$$

To infer the conclusion clause the reaction representing the rule R2 is performed. For $2 \div 3$ drops solution of $AgNO_3$ we add 2-mole $(NH_3 {}^* H_2 0)$. Tawny sediment is observed as a product of the reaction which testifies that our hypothesis is true.

Example 2

For the same knowledge base as in Example 1 derive all possible conclusions when the following facts are known: P, Q, A. According to Table 1 three reactions have been performed

$$AgNO_3 + HCl \rightarrow AgCl_{(\downarrow)} + HNO_3$$
$$AgNO_3 + (NH_3 {}^* H_2 0) \rightarrow AgOH_{(\downarrow)} + NH_4 NO_3$$
$$AgCl_{(\downarrow)} + AgOH_{(\downarrow)} \rightarrow \text{multicomponent mixture}$$

The following products have been identified: black, caseous sediment for the reaction R1, tawny sediment for the reaction R2, little brown sediment for the reaction R3. It means that the following facts have been proved: S, T, U.

Example 3

Consider the knowledge base 2. Suppose we know the facts A, B, C, E, H, L, M. Derive all possible conclusions from this knowledge base. We prepare reactions to infer conclusions from known premise clauses. In accordance with Table 2 all the rules R5 - R8 can be fired. The following reactions have been performed.

$$2AsO_3^{3-} + 5H_2S + 6H^+ \rightarrow As_2S_5 (\downarrow) + 8H_2O$$

For 2÷3 drops of a researched solution we add 2÷3 drops of AKT (amide of tioacetic acid) and 8÷10 drops of concentrated HCl solution. We heat the mixture for 8 minutes. Yellow sediment is precipitating. Thus the conclusion clause D is derived.

$$3AsO_3^{3-} + K_2Cr_2O_7 + 8H^+ \rightarrow 3AsO_4^{3-} + 4H_2O + 2Cr^{3+}$$

To fire the rule R6 for 1 cm³ 0,05-mole solution of $K_2Cr_2O_7$ we add 1÷2 drops of the researched solution and 1÷2 drops of concentrated H_2SO_4. The colour of solution is changing from pink to green, which testifies that the conclusion clause G is true.

$$3As_2S_3 + 28NO_3^- + 4H_2O + 4H^+ \rightarrow 6H_2AsO_4^- + 9SO_4^{2-} + 28NO_{(\uparrow)}$$

To fire the rule R7 for 3÷4 drops of the researched solution we add 4÷5 drops of concentrated solution HCl and 2÷3 drops of AKT. We heat the mixture for 8 minutes. Yellow sediment is dissolving and we can see air bubbles (from NO). Hence, the conclusion clauses I and K are proved.

$$2Cu^{2+} + 5OH^- + HAsO_3^{2-} \rightarrow Cu_2O_{(\downarrow)} + AsO_4^{3-} + 3H_2O$$

Finally, to fire the rule R8 for 3÷4 drops of the researched solution we add about 0.05-mole of $CuSO_4$ solution, and 1÷2 drops of 2-mole NaOH solution. For precipitated sediment we add NaOH, until dissolving it. Blue solution is created, and we heat it. The sediment's colour is red. In such a way we identify the conclusion clause N. Thus, the following conclusion clauses have been proved: D, G, I, K, N.

6 Conclusions

We have demonstrated that two kinds of inference problems can be solved by using the rules listed in Tables 1, 2 where exemplary sets of reactions are provided. Performing a chemical reaction is equivalent to firing a rule. As follows from the tables the sets of considered rules are not numerous, therefore the scope of problems to be solved by such small knowledge bases is restricted. A search for larger knowledge bases has been carried out. However by using methods of qualitative chemistry it is difficult to find to this purpose adequate sets of rules because simple molecules do not provide high degree of selectivity of chemical reaction and therefore can not serve as an information carriers.

Larger knowledge bases suitable for molecular inference can be achieved by using more complicated molecules. They should have two or several well distinguishable states such that transition from state to state is controllable by external factors such as optical, thermal, electrical. Generally, molecules of organic chemistry are more adequate for carrying information. For example polymers consist of repeated structural units called monomers. Some polymers are linear and therefore they may be considered as strings over an alphabet of monomers. To this class belong proteins which are linear polymers based on twenty amino acid monomers. In this paper we have restricted, however, our considerations to inorganic compounds. It has been shown that simple unidirectional reactions may be utilised to perform the forward chaining although one might think of further exploiting this approach for example to 'train' up a knowledge base, etc.

Acknowledgments

This work was supported by the Rector's grant PATIA. We would like to thank professors A. Jończyk, W. Fabianowski and G. Rokicki for valuable discussions and help in arranging the experiments.

References

1. Mulawka J.J.: Expert Systems (in Polish). WNT Warsaw (1996)
2. Hayes-Roth F., Waterman D.A., Lenart D.B.: Building Expert Systems. Adison-Wesley, New York (1983)
3. Nebendahl E.: Expert Systems. Wiley J. and Sons , Inc. Berlin (1988)
4. Węgleński P.: Molecular Genetics (in Polish). PWN Warsaw (1996)
5. Mahaney S., Royer J., and Simon J.: Biological Computing. To appear in the Complexity Retrospective II.
6. Stolarczyk L.Z., Piela L.: Hypothetical Memory Effect in Chains of Donor - Acceptor Complexes. Chemical Physics 85 (1984) 451 - 460
7. Nowaczek W., Piela L., Stolarczyk L.Z.: Low - Energy Metastable Electronic States of Donor - Acceptor Oligomers. Advanced Materials for Optics and Electronics, vol. 6 (1996) 301 - 306
8. Atwood J. L. et al.: Comprehensive Supramolecular Chemistry. Volume 10 Supramolecular Technology, Pergamon Elsevier Science Ltd. (1996)
9. Rudkevich D. M., Shivanyuk A. N., Brzózka et al.: A Self - Assembled Bifunctional Receptor. Angew. Chem. Int. Ed. Engl. 34 No. 19 (1995) 2124 - 2126
10. Adleman L.M.: Molecular Computation of Solution to Combinatorial Problems. Science, vol. 266 (1994) 1021 - 1024
11. Adleman L.M.: On Constructing a Molecular Computer, Manuscript (1995)
12. Mulawka J.J.: Molecular Computing Promise for New Generation of Computers. Polish-Czech-Hungarian Workshop on Circuit Theory, Signal Processing and Appl., Budapest Hungary (1997) 94 - 99
13. Mulawka J.J., Borsuk P.,Węgleński P.: Implementation of the Inference Engine Based on Molecular Computing Technique. Proc. IEEE Int. Conf. on Evolutionary Computation (ICEC '98) Anchorage USA (1998) 493 - 498
14. Mulawka J.J., Borsuk P., Węgleński P.: A Method of Deduction via Backward Chaining (in Polish). Submitted as patent nr P 322-076, Warsaw (1997)
15. Ogihara M. and Ray, A.: The Minimum DNA Computation Model and Its Computational Power. Technical Report TR 672, University of Rochester, Singapur (1998)
16. M. Amos and P. E. Dunne. DNA Simulation of Boolean Circuits. Technical Report CTAG-97009, Department of Computer Science, University of Liverpool UK (1997)

Author Index

Springer
and the
environment

At Springer we firmly believe that an international science publisher has a special obligation to the environment, and our corporate policies consistently reflect this conviction.

We also expect our business partners – paper mills, printers, packaging manufacturers, etc. – to commit themselves to using materials and production processes that do not harm the environment. The paper in this book is made from low- or no-chlorine pulp and is acid free, in conformance with international standards for paper permanency.

 Springer

Lecture Notes in Computer Science

For information about Vols. 1–1397

please contact your bookseller or Springer-Verlag